Uranium Processing and Properties

Jonathan S. Morrell • Mark J. Jackson
Editors

Uranium Processing and Properties

Springer

Editors
Jonathan S. Morrell
Compatibility and Surveillance Section
Y-12 National Security Complex
Oak Ridge, TN
USA

Mark J. Jackson
Center for Advanced Manufacturing
Purdue University
West Lafayette, IN
USA

ISBN 978-1-4614-7590-3 ISBN 978-1-4614-7591-0 (eBook)
DOI 10.1007/978-1-4614-7591-0
Springer New York Heidelberg Dordrecht London

Library of Congress Control Number: 2013943696

Printed on acid-free paper

Springer is part of Springer Science+Business Media (www.springer.com)

Preface

The processing and properties of uranium and its alloys are the subject of much scrutiny at the present time owing to a resurgence of activity in the field of nuclear technology. This book is a summary of work prepared by experts in the field of uranium, with some special submissions based on presentations and papers included at the "Processing of Specialty Metals with Emphasis on Depleted Uranium Users' Conferences" that were sponsored by the American Society of Materials, Manufacturing Sciences Corporation, and the Y-12 National Security Complex at Oak Ridge, Tennessee. These conferences occurred in 2004, 2007, and 2010 with contributors from academia, industry, and government including the Department of Energy—National Nuclear Security Administration laboratories and facilities, and the Atomic Weapons Establishment at Aldermaston, United Kingdom. The chapters provide an authoritative text on the current developments in the field of processing and properties of uranium.

Chapter 1 focuses on the physical, chemical, mechanical, and metallurgical properties of uranium metal; Chap. 2 focuses on several casting and melting techniques; Chaps. 3 and 4 cover machining with coated cutting tools and the grinding parameters of uranium, respectively; Chap. 5 deals with aqueous extraction techniques and processes with emphasis on the chemical principles involved; Chaps. 6 and 7 focus on uranium corrosion with specific detail on the phase relationships and kinetic processes describing the pure uranium–hydrogen binary system and the ternary processes involving oxygen, respectively; Chap. 8 focuses on the fundamentals supporting nondestructive evaluation techniques as well as the methods themselves which are applicable to uranium systems; lastly, Chap. 9 covers developments of new medical isotope production using low-enriched uranium fuel.

The structure of the book is based on the materials provided by many colleagues, and the author wishes to thank the contributors of this book for helping construct a current source of knowledge and information on uranium processing and properties as well as for granting the editors permission to use such material. The editors also acknowledge the help and support of Merry Stuber and her colleagues at Springer publishers for help in preparing the manuscript in a timely manner.

Oak Ridge, TN, USA Jonathan S. Morrell
West Lafayette, IN, USA Mark J. Jackson

Contents

Contributors

Rahul Bhola Department of Metallurgical and Materials Engineering, Colorado School of Mines, Golden, CO, USA

Kenneth H. Eckelmeyer 6 Valley View Court, Placitas, NM, USA

Srisharan G. Govindarajan Mechanical and Aerospace Engineering, University of Missouri-Columbia, Columbia, MO, USA

Nathan R. Gubel Department of Metallurgical and Materials Engineering, Colorado School of Mines, Golden, CO, USA

Rodney G. Handy Department of Engineering Technology & Construction Management, University of North Carolina-Charlotte, Charlotte, NC, USA

Cameron Howard Department of Metallurgical and Materials Engineering, Colorado School of Mines, Golden, CO, USA

Mark J. Jackson Center for Advanced Manufacturing, Purdue University, West Lafayette, IN, USA

Kimberly N. Johnson University of Tennessee, Knoxville, TN, USA

Kamalu Koenig Structural Integrity Associates, Inc., Centennial, CO, USA

Brajendra Mishra Department of Metallurgical and Materials Engineering, Colorado School of Mines, Golden, CO, USA

Jonathan S. Morrell Y-12 National Security Complex, Oak Ridge, TN, USA

David L. Olson Department of Metallurgical and Materials Engineering, Colorado School of Mines, Golden, CO, USA

Jonathan Poncelow Department of Metallurgical and Materials Engineering, Colorado School of Mines, Golden, CO, USA

G. Louis Powell Y-12 National Security Complex, Oak Ridge, TN, USA

Edward B. Ripley Y-12 National Security Complex, Oak Ridge, TN, USA

Grant M. Robinson Center for Advanced Manufacturing, Purdue University, West Lafayette, IN, USA

Gary L. Solbrekken Mechanical and Aerospace Engineering, University of Missouri-Columbia, Columbia, MO, USA

Kyler K. Turner Mechanical and Aerospace Engineering, University of Missouri-Columbia, Columbia, MO, USA

Craig VanHorn URS Corporation, Denver, CO, USA

Michael D. Whitfield Center for Advanced Manufacturing, Purdue University, West Lafayette, IN, USA

Chapter 1
Introduction to Uranium

Nathan R. Gubel, Kenneth H. Eckelmeyer, Kimberly N. Johnson, Mark J. Jackson, and Jonathan S. Morrell

Abstract Historical studies of uranium were attributed to its distinction as the heaviest element in nature and as the terminus of the classical periodic table. Presently, the mechanical, physical, chemical, and metallurgical properties of uranium metal are of great importance in nuclear fuels, in weapons, and as a precursor to uranium oxide fuel. New reactor designs, including metal-fueled fast reactors, require the continued research and understanding of uranium material science. All actinide metals exhibit unusual physical properties, as such; the mechanical properties of uranium are highly dependent upon fabrication and heat treatment history as well as impurity content. Many of the difficulties that arise stem from the anisotropic nature of the room-temperature (α) phase of uranium. The majority of nuclear fuel elements are manufactured from machined and heat-treated bars of cast uranium metal. Powder metallurgy practices are not practical due to irregular sintering characteristics and the pyrophoric nature of fine uranium particles. As with other metals, the tensile properties of uranium can be improved

N.R. Gubel (✉)
Department of Metallurgical and Materials Engineering, Colorado School of Mines, 1500 Illinois Street, Golden, CO 80401, USA
e-mail: ngubel@mines.edu

K.H. Eckelmeyer
6 Valley View Court, Placitas, NM 87043, USA
e-mail: keckelmeyer@comcast.net

K.N. Johnson
University of Tennessee, 908 Trent Lane, Knoxville, TN 37922, USA
e-mail: knjohnson629@gmail.com

M.J. Jackson
Center for Advanced Manufacturing, Purdue University, West Lafayette, IN, USA
e-mail: jacksonmj04@yahoo.com

J.S. Morrell
Y-12 National Security Complex, 1 Bear Creek Road, MS #8097, Oak Ridge, TN 37831, USA
e-mail: morrelljs@y12.doe.gov

J.S. Morrell and M.J. Jackson (eds.), *Uranium Processing and Properties*,
DOI 10.1007/978-1-4614-7591-0_1, © Springer Science+Business Media New York 2013

with alloying additions although options are inhibited by the low solubility of other elements in uranium. The study of uranium processing is an ongoing effort due to the inherent issues present in nuclear reactors used for electricity production; nuclear weapon production and dismantlement; the treatment, recycling, and storage of nuclear waste; and the cleanup of Cold War nuclear material production sites. These are causes of acute global interest, in all of which uranium is intimately involved.

Keywords Uranium metallurgy • Uranium alloys • Uranium metal • Mechanical properties • Microstructure • Martensite • Corrosion • Uranium oxide • Uranium hydride • Oxidation • Uranium decay • Uranium production • Stress corrosion cracking

1.1 Introduction

Uranium is found in modest concentrations in a wide variety of rocks, soils, and salt water. Approximately 0.0004 % of the earth's crust consists of naturally occurring uranium. This makes uranium about 1/20th as abundant as the common metals copper and nickel, approximately equally abundant as tantalum, and nearly 1,000 times more abundant than gold.

Several isotopes are present in naturally occurring uranium, as shown in Table 1.1. Of these, U^{238} strongly predominates at 99.27 %, followed by U^{235} at only 0.72 %. Both U^{238} and U^{235} are mildly radioactive, with half-lives of 10^9 and 10^8 years, respectively. The levels and types of radioactivity given off are relatively mild and non-penetrating, making external exposure to bulk uranium fairly safe. Inhalation or ingestion of finely divided uranium is more dangerous due to a combination of radioactivity and chemical toxicity. In general, uranium and lead are considered to be roughly comparable health hazards.

U^{235} is notable because it is fissionable by slow neutrons. Each fission event gives off a large amount of energy, as well as additional neutrons. The release of additional neutrons creates the possibility of self-sustaining chain reaction fission, thus providing the basis for the atomic bomb (where the chain reaction occurs in an uncontrolled fashion, releasing a huge amount of energy over a very short time), and nuclear power generation (where the chain reaction is carefully regulated and energy is released in a controlled manner, heating water to drive the turbines that power electrical generators). U^{238}, on the other hand, is only fissionable by very

Table 1.1 Naturally occurring uranium isotopes[a]

Mass	Percentage	Half-life (years)
238	99.27	10^9
235[b]	0.72	10^8
234	0.006	10^5

[a]Approximately 0.0004 % of earth's crust (~1,000 times more plentiful than gold)
[b]Fissions with slow neutrons

high energy neutrons. Because of this, U^{235} is the "high-equity isotope" for which uranium is mined and processed.

Isotopic separation is carried out by a number of processes. Magnetic separation (based on the same principle as mass spectroscopy) was used in the early 1940s, but has been abandoned in favor of gaseous diffusion and centrifugal separation which are faster and more efficient processes. Laser isotope separation was investigated more recently, and it is being developed as a production capability. All of the common separation techniques are based on very small differences between U^{235} and U^{238}, so they require a long cascade of low-efficiency individual separation steps. The product of the isotopic separation process is enriched uranium, and the by-product (depleted uranium) contains ~0.2 % U^{235} (i.e., ~99.8 % U^{238}).

1.2 Uranium Production

Uranium ores vary in chemical complexity from the relatively simple pitchblende ores, which are accompanied by perhaps ten other minerals, to extraordinarily complex uranium-bearing titanites, niobates, and tantalates containing rare earths and many other metallic elements.

Some pitchblende ore compounds may have in excess of 40 elements present from which uranium must be isolated. Most uranium deposits vary in composition, resulting in a constant variation in the composition of the starting materials. Such variations tend to be minimized by stockpiling methods. Because of this fact there have been many ad hoc procedures incorporated to meet special chemical situations all over the world. Highly specialized methods have no place in this publication. General features common to most extraction procedures will, however, be relevant. All common methods will include the following steps: (1) pre-concentration of the ore; (2) roasting or calcination to remove clays and carbonaceous materials, e.g., to increase solubility and improve the extraction; (3) a leaching operation to convert the uranium into an aqueous form; and (4) recovery of the uranium from the loaded leach liquors by ion exchange, direct precipitation, or solvent extraction. A brief summary of some special methods used for recovery of by-product uranium is mentioned toward the end of this section. The product of these operations is a high-grade concentrate, which is usually further purified at a site other than the uranium mill.

The element uranium is strongly electropositive, resembling aluminum and magnesium in this respect; consequently uranium metal cannot be prepared by reduction with hydrogen. Uranium metal has been prepared in a number of ways: reduction of uranium oxide with strongly electropositive elements, such as calcium, electrodeposition from molten salt baths (direct electrolytic reduction), thermal decomposition, decomposition of uranium halides (van Arkel de Boer "hot wire" method), and reduction of uranium halides (UCl_3, UCl_4, UF_4) with electropositive metals (Li, Na, Mg, Ca, Ba) [1]. Uranium metal is manufactured by the metallothermic reduction of uranium tetrafluoride (UF_4, often called green salt)

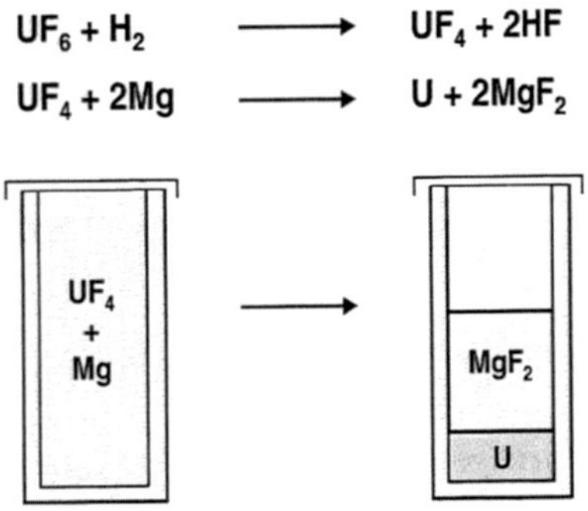

Fig. 1.1 UF_4 reduction process

with a reactive metal, usually magnesium. The enriched or the depleted UF_6 product of the isotopic separation process can either be converted to oxide (enriched oxide pellets for assembly into nuclear fuel rods) or reduced to metallic uranium, as illustrated in Fig. 1.1. The initial reduction step involves reaction of UF_6 with hydrogen to form UF_4. The UF_4 is then reduced to metallic uranium by elevated temperature reaction with magnesium or other alkali or alkaline earth metals. This produces liquid uranium covered with a blanket of MgF_2 slag. After the reaction has gone to completion the reaction vessel cools and the uranium and slag solidify. Typical production runs produce anywhere from approximately 4 kg to greater than 50 kg of metal product in the form of a fused button depending on enrichment levels.

The resulting uranium "derbies" are removed and cleaned. Casting is done in vacuum-induction furnaces using graphite molds and crucibles. The molds are coated with a thin ceramic layer (typically rare-earth oxide) to minimize the carbon contamination of the cast products. They can then be vacuum remelted, cast into shapes of interest, or formed by a variety of standard metal working processes. Alloying elements are sometimes added during the remelting process for purposes to be discussed later. Large-scale production of uranium metal requires elevated temperature where the high reactivity of uranium with most common refractory materials and metals makes the selection of container vessels a difficult problem.

Uranium metal can be melted and fabricated using the same methods that are employed with most other metals. It can be hot worked by processes such as rolling, forging, and extrusion; cold worked by rolling, swaging, or a variety of sheet metal forming processes; and welded and machined by most standard methods. It is very prone to oxidation, however, so melting, welding, and heat treating are usually done in inert environments or in vacuum. Similarly, finely divided uranium is pyrophoric, requiring special precautions to be taken in machining. The hazards associated with ingestion or inhalation also require special precautions to be taken in work areas where finely divided uranium metal or oxides are likely to be present.

Enriched uranium metal is used primarily for nuclear weapon and energy applications. Depleted uranium (DU) is used for applications requiring very high density, such as counterweights, shields, and kinetic energy penetrators (antitank rounds).

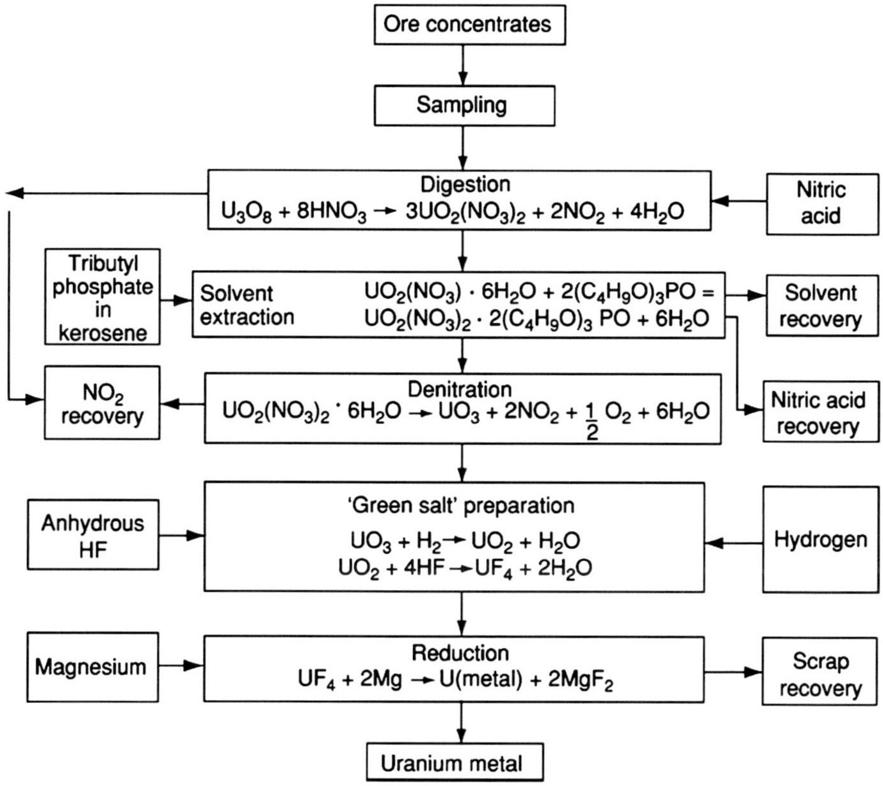

Fig. 1.2 Flow sheet for the production of uranium metal by reduction of UF_4 with magnesium [1]

The density of unalloyed uranium is 19.1 g/cm^3. This is more than twice the density of steel, and 68 % denser than lead. Among ductile materials, only a few precious metals, such as gold and platinum, exhibit comparably high densities. Tungsten is similar in density to uranium, but is brittle in elemental form. Tungsten powders are frequently liquid phase sintered with ductile metal binders. The resulting "tungsten heavy metal" composites have somewhat lower densities, but more useful engineering properties, and compete with DU for kinetic energy penetrator applications.

Chemical recovery of by-product uranium is necessary for both economic reasons and to maintain uranium in a suitable form. The major contributor to the waste stream is uranium metal fabrication operations that generate uranium scrap. This involves the casting, machining, rolling, and forming operations.

The chemical recovery system involves treatment of the low-equity scrap (Fig. 1.2), and it is generally derived from combustible materials (that are subsequently converted to contaminated uranyl nitrate solutions by means of several unit operations including dissolution, acidification, and evaporation). The low-equity scrap is typically processed for enriched uranium and consists of finely divided and highly oxidized scrap (machine turnings or skull material from castings).

The uranyl nitrate produced from low-equity scrap is purified in one or two stages depending on the setup. If two stages are used, the first is called primary extraction and uses dibutyl carbitol or other organics as the extraction agent. Dibutyl carbitol will extract nearly all of the enriched uranium from the process stream, but will also extract some contaminants. The raffinate from the primary extraction typically has a minimum level of uranium and can be sent directly to waste treatment. The second stage, known as the secondary extraction system, uses tri-butyl phosphate in kerosene as the extracting agent. The tri-butyl phosphate is highly selective toward uranium and rejects almost all contaminants, but leads to significant uranium losses in the raffinate that must be recycled back to primary extraction.

High-equity scrap generally consists of uranium oxide generated in the skull of the casting operations. The skull oxide is dissolved in nitric acid and sent directly to secondary extraction. The product leaving secondary extraction is evaporated to high-purity molten uranyl nitrate. The molten uranyl nitrate is heated to a trough-type de-nitrator and thermally decomposed to uranium trioxide. The uranium trioxide is transferred to the oxide conversion facility that consists of two fluidized beds. The first bed converts the uranium trioxide to uranium dioxide with hydrogen gas. The second fluid bed converts the uranium dioxide to UF_4 with anhydrous hydrogen fluoride gas. The green salt is subsequently reduced to uranium metal by a reactive metal. The product is a high-purity uranium button. The by-product is a slag consisting primarily of contaminated magnesium fluoride. This by-product is crushed, calcined, and dissolved in acidified aluminum nitrate solution and processed through primary extraction.

1.3 Electronic Structure of Uranium

The actinide series of elements has a complex electronic relationship wherein the 5f, 6d, 7s, and 7p orbitals are all at relatively comparable energies and electron shifting from the 5f to 6d can occur spontaneously. The electronic orbitals also overlap spatially, implying that actinide bonding can involve any or all of them. Regularly determining whether the complex bonding is covalent or ionic is impossible.

Differing from lanthanide chemistry is the contribution of covalent-hybrid bonding involving the 5f electrons. Relative to the 5s and 5p orbitals, the 5f orbitals have a much greater spatial extension relative to the 6s and 6p orbitals than that of the 4f to the 5s and 5p orbitals. Hybrid bonding has all been shown experimentally, most often involving the 5f overlap with fluorine ions and the covalent-ionic bonding that occurs to form UF_3 and UF_6 compounds [2].

Uranium is very chemically reactive and bonds directly with most elements. Given the reactiveness of uranium, combined with the overlap of electronic bonding modes, a wide range of intermetallic compounds are formed while extensive solid solutions are unusual. Without delving into further detail, the knowledge of this situation that occurs with uranium and other actinides sheds light into the complexity of uranium kinetics and compatibility studies.

1.4 Uranium Metallurgy

The light actinide metals (Th–Pu) have always occupied unique placement in the periodic table because of their wandering 5f electrons and intricate crystallographic phases at high pressures and high temperatures [2]. The contributions of 5f electrons to bonding give rise to low-symmetry crystal structures and a narrow 5f-band related to the Fermi-energy level.

Uranium metal has been the subject of intense experimental and theoretical investigation. Its crystallography and properties have been characterized to pressures as high as 100 GPa and temperatures up to 1,500 K. Solid uranium exhibits three polymorphic forms: gamma-phase (body-centered cubic (BCC)) above 775 °C, beta-phase (body-centered tetragonal) between 668 and 775 °C, and alpha-phase (orthorhombic) below 668 °C (see Table 1.2). The mechanical properties of these three phases are described in Table 1.3 of the following section. Gamma-phase is soft and ductile, whereas beta-phase is brittle. The properties of alpha-phase vary substantially with temperature, as will be discussed in more detail.

Table 1.2 Allotropes of uranium including ranges of temperature stability and lattice arrangements

Allotrope	Temperature range of stability (°C)	Space lattice and space group	Unit cell dimensions (Å)	Atoms per unit cell
α	<668	Orthorhombic	$a = 2.854$, $b = 5.865$, $c = 4.955$	4
β	668–775	Tetragonal P42 nm or P42 mnm	$a = 5.656$, $b = c = 10.76$	30
γ	775–1,130 °C (T_m)	Body-centered cubic	$a = 3.524$	2

Table 1.3 Uranium allotropes and respective standard mechanical properties

Allotrope	Structure	Temperature range (°C)	Mechanical behavior
		<200	Decreasing ductility, decreasing slip, increasing twinning, ductile-to-brittle transition
α	Orthorhombic	200–400	Ductile, deforms primarily by slip, strain hardens, does not recrystallize
		>400	Soft and ductile, deforms readily by slip, dynamically recrystallizes
β	Tetragonal	668–775	Brittle
γ	Body-centered cubic	764–1,130	Soft and ductile, deforms readily by slip, dynamically recrystallizes

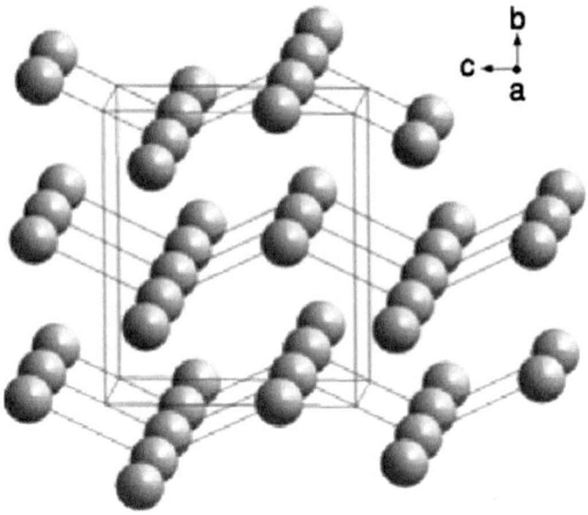

Fig. 1.3 The structure of α-uranium is a layered structure with *ac*-layers perpendicular to the *b*-axis with clear strongly bonded and corrugated (010) planes. The uranium–uranium distances in the ac layer are (2.80 ± 0.05) Å and between the layers 3.26 Å [1]

The α-structure consists of corrugated sheets of atoms, parallel to the *ac*-plane and perpendicular to the *b*-axis. Within the sheets the atoms are tightly bonded, whereas the forces between atoms in adjacent sheets are relatively much weaker (Fig. 1.3). This arrangement is highly anisotropic and resembles the layer structures of arsenic, antimony, and bismuth.

Both the interstitial sites in α- and β-U phases are too small to be occupied by interstitials. These facts are in agreement with the lack of primary solid solubility of the interstitial impurity atoms carbon, nitrogen, and oxygen in the α- and β-U phases and the very limited substitution solid solubility of certain transition metals in β-U [3]. The limited solubility tends to invoke complex formation; if suitable complexes exist, these could provide interstitial positions for the 3d impurity atoms such as iron, manganese, nickel, and cobalt in the α- and β-U phases. Because of strong U–M bonding both in α-U and β-U, the presence of solute atoms M will transform the geometry of their neighboring uranium atoms into complexes which lead to anisotropy.

The physical properties of the α-phase are a reflection of its structure, e.g., the strongly anisotropic coefficient of thermal expansion. The large anisotropy is obvious at room temperature and increases with temperature [4]. The average values of the thermal expansion coefficient over the temperature range 25–325 °C are 26.5, −2.4, and 23.9×10^{-6}, respectively, along *a*, *b*, and *c*. A chemical consequence of the unique orthorhombic structure of α-uranium is that the formation of solid solutions with metals of the common structure types is severely restricted [1, 5].

The β-phase of uranium exists between 668 and 775 °C; it has a complex structure with six crystallographically independent atoms in the tetragonal unit cell [6]. The space group is P42/mnm, P42nm, or P4n2, with unit cell parameters $a = 5.656$ Å and $b = c = 10.759$ Å. The tetragonal lattice is a stacked layer structure with layers parallel to the *ab* plane of the unit cell at c/4, c/2, and 3c/4.

Additional high-precision measurements are required to solve the structure completely [1, 6].

The γ-phase of uranium is formed at temperatures above 775 °C; it has a BCC structure with the cell parameter $a = 3.524$ Å. The gamma-phase can be stabilized to room temperature by moderate alloying additions, for example 10 wt% Mo, thus facilitating characterization of the gamma-phase in early studies.

1.4.1 Mechanical Properties of Unalloyed Uranium

Alpha-uranium might best be described as a hybrid between a metal and a covalently bonded solid. In face-centered cubic metals each atom is surrounded by 12 equidistant nearest neighbors. This represents the arrangement in which "spherical atoms" can be most closely packed, and is most suggestive of nondirectional "central force field bonding" associated with "ideal metallic behavior."

Accordingly, FCC metals (such as aluminum and copper) are typically ductile at all temperatures and exhibit very low electrical resistivity. In BCC metals, on the other hand, bonding is somewhat more directional, indicating less ideal metallic behavior. Each atom is surrounded by eight nearest neighbors, and then four next-nearest neighbors at somewhat greater distances. Comparing FCC and BCC iron, it can be seen that the eight-nearest neighbors are actually somewhat closer in the BCC arrangement than in FCC, while the four next-nearest neighbors are considerably more distant. The "less ideal metallic bonding" characteristic of BCC metals (such as iron and many of the refractory metals) typically results in higher strength, a ductile-to-brittle transition at lower temperatures, and somewhat higher electrical resistivity. Contrary to the situation with titanium and zirconium, the introduction of small amounts of oxygen or nitrogen does not have an adverse effect on the mechanical properties of the metal.

Alpha-uranium takes this progression one step further. In alpha-uranium the atoms are arranged such that each atom has neighbors at four different distances. Four atoms are even closer than in BCC gamma uranium, whereas eight are farther away. As mentioned previously, the most closely spaced atoms are arranged in "tightly bonded" corrugated planes, with relatively large separations characteristic of much "weaker bonds" between adjacent corrugated planes, as illustrated in Fig. 1.3. This is indicative of a greater component of directional covalent bonding, rather than ideal metallic bonding. Consistent with this, alpha-uranium exhibits still higher electrical resistivity than BCC metals. Alpha-uranium also exhibits relatively high strength and brittle behavior at low temperatures. Fortunately, however, the ductile-to-brittle transition in alpha-uranium is slightly below room temperature, making it a useful engineering material.

The mechanical properties of alpha-uranium vary substantially with temperature, as shown in Fig. 1.4. The tightly bonded corrugated basal planes in alpha-uranium result in a very limited number of slip systems. In fact, room-temperature slip occurs only on these corrugated basal planes parallel to the direction of the

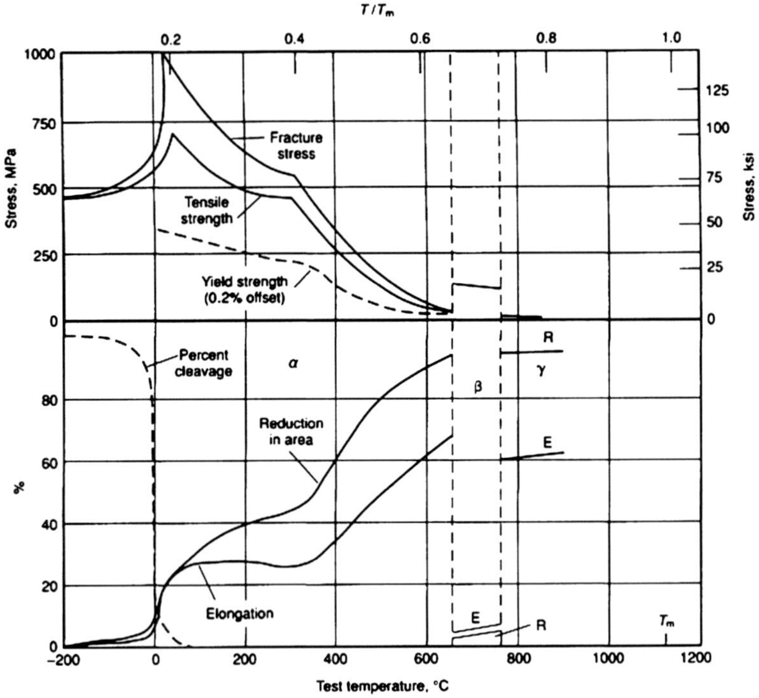

Fig. 1.4 Effect of temperature on the tensile properties of unalloyed uranium [7]

corrugations (the (010)[100] system). This singular slip system would not support macroscopic ductility were it not supplemented by a multiplicity of twinning systems. Combined, however, these provide a modest level of room-temperature ductility.

As temperature increases to ~250 °C and above, additional slip systems become active and the material becomes softer and much more ductile. Hence, while uranium plates can be cold rolled, warm rolling in the 250–350 °C range is much easier. Material that is warm rolled in this temperature range retains elongated grains and substantial work hardening. Previously cold or warm worked plates are sometimes recrystallized at 500–600 °C, but frequently the material is left in the worked condition, as will be discussed later.

Dynamic recrystallization begins in the vicinity of 400 °C, but varies substantially with rolling temperature and reduction, as well as the purity of the material. Ingots are typically heated to ~630 °C in molten salt in preparation for hot rolling. The molten salt prevents excessive surface oxidation, as would occur if the material were heated to this temperature in air. Substantial reductions are possible before the temperature drops below the dynamic recrystallization temperature. Plates are typically reheated in salt if additional amounts of hot working are needed.

Table 1.4 Room-temperature tensile properties of annealed uranium

Form	Heat treatment	YS (MPa)	UTS (MPa)	Elong (%)
Cast	None	200	450	6
Cast	Hydrogen outgassed at 630 °C	185	560	13
Cast	Hydrogen outgassed and beta grain refined at 720 °C	295	700	22
600 °C rolled	Preheated in salt (no hydrogen outgassing)	270	575	12
600 °C rolled	Hydrogen outgassed at 630 °C	270	720	31
600 °C rolled 300 °C finished	Hydrogen outgassed at 630 °C and 550 °C (final)	220	750	49

At temperatures below ~0 °C slip becomes more difficult, even on the preferred (010)[100] system. A transition from ductile to brittle behavior accompanies this decrease in slip and increase in twinning. As in steel, the ductile-to-brittle transition temperature is strongly influenced by microstructure and impurities. Grain refinement suppresses the transition temperature; hence fine-grained wrought materials typically exhibit much better room-temperature ductility than coarse-grained castings. The presence of even small amounts (<1 ppm by weight) of dissolved hydrogen increases the transition temperature, thus decreasing room-temperature ductility. Preheating in molten salt typically introduces hydrogen, so subsequent vacuum outgassing (at 350–630 °C) is often employed to reduce hydrogen content, thereby enhancing room-temperature ductility. Warm working makes the material more tolerant of hydrogen (apparently by providing dislocation hydrogen sinks), thus suppressing the transition temperature and enhancing room-temperature ductility.

Table 1.4 shows typical room-temperature tensile properties of annealed uranium following various types of processing. As-cast material typically exhibits very large grains and contains 1–2 ppm hydrogen. These result in nonuniform deformation and low tensile ductility. Castings are often beta heat treated to refine the alpha grain structure and increase ductility, particularly if these treatments are done in vacuum to also reduce hydrogen content. Alpha-worked and recrystallized uranium contains much finer grains and exhibits substantially higher ductility, particularly when residual hydrogen has been removed by vacuum outgassing. Plates that have been extensively warm rolled exhibit yield and ultimate strengths as high as 650 and 1,200 MPa, respectively [8]. Elongations of 15–20 % can be retained at these warm-worked strength levels, even without hydrogen outgassing. Recrystallization of these previously warm-worked materials actually reduces both strength and ductility unless hydrogen is also removed during the recrystallization anneal.

While unalloyed uranium exhibits yield and ultimate strengths similar to those of mild steel, uranium is substantially less tough. In particular, uranium and uranium alloys exhibit relatively low toughness in tests using notched or pre-cracked samples. There is some indication that this is due to the triaxial stresses in such samples shifting the ductile-to-brittle transition to higher temperatures. In any event, room-temperature Charpy V-notch energies for unalloyed uranium are typically in the 10–25 J range, whereas 20 J is often considered a bare minimum for structural steels. Uranium alloys heat treated to 950 MPa yield strength have a KIC value in the vicinity of 40 MPa $m^{1/2}$, considerably lower than high-strength steels at equivalent strength levels.

1.5 Uranium Alloy Metallurgy

Relatively few elements are metallurgically similar to uranium. Many exhibit liquid-phase immiscibility. Others are soluble in liquid uranium, but form only intermetallic compounds on solidification. Only a few exhibit substantial solid solubility at high temperature: Cr, Mo, Nb, Pu, Re, Rh, Ti, V, and Zr. All of these but Cr, Nb, and V form intermetallic compounds with uranium.

Uranium is often alloyed to improve its very limited resistance to oxidation and corrosion, as well as to increase its strength. As shown in Table 1.5, alloying additions such as titanium, niobium, molybdenum, zirconium, and vanadium are extensively soluble in the high-temperature gamma-phase, much less soluble in the intermediate-temperature beta-phase, and essentially insoluble in the low-temperature alpha-phase. Typical phase diagrams are shown in Fig. 1.5.

Table 1.5 Alloying elements solubility at high temperatures [9]

Phase/structure	Elemental solubility
α (Orthorhombic)	<0.3 %: all
β (Tetragonal)	<1 %: all
γ (BCC)	20–100 %: Mo, Nb, Ti, Zr
	2–10 %: Au, Cr, Pd, Pt, Re, Rh, Ru, V

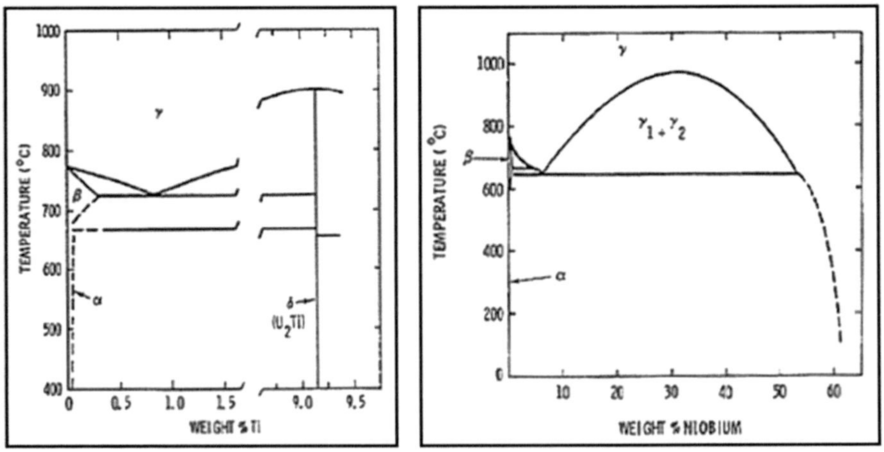

$$\gamma_{0.8\%Ti} \longrightarrow \beta_{0.2\%Ti} + U_2Ti \longrightarrow \alpha_{<0.1\%Ti} + U_2Ti$$

Fig. 1.5 Phase diagrams of uranium with decomposition of gamma-phase [10, 11]

$$\gamma_{0.8\%Ti} \longrightarrow \beta_{0.2\%Ti} + U_2Ti \longrightarrow \alpha_{<0.1\%Ti} + U_2Ti$$

- ***Oxidation and corrosion resistance: similar to unalloyed uranium (little Ti in solid solution)***
- ***Mechanical properties:***

	YS (MPa)	UTS (MPa)	Elong (%)
Unalloyed	220 – 270	575 - 720	12 - 30
U-0.8%Ti	520	1070	10

Fig. 1.6 Slow cooling results in diffusional transformations to two-phase microstructures. Brightfield micrograph of U-0.75Ti cooled from 800 °C at <1 °C/s showing light-etching acicular martensite and darker etching $\alpha + U_2Ti$ produced by the $\gamma \rightarrow \alpha + U_2Ti$ reaction in the interstices between the martensite plates. Shown at 100× [9]

Titanium is most commonly used to increase strength, niobium and molybdenum to increase oxidation and corrosion resistance, and vanadium to refine alpha grain size in castings. Silicon, iron, and carbon are commonly present in concentrations of a few tens to a few hundreds of ppm, but are usually considered impurities, rather than deliberate alloying additions.

Uranium alloys are typically vacuum solution heat treated in the gamma range (~800–850 °C) to dissolve the alloying elements and remove hydrogen. The cooling rate from the solution treatment temperature is a critical parameter that determines subsequent microstructures and properties.

Slow cooling enables the gamma-phase to undergo diffusional decomposition, as described in Fig. 1.6. Hence, slowly cooled alloys typically exhibit two-phase microstructures in which the alloying elements are concentrated in the second phases, leaving the alpha-phase virtually devoid of alloying. The presence of the harder second phases results in modest increases in strength and decreases in ductility. However, oxidation and corrosion resistance remain similar to unalloyed uranium because the alloying elements are not retained in solid solution in the alpha-phase.

Figure 1.7 shows how quenching enables diffusional decomposition of the gamma-phase to be suppressed, thereby producing a variety of metastable microstructures that are supersaturated with alloying elements and exhibit a wide range of useful properties. Figure 1.8 summarizes the changes in microstructure, crystal structure, and strength that occur with increasing alloy content, regardless of whether alloying is done with titanium, niobium, molybdenum, or zirconium. However, the concentrations corresponding to the various transitions vary somewhat for different alloying elements.

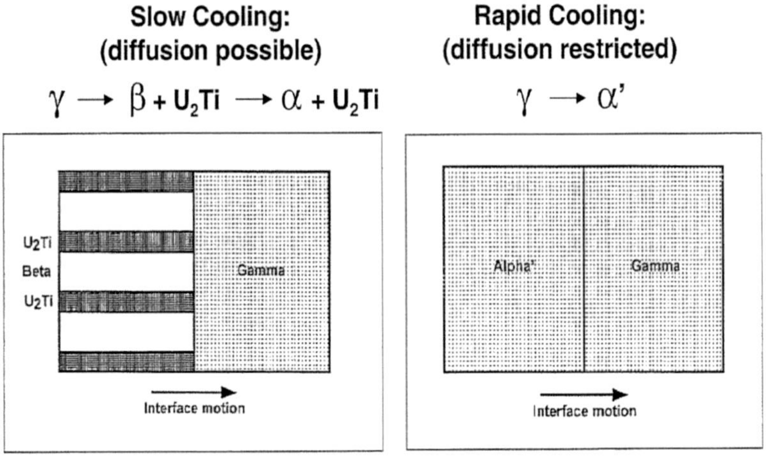

Fig. 1.7 Quenching effects on diffusion that promote martensitic transformation

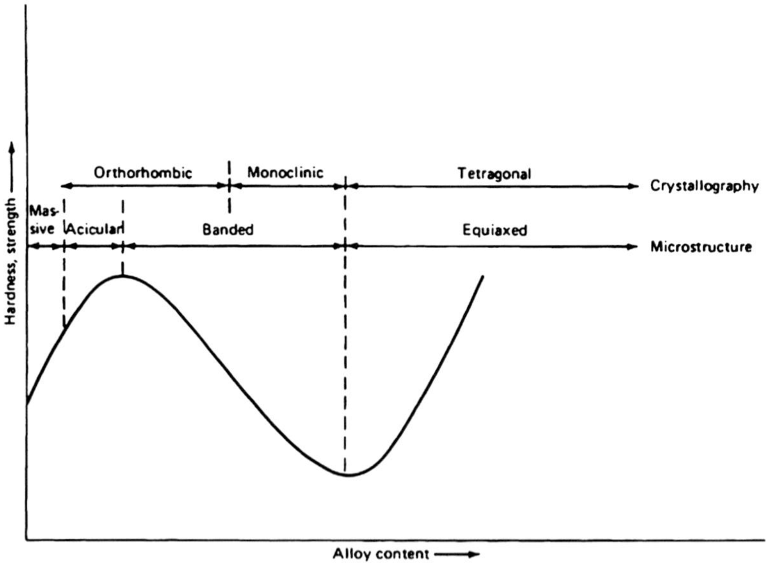

Fig. 1.8 Alloy content influencing microstructure, crystal structure, and properties [9]

Very dilute alloys (<0.3 wt% alloying addition) typically undergo the equilibrium sequence of phase changes on quenching from the gamma-phase. However, the combined rapid cooling 50 % and alloying additions enable the transformations to be suppressed to lower temperatures. This is sometimes exploited by using small vanadium additions to reduce alpha grain size in castings, as illustrated in Fig. 1.9.

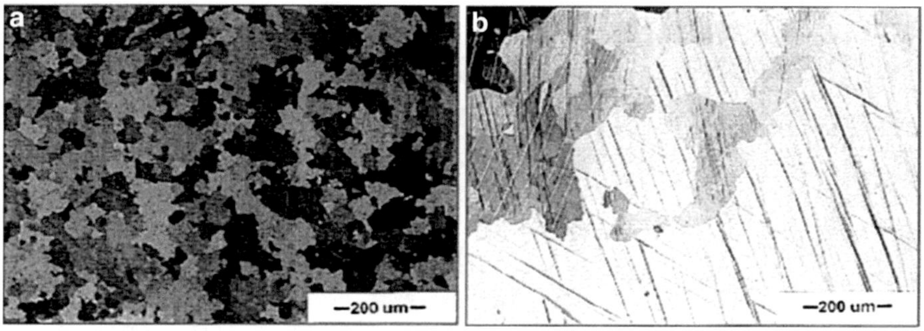

Fig. 1.9 Transformation of very dilute alloys at low temperatures. (**a**) U-0.27 %V, (**b**) unalloyed uranium. This transformation can be exploited to refine grain size in castings

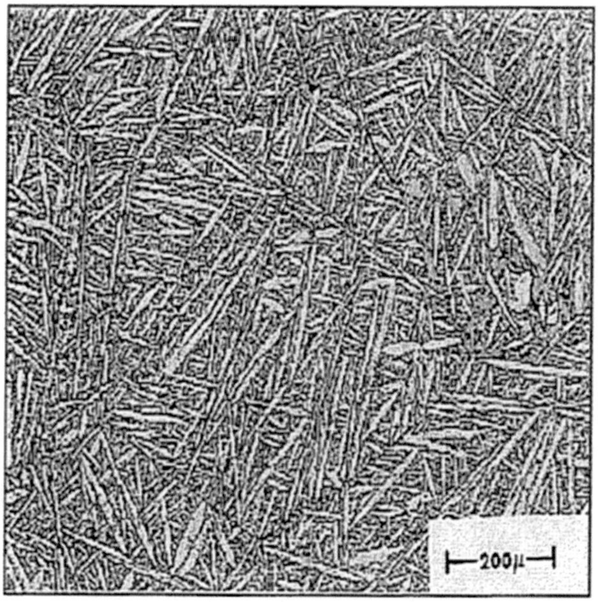

Fig. 1.10 Acicular martensite in U-0.8 % Ti rapidly quenched from 800 °C

In more concentrated alloys the equilibrium phase changes can be completely avoided by quenching. This results in a variety of supersaturated solid solutions being retained at room temperature. In alloys of intermediate concentration the gamma-phase transforms martensitically to one of several supersaturated variants of the alpha-phase. These are analogous to the low-carbon martensite in as-quenched maraging steel.

Quenched alloys containing approximately 0.5–1.5 wt% alloying addition undergo diffusion-less martensitic transformation to acicular martensite, as shown in Fig. 1.10. The alloy atoms retained in solid solution provide a substantial

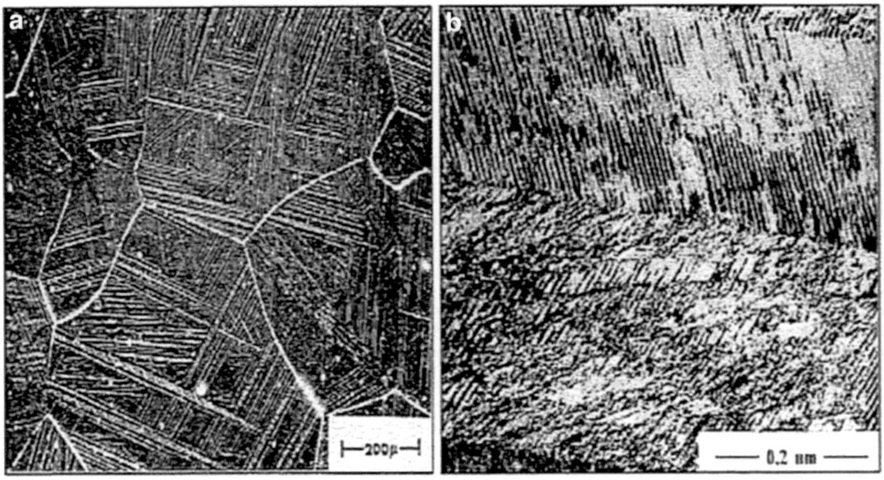

Fig. 1.11 Intermediate concentrations creating martinisitc twin-related parallel plates. (**a**) U-6 % Nb observed with polarized light, (**b**) same sample observed with TEM [10]

component of solid solution strengthening that increases with increasing alloy concentration. Unlike iron–carbon martensite, where the carbon atoms occupy interstitial sites and cause dramatic increases in strength and decreases in ductility, the alloying atoms in uranium martensites occupy substitutional sites. As a result they cause more moderate strengthening and permit good ductility to be retained. For example, as-quenched U-0.75 % Ti has a yield strength of ~650 MPa, three times that of unalloyed uranium, but exhibits 30 % tensile elongation and 50 % reduction in area.

Quenched alloys containing somewhat higher alloy concentrations (for example, 2–6 wt% niobium) also transform martensitically, but the morphology of the martensite is substantially different. These martensites consist of packets of twin-related parallel plates. The individual plates also contain a profusion of internal twins. This microstructure is shown at low and high magnifications in Fig. 1.11. These are frequently termed "banded martensites" to distinguish them from the "acicular martensite" characteristic of lower alloy contents. Complicating matters still more, the lattice angle gradually departs from 90°, causing a slight crystal structure change from orthorhombic to monoclinic.

The morphology of the banded martensites is similar to thermoelastic martensites that was observed in a number of nonferrous systems. Initial plastic deformation in these structures occurs by twin reorientation, rather than slip. When stresses are applied, favorably oriented twin variants grow at the expense of unfavorably oriented ones, as illustrated in Fig. 1.12. The curious decrease in hardness and strength with increasing alloy content is apparently caused by increasing twin boundary mobility, although the details of this are not completely understood. As with other thermoelastic martensites, a shape memory effect occurs in banded uranium martensites, particularly those near the hardness minimum.

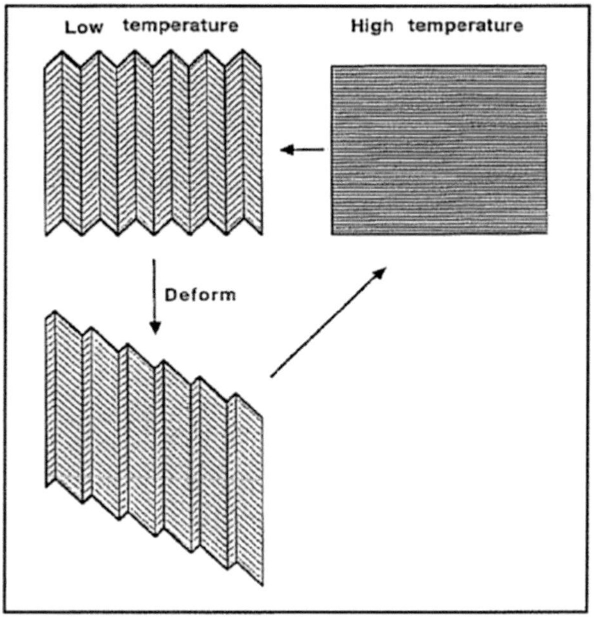

Fig. 1.12 Twin boundary mobility and shape-memory effect

The plastic deformation associated with twin reorientation (approximately 5 % tensile strain in U-6 % Nb) can be completely recovered by heating to the temperature at which the alpha-martensite reverse transforms to the parent phase (a variant of gamma-phase).

Alloys of still higher concentration (above ~6.5 wt% Nb or ~5.5 wt% Mo) retain metastable variants of the gamma-phase when quenched. The martensite start and finish temperatures decrease with increasing alloy content. If the M_s is suppressed to below room temperature, supersaturated variants of the gamma-phase are retained at room temperature. These are analogous to austenitic stainless steels. The primary microstructural features are typically the gamma-phase grain boundaries, as shown in Fig. 1.13. Under some conditions, however, parallel substructural features appear within the gamma grains. These correspond to domains within ordered variations of gamma-phase.

Alloys must be cooled at some critical rate or faster to avoid diffusional transformation and obtain these supersaturated structures, as illustrated in Fig. 1.14. The critical cooling rates vary with alloy content [12]. Relatively dilute alloys require rapid quenching. This restricts the section thicknesses in which they can be effectively heat treated. For example, cooling rates of ~100 °C/s or higher are required to obtain near-fully martensitic microstructures in U-0.75 wt% Ti—this limits the use of this alloy to section thicknesses of ~3 cm or less. Furthermore, increasing the alloy content in relatively dilute alloys actually increases quench rate sensitivity. This occurs because the M_s temperatures in dilute alloys are above the knee of the C-curve for diffusional decomposition (typically ~530 °C). The "critical quenching rate" in these alloys is defined by the intersection of the C-curve with the M_s. Increasing alloy content typically pushes the C-curve to longer

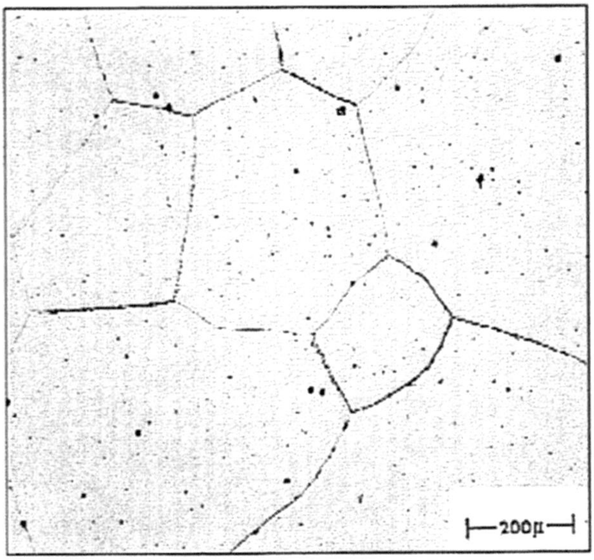

Fig. 1.13 Concentrated alloy consisting of U-10 % Mo showing the typical gamma-phase grain boundaries

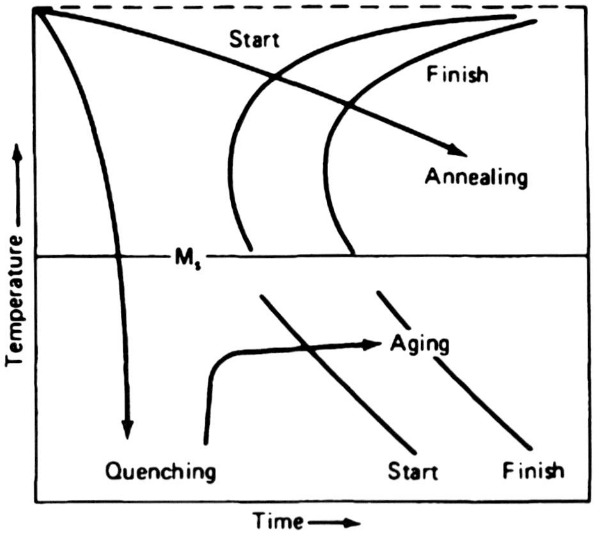

Fig. 1.14 Generalized time–temperature-transformation diagram showing heat treatments employed with uranium alloys. Increasing alloy content shifts diffusional transformations to longer times and decreases martensite transformation temperature. Quenching results in diffusion-less transformation of γ-phase to supersaturated martensites, which can be subsequently age hardened. M_s martensite start temperature [9]

times, but also lowers the M_s temperature. The decrease in M_s pushes the "critical intersection" to a lower temperature on the C-curve where diffusional decomposition occurs in shorter times. The net effect is that higher quenching rates are required to suppress diffusional decomposition prior to the onset of martensite formation, as shown for U-0.6 to 1.5 wt% Ti alloys in Fig. 1.15.

More concentrated alloys require less severe quenching, so they can be processed in thicker sections. For example, U-6 wt% Nb can be cooled at rates as low as ~10 °C/s and still get fully martensitic microstructures. U-10 wt% Mo is so quench-rate insensitive that it does not require artificial quenching—supersaturated

Dilute alloys
(Ms above knee of C-curve):

- Decrease in Ms dominates
- Critical quench rate increases

% Ti	Cooling Rate for 50% Martensite (°C/sec)
0.60	10
0.75	30
1.0	41
1.5	>45

Concentrated alloys
(Ms below knee of curve):

- Increased time for diffusional transformations dominates
- Critical quench rate decreases

%Nb	Cooling Rate for 100% Martensite (°C/sec)
2.3	>45
4.5	14
6.0	8

Fig. 1.15 Alloy content effects with dilute and concentrated alloys [10, 11]

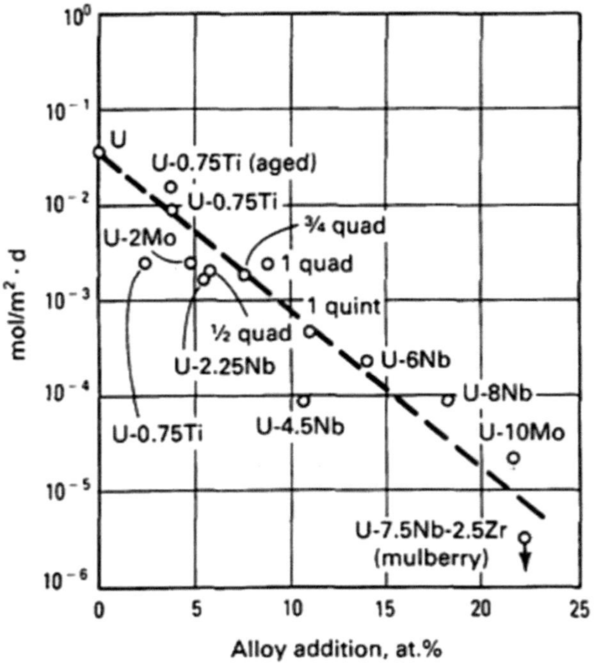

Fig. 1.16 Oxidation and corrosion resistance increase with alloying addition in solid solution for selected uranium alloys [13]

variants of the gamma-phase are retained even after air-cooling. Furthermore, once alloy content is sufficiently high to suppress the M_s temperature below the knee of the *C*-curve, the *C*-curve alone defines the critical quenching rate. As a result, further increases in alloy content now decrease quench rate sensitivity, as shown for U-2.3 to 6.0 wt% Nb alloys in Fig. 1.16. This is similar to the effect typically observed in alloy steels.

Residual stresses frequently pose a problem in alloys that must be rapidly quenched. These stresses are introduced due to substantial thermal gradients that occur during quenching. The magnitude of the problems they cause increases with quench severity and the geometric complexity of the parts being quenched. If sufficiently high and/or complex, these residual stresses can result in distortion during subsequent machining and/or delayed cracking (as will be discussed in more detail later). Relatively dilute alloys, such as U-0.75 wt% Ti, tend to be particularly susceptible to such problems because they require rapid quenching and have high as-quenched yield strengths that can support high residual stresses. On the other hand, U-6 wt% Nb has such a low as-quenched yield strength that it cannot support large residual stresses, even if rapidly quenched—as a result it is fairly immune to residual stress problems. Highly concentrated alloys, such as U-10 wt% Mo, have sufficient yield strengths to support large residual stresses, but are quench rate insensitive that they can frequently be air-cooled—quenching of such alloys unnecessarily invites delayed cracking and should be avoided.

The weldability of uranium alloys is closely related to the issues of quench rate sensitivity and residual stress magnitude. Dilute alloys, such as U-0.75 wt% Ti, are very difficult to weld for two reasons. First, their high as-quenched yield strengths permit high residual stresses to develop in the vicinity of the welds. Second, the cooling rates following welding are typically subcritical, thus permitting the formation of brittle non-martensitic microconstituents in the weld and heat-affected zones. This combination of high residual stresses and undesirable microstructures virtually assures that cold cracking will occur either immediately or worse yet, delayed. Conversely, alloys near the hardness minimum, such as U-6 wt% Nb, are relatively easy to weld. Low quench rate sensitivity typically permits fully martensitic microstructures to form, and their low as-quenched yield strength prevents the development of high residual stresses. Highly concentrated alloys, such as U-10 wt% Mo, are again more difficult to weld. Low quench rate sensitivity prevents undesirable diffusional decomposition, but substantial thermal gradients are unavoidable in the vicinity of the weld and the relatively high yield strengths permit the development of high residual stresses that promote delayed cracking.

Significant changes in chemical properties also occur as alloying elements are retained in supersaturated solution. First, general corrosion and oxidation resistance increase. To a first approximation, corrosion and oxidation resistance increase monotonically with increasing alloy content, as shown in Fig. 1.16. This figure shows the rate at which hydrogen is produced by the reaction of water with the metal surface—this is a frequently used measure of general corrosion resistance in uranium alloys. Relatively concentrated alloys, such as U-6 wt% Nb and U-10 wt % Mo, are often used in applications requiring good corrosion resistance.

Second, the species that promote environmental embrittlement and cracking change with increasing alloy content, as shown in Fig. 1.17. Embrittlement in what would normally be considered very benign environments is a significant problem in engineering applications of high-strength uranium alloys. Unalloyed uranium and dilute uranium alloys are embrittled primarily by internal hydrogen and by atmospheric moisture. Since hydrogen is formed when moisture reacts with

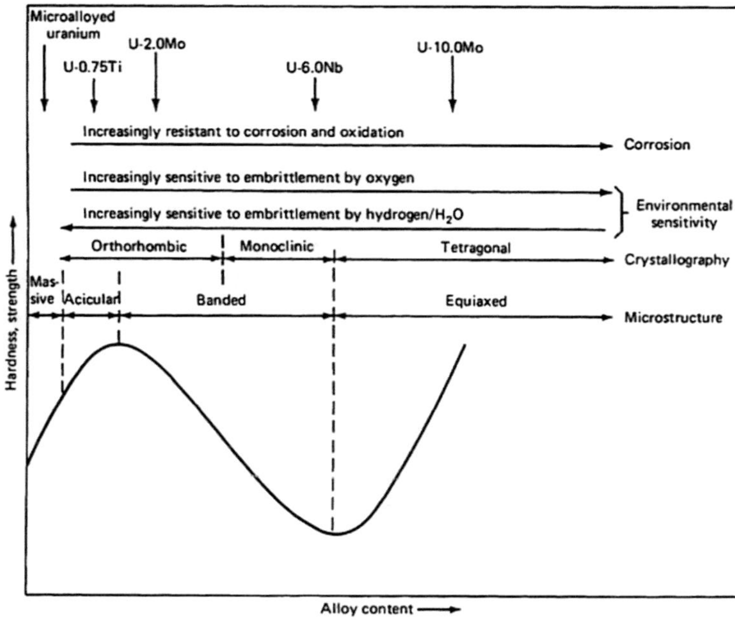

Fig. 1.17 Effects of alloy composition on microstructure, crystal structure, and properties of quenched uranium alloys [9]

the uranium surface, some researchers believe that these two effects are manifestations of the same underlying phenomena. Others, however, believe that they are distinct. Regardless of mechanistic details, their macroscopic results are similar. In unalloyed uranium this takes the form of reduced tensile ductility in humid environments. In higher strength alloys, such as U-0.75 wt% Ti, time-dependent crack formation and growth can also occur in moist air. The quenching-induced residual stresses present in U-0.75 wt% Ti frequently provide sufficient stress, thus resulting in delayed cracking in complex geometry parts or welded assemblies.

Concentrated alloys, on the other hand, are much less sensitive to either internal hydrogen or atmospheric water. Instead, environmentally assisted crack growth is promoted by atmospheric oxygen. Under some circumstances, the wedging effect produced by the low-density oxide forming on the freshly exposed uranium surfaces near the crack tip provides sufficient stress to drive continued crack growth, even in the absence of residual or applied stresses. The presence of atmospheric moisture actually reduces cracking problems in these concentrated alloys, apparently because moisture condenses in the cracks, thus limiting crack tip exposure to atmospheric oxygen.

In addition to changes in microstructure, crystallography, general corrosion resistance, and environmental sensitivity, the supersaturated microstructures obtained by quenching are also amenable to age hardening. This can be exploited

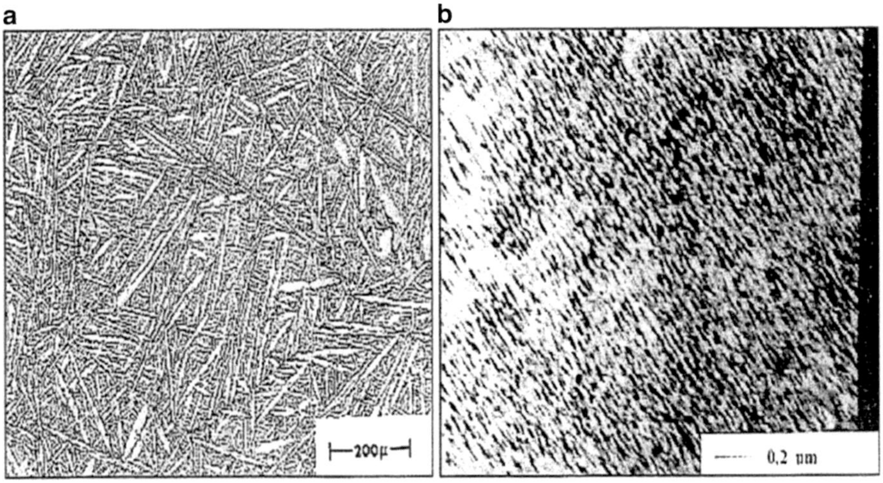

Fig. 1.18 Age-hardened U-0.75 % Ti. (**a**) Observed via optical metallograph, (**b**) same sample observed with TEM

to obtain very-high-strength uranium, analogous to age-hardened maraging steels. Age hardening occurs when supersaturated microstructures are heat treated at temperatures up to ~450 °C, resulting in increasing strength and decreasing ductility. Age-hardened alloys retain their enhanced corrosion resistance, as the alloying additions remain mostly in solid solution. Age hardening in U-0.75 wt% Ti occurs by the formation of very fine precipitates. Figure 1.18 shows a light micrograph of age-hardened U-0.75 wt% Ti—the appearance at this magnification is identical to that of the as-quenched martensite. Transmission electron microscopy reveals the fine precipitates of U_2Ti that result in strengthening. Overaging occurs by cellular decomposition at temperatures above ~450 °C, as shown in Fig. 1.19. This results in substantial decreases in strength. Corrosion resistance also decreases on overaging, as the alloying additions are no longer retained in solid solution. The degree of strengthening that can be obtained by age hardening increases with increasing alloy content, as shown in Fig. 1.20. The properties that can be obtained by various degrees of age hardening of U-0.75 wt% Ti are shown and compared with those of unalloyed uranium in Table 1.6.

More concentrated alloys, such as U-6 wt% Nb and U-10 wt% Mo, can also be age hardened. The phenomenology is similar—increasing strength at low aging temperatures due to very-fine-scale microstructural changes, and overaging by cellular decomposition at higher aging temperatures. The microstructural details of the aging phenomena appear to be somewhat different in these non-acicular microstructures, however. Discrete precipitate particles do not seem to form during aging. Instead, transmission electron microscopy reveals diffuse contrast effects that develop as aging proceeds. Recent work appears to confirm that these correspond to spinodal decomposition [15]. This apparently accounts for the age hardening that occurs in these more concentrated alloys.

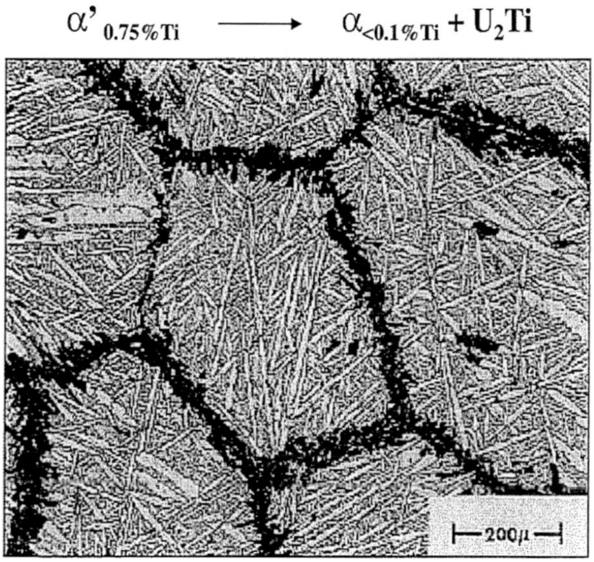

Fig. 1.19 Overaging of U_2Ti, resulting in cellular decomposition [11]

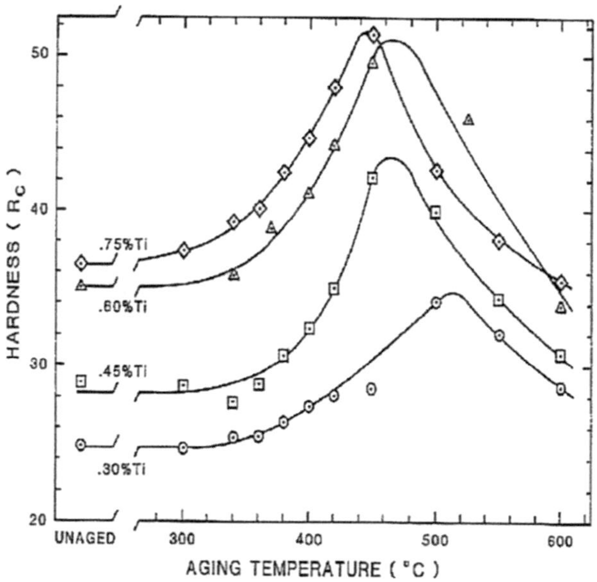

Fig. 1.20 Age hardening of various U-Ti alloys ranging from 0.30 to 0.75 % Ti varied with temperature [14]

Table 1.6 Age hardening comparisons with strength and ductility measurements

Material/heat treatment	YS (MPa)	UTS (MPa)	Elong (%)	Corrosion resistance
Unalloyed, α-rolled	270	720	31	Poor
U-0.75 % Ti, as quenched	650	1,310	31	Moderate
U-0.75 % Ti, partially aged	965	1,565	19	Moderate
U-0.75 % Ti, fully aged	1,215	1,660	2	Moderate
U-0.75 % Ti, overaged	585	1,140	10	Poor

Table 1.7 Typical mechanical properties of uranium and uranium alloys

Material and treatment	YS (MPa)	UTS (MPa)	Elong (%)	Corrosion resistance
As-cast U	200	450	6	Poor
Beta refined and H-outgassed	295	700	22	Poor
Annealed U-plate	270	720	31	Poor
Warm rolled U-plate	650	1,200	16	Poor
U-0.75 % Ti Q and aged	965	1,565	19	Fair
U-6 % Nb quenched	160	825	34	Excellent
U-10 % Mo air cooled	900	930	9	Excellent

Quenching-induced residual stresses are partially relieved during aging, but the sensitivity to environmental cracking is also increased. As a result, susceptibility environmental cracking driven by quenching stresses actually increases when alloys are age hardened. Overaging results in the complete elimination of residual stresses. Unfortunately, relatively coarse particles of U_2Ti tend to form along the prior martensite plate boundaries during overaging of U-0.75 wt% Ti—these result in relatively low ductility and toughness in the overaged condition. Alloys with banded martensitic microstructures, however, appear to be free from this problem, and exhibit better combinations of strength and ductility in substantially overaged conditions.

Table 1.7 provides a brief summary of the properties of unalloyed uranium and commonly used uranium alloys in a number of conditions.

1.6 Nuclear Properties of Uranium

Table 1.8 gives a compilation of the nuclear properties of all known isotopes of uranium. Only three of these isotopes occur naturally, U^{238} (99.275 % abundance), U^{235} (0.720 % abundance), and U^{234} (0.0005 % abundance). The U^{238} and the U^{234} isotopes are in secular equilibrium, the phenomena that the amount of U^{234} being formed by the radioactive decay of U^{238} at the same rate it is disappearing by alpha decay. Tables 1.9 and 1.10 show the radioactive decay schemes for U^{235} and U^{238}, respectively.

1.7 Solid-State Uranium Corrosion

Typically, trivalent oxidation states are the most stable for all lanthanides. This oxidation state may be the case for some actinides, but six is the most stable state for uranium. This behavior can be explained by the fact that the energies of the 5f and 6d levels are close together, and the chemical binding energies have the same order of magnitude. As a consequence, the electronic configuration and the oxidation

Table 1.8 Nuclear properties of uranium isotopes [16–18]

Mass number	Half-life	Mode of decay	Q value (MeV)	Method of production
217	16 ms	α	8.160	$^{182}W(^{46}Ar,5n)$
218	6 ms	α	8.786	$^{197}Au(^{27}Al,6n)$
219	42 ms	α	9.860	$^{197}Au(^{27}Al,x)$
222	1.4 μs	α	9.500	$W(^{40}Ar,xn)$
223	15 μs	α	8.941	$^{208}Pb(^{20}Ne,5n)$
224	0.94 ms	α	8.619	$^{208}Pb(^{20}Ne,4n)$
225	69 ms	α	8.014	$^{208}Pb(^{22}Ne,5n)$
226	0.35 s	α	7.715	$^{232}Th(\alpha,10n)$
227	1.1 months	α	7.211	$^{232}Th(\alpha,9n)$
				$^{208}Pb(^{22}Ne,3n)$
228	9.1 months	$\alpha > 95\ \%$	6.804	$^{232}Th(\alpha,8n)$
		$EC < 5\ \%$		
229	58 months	EC—80 %	6.475	$^{230}Th(^{3}He,4n)$
		α—20 %		$^{232}Th(\alpha,7n)$
230	20.8 days	α	5.992	^{230}Pa daughter
				$^{231}Pa(d,3n)$
231	4.2 days	α	5.577	$^{230}Th(\alpha,3n)$
		ε	0.382	$^{231}Pa(d,2n)$
232	68.9 years	α	5.413	$^{232}Th(\alpha,4n)$
233	1.592 E^5 years	α	4.908	^{233}Pa daughter
234	2.455 E^5 years	α	4.859	Nature
235	7.038 E^8 years	α	4.678	Nature
236	2.342 E^7 years	α	4.573	$^{235}U(n,\gamma)$
237	6.75 days	β^-	0.518	$^{236}U(n,\gamma)$
				^{241}Pu daughter
238	4.468 E^9 years	α	4.269	Nature
239	23.45 months	β^-	1.263	$^{238}U(n,\gamma)$
240	14.1 h	β^-	0.400	^{244}Pu daughter
242	16.8 months	β^-	1.200	$^{244}Pu(n,2pn)$

states are strongly affected by the chemical environment, magnetic and electric fields, temperature, and pressure. As a result, both 5f and 6d electrons are assumed to join in on the formation of a chemical bond [19].

A large number of stable oxide phases occur. Most exist over a range of stoichiometries. Karkhanavala [20] and Willis [21] long ago demonstrated that the range of non-stoichiometric compounds of UO_2 are created by the penetration and accommodation of excess oxygen in channels created between cations along the $\langle 111 \rangle$ and $\langle 110 \rangle$ directions. UO_2 will naturally convert to U_3O_8 at ambient temperatures in the presence of oxygen.

The electrical properties of the uranium oxide behave similar to p-type or n-type semiconductors. In particular, hyperstoichiometric UO_2 is a p-type extrinsic semiconductor up to a temperature of about 1,100 K. Due to deviations from stoichiometric composition, the electrical conduction arises from positive holes. Oxide films may or may not be oriented when growing on textured surfaces. Orientation

Table 1.9 Nuclear decay chain and properties for ^{235}U [16, 17]

Nuclide	Half-life	α *Q*-value (MeV)	β *Q*-value (MeV)	Product of decay
^{239}Pu	2.41 E^4 years	5.244		^{235}U
^{235}U	7.04 E^8 years	4.678		^{231}Th
^{231}Th	25.52 h		0.391	^{231}Pa
^{231}Pa	3.27 E^4 years	5.150		^{227}Ac
^{227}Ac	21.7 years	5.042 (1.38 %)		^{227}Th
^{227}Ac	21.7 years		0.045 (98.62 %)	^{223}Fr
^{227}Th	18.68 days	6.147		^{223}Ra
^{223}Fr	22 months		1.149 (99.9 %)	^{223}Ra
^{223}Fr	22 months	5.340 (6.0E^{-3} %)		^{219}At
^{223}Ra	11.43 days	5.979		^{219}Rn
^{219}At	11.43 days	6.275 (97 %)		^{215}Bi
^{219}At	11.43 days		1.700 (3 %)	^{219}Rn
^{219}Rn	3.96 s	6.946		^{215}Po
^{215}Bi	7.6 months		2.250	^{215}Po
^{215}Po	1.78 ms	7.527 (99.999 %)		^{211}Pb
^{215}Po	1.78 ms		0.715 (0.0002 %)	^{215}At
^{215}At	0.1 ms	8.178		^{211}Bi
^{211}Pb	36.1 months		1.367	^{211}Bi
^{211}Bi	2.14 months	6.751 (99.72 %)		^{207}Tl
^{211}Bi	2.14 months		0.575 (0.28 %)	^{211}Po
^{211}Po	0.52 s	7.595		^{207}Pb
^{207}Tl	4.79 months		1.418	^{207}Pb
^{207}Pb	Stable			

Table 1.10 Nuclear decay chain for ^{238}U [16, 17]

Nuclide	Half-life	α *Q*-value (MeV)	β *Q*-value (MeV)	Product of decay
^{238}U	4.468 E^9 years	4.270		^{234}Th
^{234}Th	24.1 days		0.273	^{234m}Pa
^{234m}Pa	1.16 months		2.271 (99.84 %)	^{234}U
^{234m}Pa (IT)	1.16 months		0.074 (0.16 %)	^{234}Pa
^{234}Pa	6.70 h	2.197		^{234}U
^{234}U	2.455 E^5 years	4.859		^{230}Th
^{230}Th	7.538 E^4 years	4.770		^{226}Ra
^{226}Ra	1,602 years	4.871		^{222}Rn
^{222}Rn	3.823 days	5.590		^{218}Po
^{218}Po	3.10 months	6.115 (99.98 %)		^{214}Pb
^{218}Po	3.10 months		0.265 (0.02 %)	^{218}At
^{218}At	1.5 s	6.874 (99.90 %)		^{214}Bi
^{218}At	1.5 s		2.883 (0.10 %)	^{218}Rn
^{218}Rn	35 ms	7.263		^{214}Po
^{214}Pb	26.8 months		1.024	^{214}Bi
^{214}Bi	19.9 months		3.272 (99.98 %)	^{214}Po
^{214}Bi	19.9 months	5.617 (0.02 %)		^{210}Tl
^{214}Po	164 μs	7.883		^{210}Pb
^{210}Tl	22.3 years		5.484	^{210}Pb
^{210}Pb	22.3 years		0.064	^{210}Bi
^{210}Bi	5.01 days		1.426 (99.9999 %)	^{210}Po
^{210}Bi	5.01 days	5.982 (0.0001 %)		^{206}Tl
^{210}Po	138.4 days	5.407		^{206}Pb
^{206}Tl	4.199 months		1.533	^{206}Pb
^{206}Pb	Stable			

may occur when the inherent electrical properties of an oxide given stoichiometry are affected due to the variation in surface energies of different crystallographic oxide planes [19]. As a thin film of oxide thickens on the surface of any of the uranium, compressive stresses are set up between the oxide and the metal because UO_2 occupies 1.97 times as much volume as the metal on a mole-for-mole basis. Cracks will then occur once a critical thickness is reached or the film becomes noncoherent. A coherent, very thin film remains next to the metal under these conditions. As a consequence of the electronic properties of the thin film, the oxidation kinetics is controlled as well as the migration of the ionic or the electronic species. U_4O_9 and U_3O_7 appear from the UO_2 lattice absorbing oxygen, which decreased the electrical conductivity. This decrease in electrical conductivity changes the material from p-type to n-type conduction. U_3O_8 is thought to be an n-type excess metal semiconductor with anion vacancies [22].

1.8 Uranium Oxide

The complexity of uranium compounds, especially that of uranium oxides, is due to the ability of uranium to wholly and simultaneously exist in numerous valence states. At least 13 uranium oxides are known to exist, most notably UO_2, U_3O_8, and UO_3. U_3O_8 is the most stable uranium oxide under ambient conditions. The crystal structures of uranium oxides in the O/U composition range of $2.00 \leq O/U \leq 2.375$, which consist of UO_2 and polymorphs of U_4O_9 and U_3O_7, are very similar to the recognizable fluorite crystal structure. In contrast, the crystal structures of U_3O_8 and many UO_3 phases comprise layered structures, distinguishable from the presence of UO_2^{2+} uranyl groups arranged normal to the layered planes [23]. The phase diagram of the U–O system is shown in Fig. 1.21.

Uranium dioxide is the most studied due to its use as fuel in nuclear power generation. The room-temperature face-centered cubic fluorite crystal structure of UO_2 contains uranium atoms at the face-centered sites, while oxygen atoms occupy the corners of a cube centered inside the unit cell structure. The relatively open structure allows the anionic oxygen atoms to migrate toward the holes, resulting in inconsistent motion of the oxygen atoms [24]. In addition, the holes allow for the additional oxygen atoms to occupy excess lattice room up to a hyperstoichiometric composition of $UO_{2.25}$ without altering the cubic structure.

UO_2 is created by hydrogen reduction of UO_3 or U_3O_8 at a temperature range of 800–1,100 °C [25]. High-purity H_2 gas must be used as O_2 impurities tend to create hyperstoichiometric UO_{2+x} upon cooling below 300 °C. Oxygen diffusion mechanisms change with temperature for stoichiometric UO_2. At low temperatures, thermal oxygen vacancies are the primary contributors to oxygen mobility, whereas thermal interstitial oxygen atoms dominate at high temperatures. In the temperature range of 800–1,800 °C, both defects combine to create significant atomic mobility [26]. UO_2 pellets produced for reactor fuel incorporate this enhanced mobility of oxygen atoms at elevated temperatures approaching 1,700 °C to vastly increase

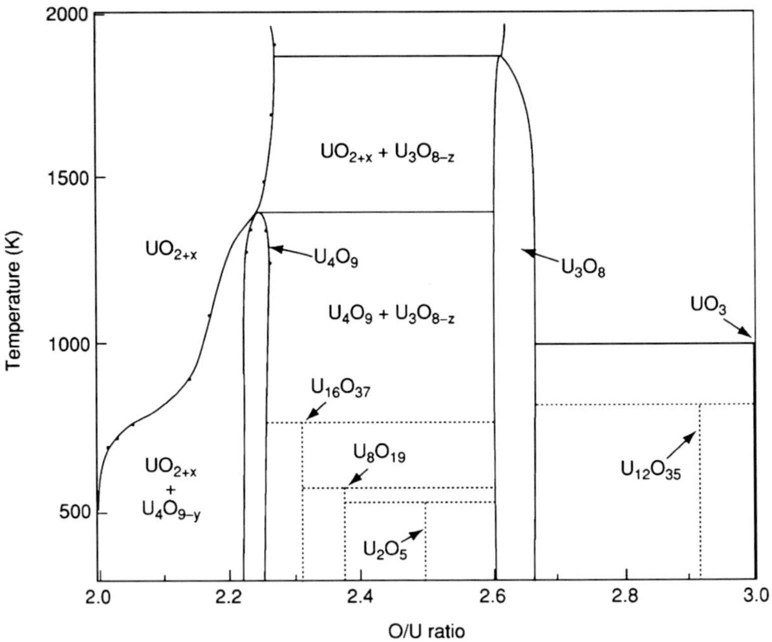

Fig. 1.21 Phase diagram of the U–O system in the region $2.0 \leq O/U \leq 3.0$ [1]

purity and foster near-theoretically dense fuel pellets. When the particle sizes are 0.2–0.3 μm or larger, UO_2 is fairly resistant to oxidation. Highly dense, sintered UO_2 pellets are not oxidized for years due to protection by slightly oxidized thin surface film formed on the large UO_2 grains.

The most stable uranium oxide form, U_3O_8, has two common polymorphs (α and β). The most stable form is α-U_3O_8, an orthorhombic crystal structure where oxygen atoms are all positioned at corners of a pentagonal bipyramid around the central uranium atom. α-U_3O_8 is prepared by the oxidation of UO_2 in air above 800 °C followed by slow cooling. β-U_3O_8 is created by heating α-U_3O_8 above 1,350 °C in air or oxygen, followed by slow cooling over the course of approximately 12–14 days (100 K/day) [23, 27]. β-U_3O_8 reveals a hexagonal, two uranium coordination situations that form two distinct uranium chains along the *c*-axis. Although both phases are closely related, there is a distinct difference in the atom arrangement. In the β-U_3O_8 phase the uranium atoms occupy a single threefold position, whereas at ambient temperatures (α-U_3O_8) they are located at twofold and fourfold positions. This change in crystal structure allows the uranium atoms to have different localized charges. Two other polymorphs of U_3O_8 exist but are only stable during special conditions.

UO_3 is capable of existing in seven unique structures, both crystalline and amorphous: A, α, β, γ, δ, ε, and ζ-UO_3. Their preparation processes are illustrated in Fig. 1.22 as specific details of each polymorph will not be elaborated upon here.

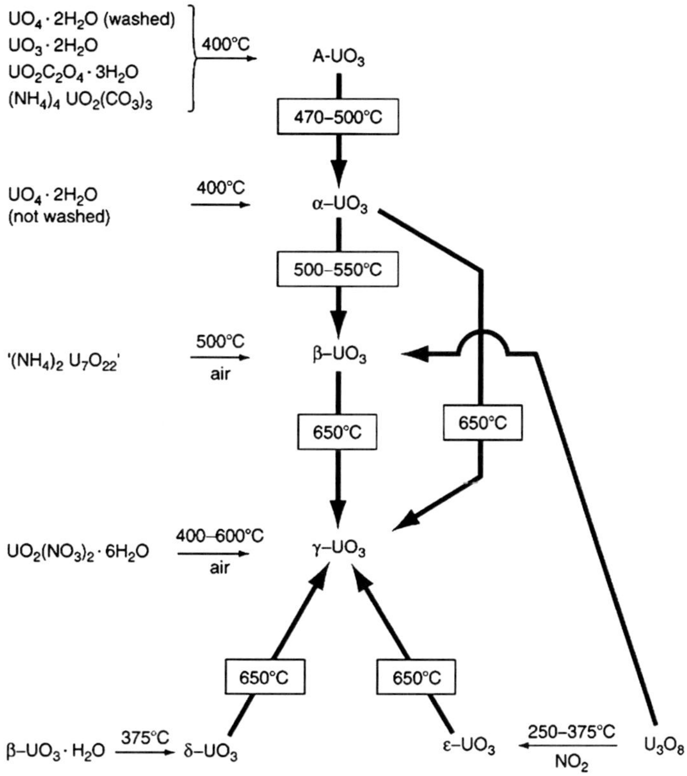

Fig. 1.22 Uranium trioxide polymorphs and means of preparation [1]

Worth mentioning, the amorphous oxide phase is referred to as A-UO_3 and is similar to the structure of uranium-deficient α-U_3O_8, where the absence of uranium atoms is counteracted by the reduction of some U–O bond distances by as much as 25 % [28].

1.9 Uranium Hydride

Uranium hydride is commonly used as a starting material in many uranium production reactions, including preparation of finely divided uranium metal. Hydrogen, deuterium, and tritium gases are regularly stockpiled as the hydride forms: UH_3, UD_3, and UT_3. These gases are liberated when these compounds are heated above their decomposition temperatures.

Uranium hydride is a very reactive compound and its reactivity resembles that of finely divided uranium metal. Uranium hydride ignites spontaneously in air, but

Table 1.11 Commonly utilized uranium hydride reactions

Reagent	Reaction temperature (°C)	Product
O_2	Ignites at room temperature	U_3O_8
H_2O	350	UO_2
H_2S	400–500	US_2
N_2	250	U_2N_3
NH_3	250	U_2N_3
PH_3	400	UP
Cl_2	250	UCl_4
CCl_4	250 (possibility of explosion at 25 °C)	UCl_4
HCl	250–300	UCl_3
HF	200–400	UF_4
Br_2	300–350	UBr_4
HBr	300	UBr_3
CO_2	300	UO_2

gradual oxidation at room temperature at low oxygen partial pressures tends to result in the creation of a protective oxide film on the surface of the hydride, preventing ignition of the hydride. Small quantities of UH_3 react slowly with water, but larger samples react violently and generate large amounts of exothermic heat.

Two crystalline phases of uranium trihydride exist and are known as α-UH_3 and β-UH_3. The first uranium hydride phase to be discovered was the β-form of the trihydride which one can rapidly form if uranium is heated above 250 °C in the presence of hydrogen gas [29]. Neutron diffraction of UD_3 reveals a cubic structure where each hydrogen atom lies in a distorted tetrahedron, surrounded by four uranium atoms at a constant distance of 2.32 Å. There are two kinds of uranium atoms in the UH_3 lattice, each surrounded by 12 hydrogen atoms. Uranium atoms of the first type are surrounded at the corners of an icosahedron by hydrogen atoms, while the uranium atoms of the second type are bounded by groups of three hydrogen atoms, with each forming the face of a different icosahedron. The absence of metallic U–U bonds indicates that the structure is held together solely by U–H interactions [30].

The metastable α-UH_3 is a simple cubic structure with a lattice parameter of 4.160 Å and has U–H atomic distances equal to those of the β-form. Reactions of hydrogen with uranium powder at low temperatures (−80 °C) produce a mixed composition of the α- and β-forms creating the greatest possible percentage of α-UH_3 without diffusive methods. Heating mixtures of known α-/β-compositions above 250 °C result in a total conversion to β-UH_3, while heating of the same mixture at 100 °C causes no difference in the diffraction pattern. Thus, the appearance of α-UH_3 is only by forming more quickly than it decomposes at low temperatures [31].

As previously mentioned, the reactivity of UH_3 is often utilized in the production of other uranium compounds since most reactions of UH_3 are thermodynamically favored, and many cases where reactions do not occur are a result of kinetic barriers [32]. Commonly utilized uranium hydride reactions are listed in Table 1.11.

1.10 Stress Corrosion Cracking

Stress corrosion cracking refers to cracking caused by the simultaneous presence of tensile stress and a specific corrosive medium. Stress corrosion cracking can be distinguished from mechanical failure by rupture at lower loads when exposed to an environment. The types of stress that can attribute to stress corrosion cracking are applied, residual, thermal, and welding. Increase in temperature increases the susceptibility to stress corrosion cracking. During uranium casting, the ingot may experience one or more forms of stress, leading to possible stress corrosion cracking. As oxidation occurs on the surface of the uranium, while cooling from casting, or at ambient temperatures, compressive stresses arise. Stress corrosion cracking may be a small factor when coupled with the chemical reactiveness of liquid uranium, but the corrosion kinetics of uranium are multifaceted. Some characteristics that stress corrosion cracking exhibits are the following [33]:

- Most of the surface is left unattacked.
- Occurs at stresses within a range of typical design stresses.
- Can be intergranular or transgranular.
- Crack is generally perpendicular to the applied stress.

With transgranular failures, the cracks propagate usually in specific crystal planes typically having low indices such as {100}, {110}, {101}, and {210} [34, 35]. Intergranular failures are more common than transgranular, but both can be present in the same system or even in the same failed part. Facing halves matching each other showing little corrosion from a recent failure indicates that fracture is induced primarily by mechanical fracture. Even though primary mechanical fracture is apparent, anodic dissolution does play a significant role.

Koger [33] states that even pure metals can stress corrosion crack, and no unifying mechanism of stress corrosion cracking exists. Jones [36] states that pure metals are more resistant to stress corrosion cracking but not immune.

Crack initiation and propagation are enhanced by any kind of stress riser such as pits, trenches, and other discontinuities. Also, breaks in passive layers initiate cracks. Once a crack has started, tip of the advancing crack has small radius and the attendant stress concentration is great. Lowering stress, eliminating critical environmental species, changing alloys, cathodically protecting, adding inhibitors to solution, and design change are possible solutions to prevent stress corrosion cracking. Design change would include the elimination of positions where small amounts of corrosive constituents can concentrate [33].

1.10.1 Stress Corrosion Cracking of Uranium Alloys

Uranium alloys are very vulnerable to environmentally enhanced cracking, and if coupled with residual stresses, can occur easily. An extensive body of literature is available in this area [33]; hence only a brief overview is presented here.

Very dilute uranium alloys, such as U-0.75 % Ti and U-2.3 % Nb, tend to be most sensitive to moisture due to surface reactions between water and uranium which produces embrittling hydrogen. More concentrate uranium alloys, such as U-10 % Mo and U-7.5 % Nb-2.5 %Zr, tend to be most sensitive to oxygen. Threshold stress intensities for stress corrosion cracking range from 5 to 40 MPa depending on composition, heat treatment, and environmental history. Under extreme conditions, the formation of large-volume corrosion products in cracks will often produce a “wedging” effect, resulting in increased tensile stresses at the crack tip. This effect often causes self-propagating stress corrosion crack growth, even in areas where macro-stresses are locally absent or compressive stresses are present. This can quickly result in total failure of unstressed components.

As is the case with other alloy systems, susceptibility to stress corrosion increases significantly with increasing material strength [37]. However, the age hardening process will decrease the threshold for stress corrosion cracking and decrease the magnitudes of the present residual stresses which provide the driving forces for crack growth.

Acknowledgements The first author, Nathan Gubel, appreciates receiving permission from Dr. Kenneth Eckelmeyer to copy verbatim his entire contribution to the 2004 ASM/MSC conference on uranium and uranium alloys into this chapter. Dr. Eckelmeyer’s contribution makes up well over 50 % of this chapter. Secondly, the chapter authors wish to thank Nate Gubel for granting permissions to reproduce portions of his thesis from Gubel, N. R. (2012) Investigation of Liquid Uranium-Ceramic Corrosion Kinetics for Metal Fuel Processing (Master’s Thesis), Colorado School of Mines, Golden, CO.

References

1. Grenthe I, Drozdzynski J (2011) Uranium. In: Morss L, Edelstein N, Fuger J (eds) The chemistry of the actinide and transactinide elements. Springer, New York
2. Eriksson O, Wills J (1992) Crystal-structure stabilities and electronic structure for the light actinides Th, Pa, and U. Phys Rev B Condens Matter 45(24):13879–13890
3. Kassner M, Peterson D (1995) Phase diagrams of binary actinide alloys. ASM, Materials Park, OH
4. Blanter M, Glazkov V, Somenkov V (2009) Anisotropy of thermal vibrations and polymorphic transformations in lanthanum and uranium. Phys Status Solidi B 246(5):1044–1049
5. Bihan TL, Heathman S, Idiri M, Lander G, Wills J, Lawson A, Lindbaum A (2003) Structural behavior of α-uranium with pressures to 100 GPa. Phys Rev B 67(13):134102
6. Donohue J, Einspahr H (1971) The structure of beta-uranium. Acta Crystallogr B 27:1740–1743
7. Taplin D (1967) The tensile properties and fracture of uranium between 200°C and 900°C. J Aust Inst Met 12:32
8. Eckelmeyer K, Dudder GB (1993) Unpublished research on mechanical properties of warm rolled uranium. Sandia National Lab and Pacific Northwest National Lab
9. Eckelmeyer K (1990) Uranium and uranium alloys. In: Davis JR (ed) Properties and selection of nonferrous alloys and special purpose materials—metals handbook, vol 2, 10th edn. ASM International, Materials Park, OH

10. Eckelmeyer KH, Romiy AD, Weirick LJ (1984) The effect of quench rate on the microstructure, mechanical properties, and corrosion behavior of U-6 Wt. Pct. Nb. Met Trans 15A:1319–1330
11. Eckelmeyer KH, Zanner FJ (1977) Quench rate sensitivity in U-0.75 wt% Ti. J Nucl Mater 67:33–41
12. Eckelmeyer KH (1997) Unpublished research on quench rate sensitivity. Sandia National Lab
13. Weirick L (1987) Corrosion of uranium and uranium alloys. In: Metals handbook-corrosion, vol 13, 9th edn. ASM
14. Ammons A (1976) Precipitation hardening in uranium rich uranium-titanium alloys. In: Burke JJ et al (eds) Physical metallurgy. Brook Hill, Chestnut Hill, MA
15. Eckelmeyer K (1997) Unpublished research on quench rate sensitivity. Sandia National Lab
16. United States National Nuclear Data Center (2012) Interactive chart of nuclides. U.S. Department of Energy, http://www.nndc.bnl.gov/
17. International Atomic Energy Agency IAEA (2012) The live chart of nuclides. http://www-nds.iaea.org/relnsd/vcharthtml/VChartHTML.html
18. Audi G, Bersillon O, Blachot J, Wapstra AH (2003) The NUBASE evaluation of nuclear and decay properties. Nucl Phys A 729:3–128
19. Colmenares C (1975) The oxidation of thorium, uranium and plutonium. Solid State Chem 9:139–239
20. Karkhanavala M (1967) Structural considerations in surface reactions with uranium oxides. Bhabha Atomic Research Centre, Bombay
21. Willis B (1963) Point defects in uranium. Atomic Energy Research Establishment, Harwell, England
22. Waber J (1956) A review of the corrosion behavior of uranium. DOE Technical Report, Los Alamos, NM
23. Loopstra BO, Cordfunke EH (1966) On the structure of alpha-UO_3. In: Rec Trav Chim Pays-Bas, vol. 85(2):135–142
24. Rouse KD, Willis BT, Pryor AW (1968) Anharmonic Contributions to Debye-Waller Factors of UO_2. In: Acta Crystallogr B 24:117–122
25. Wedemeyer H (1984) Suppl. Ser. Uranium. In: Gmelin Handbuch der Anorganischen Chemie. pp 1–64
26. Kim KC, Olander DR (1981) Oxygen Diffusion in UO_2-X. In: J Nucl Mater 102:192–199
27. Loopstra BO (1964) Neutron Diffraction Investigation of U_3O_8. In: Acta Crystallogr 17:651–654
28. Greaves C, Fender BE (1972) Structure of alpha-UO_3 by neutron and electron-diffraction. In: Acta Crystallogr B 28:3609–3614
29. Libowitz GG, Gibb TR (1957) High Pressure Dissociation Studies of the Uranium Hydrogen System. In: J Phys Chem 61(6):793–795
30. Rundle RE (1951) The hydrogen positions in uranium hydride by neutron diffraction. In: J Am Chem Soc 78:4172–4174
31. Mulford RNR, Ellinger FH, Zachariasen WH (1954) A new form of uranium hydride. In: J Am Chem Soc 76(1):297–298
32. Haschke JM (1991) Actinide hydrides. In: Meyer G, Morss LR (eds) Synthesis of lanthanide and actinide compounds. Kluwer, Dordrecht, pp 1–53
33. Koger J (1981) Overview of corrosion, corrosion protection, and stress corrosion cracking of uranium and uranium alloys. ASM metallurgical technology of uranium and uranium alloys seminar, Gatlinburg, TN, USA, 26 May 1981
34. Holden A.N (1958) Physical metallurgy of uranium. Addison-Wesley Publishers, Reading, MA
35. Koger J, Armstrong A, Ferguson JE (1975) Effect of aging on the general corrosion and stress corrosion cracking of Uranium-6 Wt% niobium alloy. DOE Technical Report
36. Jones D (1996) Principles and prevention of corrosion, 2nd edn. Prentice-Hall, Upper Saddle River, NJ, p 35
37. Eckelmeyer KH (1982) Residual stresses and stress relieving in uranium alloys. In: Kula E, Weiss V (eds) Residual stress and stress relaxation. Plenum, New York

Bibliography

Rather than citing complete references for topics that have been discussed in previous reviews, the reader desiring more detailed information should consult the following comprehensive reviews, each of which are extensively referenced:

Burke et al (eds) (1976) Physical metallurgy of uranium alloys. Brook Hill. Proceedings of the 1974 Vail Conference including papers on a wide range of uranium and uranium alloy topics, including melting, casting, forming, welding, crystal structures, phase transformations, heat treatment, mechanical behavior, fracture, corrosion, stress corrosion, hydrogen embrittlement, etc. Probably the most frequently cited source of information on uranium alloys

Eckelmeyer KH (1985) Uranium and uranium alloys. In: Metals handbook, 9th edn, vol 9, Metallography and microstructures. ASM. A review of metallographic techniques and microstructures in uranium and uranium alloys

Eckelmeyer KH (1990) Uranium and uranium alloys. In: Metals handbook, 10th edn, vol 2, Properties and selection of nonferrous alloys and special purpose materials. ASM. A review of heat treatment, microstructural development, and mechanical properties in uranium and uranium alloys

Eckelmeyer KH (1992) Environmentally assisted cracking of uranium alloys. In: Jones RH (ed) Stress corrosion cracking. ASM. The most recent review of stress corrosion cracking and hydrogen embrittlement in uranium and uranium alloys

Eckelmeyer KH (1992) Environmentally assisted cracking in uranium alloys. In: Jones RH (ed) Stress corrosion cracking. ASM. A review of information on materials performance and evaluation

Holden N (1958) Physical metallurgy of uranium. Addison-Wesley. A textbook summarizing the understanding developed mostly on unalloyed uranium during the 40s and 50s

Lillard JA, Hanrahan RJ Jr (2005) Corrosion of uranium and uranium alloys. In: ASM handbook, 10th edn, vol 13B, Corrosion: materials. A review of general corrosion and corrosion protection in uranium and uranium alloys

Magnani NJ (1976) Hydrogen embrittlement and stress corrosion cracking of uranium and uranium alloys. In: Advances in corrosion science and technology, vol 6. A thorough review of mid-50s through early-70s research on stress corrosion cracking and hydrogen embrittlement in uranium and uranium alloys

Metallurgical technology of uranium alloys, vols 1, 2, and 3. ASM, 1982. Proceedings of the 1982 Gatlinburg Conference including papers on a wide range of uranium and uranium alloy topics, including isotopic separation, smelting, melting, casting, forming, welding, heat treatment, phase transformations, mechanical behavior, corrosion, stress corrosion, etc.

Weirick LJ (1987) Corrosion of uranium and uranium alloys. In: Metals handbook, 9th edn., vol 13, Corrosion. ASM. A review of general corrosion and corrosion protection in uranium and uranium alloys

Chapter 2
Melting and Casting of Uranium

Edward B. Ripley

Abstract The art of melting and casting uranium is a fascinating subject. Uranium, which is anisotropic and allotropic, reacts readily with oxygen. When methods to melt and cast uranium are considered, those unique behaviors need to be taken into account. The proven methods that will be covered in this chapter are vacuum induction melting (VIM), vacuum arc remelt (VAR), and microwave (MW) melting and casting. Each of these methods has been demonstrated through hundreds of castings to deliver consistent and reproducible results. A thorough understanding of these methods is essential when work is being done with any reactive metal or alloy system.

Keywords Vacuum induction melting • VIM • Vacuum arc remelt • VAR • Microwave • MW • Uranium • Melting • Casting • Ohmic heating • Reactive metal • Batch process • Dissolved gas • Inert atmosphere • Vacuum • Mold • Coating • Electrode • Crucible • Insulation

2.1 Introduction

Casting molten uranium into a useable form that meets dimensional, mechanical, and chemical specifications presents a number of challenges. The first of these challenges is thermal. In order to obtain a sound casting, a certain amount of additional heat, commonly referred to as *superheat*, has to be supplied to the molten metal. Superheat is the difference between the melting temperature and the pouring temperature. Uranium melts at 1,132 °C [1]. If the molten metal is to pour and fill the mold well, the mold should be hot (800–1,000 °C is ideal) and the metal needs

E.B. Ripley (✉)
Y-12 National Security Complex, 1 Bear Creek Road, MS #8097, Oak Ridge, TN 37831, USA
e-mail: ripleyeb@y12.doe.gov

J.S. Morrell and M.J. Jackson (eds.), *Uranium Processing and Properties*,
DOI 10.1007/978-1-4614-7591-0_2,

Table 2.1 Variations in phase change upon heating and cooling

	Temperature (C°)		
	Heating	Cooling	Average
Melt point (0.3)	1,131.85	1,132.15	1,132
$\beta \rightarrow \gamma$ (5.4)	774.7	769.3	772
$\alpha \rightarrow \beta$ (9.4)	672.7	663.3	668

to be approximately 1,350–1,450 °C. This will ensure that the metal does not start to solidify before it fills the mold [2].

The second challenge is chemical. Since uranium is chemically reactive, anything that comes in contact with the molten uranium is a potential source for contamination. This includes the crucible, mold, and foundry hardware as well as the chamber atmosphere.

Third, thermodynamic challenges exist because many deleterious reactions become more thermodynamically favorable at high temperatures. Since uranium is anisotropic (having properties that are different depending on grain orientation) and allotropic (having three distinct temperature-dependent solid phases), it presents some unique challenges as it changes from its room-temperature form to its molten state.

The allotrope α-uranium has an orthorhombic structure with four atoms per unit cell. This is the structure of uranium at room temperature. For α-uranium, the dimensions are as follows: $a = 2.85$ Å, $b = 5.87$ Å, and $c = 4.95$ Å. The first phase transformation is the $\alpha \rightarrow \beta$ phase transformation, which occurs at 668 °C. The structure of β-uranium is tetragonal with 30 atoms per unit cell. For β-uranium, the dimensions are as follows: $a = 10.76$ Å and $b = 5.66$ Å. The second phase transformation is the $\beta \rightarrow \gamma$ phase transformation, which occurs at 772 °C. Having a body-centered cubic structure with two atoms per unit cell, γ-uranium has the dimension of $a = 3.52$ Å. Uranium melts and undergoes the final phase transformation from $\gamma \rightarrow$ liquid at 1,132 °C (Table 2.1).

For ultrahigh-purity uranium, the phase changes differ between heating and cooling. The magnitudes have been verified from literature; however, the average values have been adjusted to correspond with generally accepted values.

From a practical point of view, these differences in temperature are beyond most people's ability to accurately measure. The concept that the temperatures may vary slightly from the heating and cooling cycles is academically interesting. As these phase changes occur, the number of atoms per unit cell changes and the size of the unit cells also changes. These allotropic transformations manifest themselves as dimensional changes. When room-temperature α-uranium is heated and melted, it goes through three distinct transformations, but what is often more important are the transformations that occur during the solidification of molten metal within the mold.

Uranium and its alloys, as highly reactive metals, have unique requirements for processing in order to achieve an acceptable end product. Three casting methods have been demonstrated to consistently produce satisfactory results. These are vacuum induction melting (VIM), vacuum arc remelt (VAR), and microwave (MW).

2.2 Vacuum Induction Melting

Vacuum induction melting (VIM) is the oldest and currently the most widely accepted method of melting and casting reactive metals such as uranium, titanium, and many of the modern superalloys. This melting and casting technique is an extension of induction heating. The basic principles of induction were independently conceived by Michael Faraday and Joseph Henry in 1831. Since Faraday was first to publish, he is generally given credit for the discoveries that make induction melting possible. Induction heating occurs when an electrically conducting object, like a graphite crucible, is placed in a coil and a high-frequency alternating current (AC) is applied to the coil. The coil is usually made of copper and is cooled internally with water or gas. When the AC current passes through the coil, these electromagnetic fields generate eddy currents in the graphite crucible, and the resistance to these changing fields leads to heating. This is often referred to as Joule, or ohmic, heating. The metal inside the crucible is heated by both the hot crucible (by conduction) and induction field. After the metal is melted, the induction field stirs the metal and leads to improved homogeneity and uniformity in the melt.

Uranium reacts with many gaseous species and atmospheres at high temperature, and a vacuum or inert environment is generally employed. An additional benefit of casting in a casting chamber or enclosure is that this helps contain any contamination and reduce potential exposure to personnel. It is widely accepted that a vacuum aids in the removal of dissolved gases.

Uranium is highly reactive with most refractory materials, and the crucible is routinely coated with a paint or wash that is resistant to reaction by molten uranium. In addition, all of the foundry hardware, such as molds, pour rods, filters, and launders, needs to be either inert in molten uranium or thoroughly coated with a protective refractory paint or wash. Some experimental results suggest that the coating of all surfaces, not just those in contact with the molten uranium, can reduce the pickup of carbon, which can enter through the gas phase. It is important to pick coatings or paints that are thermally and thermodynamically stable with molten uranium at several hundred degrees above the melt temperature of 1,132 °C. This is because the metal needs to be completely molten and supplied with enough superheat to decrease the viscosity to a point at which the metal flows quickly and smoothly through the pouring mechanism and into the mold.

2.3 Overview of VIM Technology

Electromagnetic induction is at the heart of VIM. Electromagnetic induction is an incredible phenomenon that has found use in a number of applications. To better understand VIM, it is important to have at least a rudimentary understanding of induction. Induction is used in both the generation of electricity and the conversion

Fig. 2.1 Example of a large bottom-loading vacuum induction casting line. Stack loading and unloading occur on this first floor, and the furnace is operated from the second floor

of electricity into motion. *Faraday's law of induction* dates back to the 1830s and is the basic law governing the behavior of electromagnetism. Faraday's law is applicable to a closed-circuit conductor and states: "The induced electromotive force (EMF) in a closed circuit is equal to the time rate of change of the magnetic flux through the circuit." The converse is also true: "The EMF generated is proportional to the rate of change of the magnetic flux." Therefore, if a stationary conductor is placed in a changing magnetic field, electrical current will be induced in the conductor. This can cause circulating flows of electrons in the conductor called Foucault currents, or eddy currents. These fields have inductance and can also induce magnetic fields. Because there is an inherent resistance to these rapidly changing fields, the eddy currents generate heat in the form of Joule, or ohmic, heating [3] (Fig. 2.1).

VIM offers a number of advantages in the melting and casting of reactive metals:

- Ease of operation
- Precise control of material chemistry (achieved through small batches)
- Low losses to oxidation due to vacuum processing
- Reduction of any high-vapor-pressure contaminants
- Precise temperature control
- Removal of dissolved gases

As a heating technology, induction heating is very efficient; nearly every watt of electrical power supplied to the coil is converted into heat. Changing the frequency can directly affect the depth of penetration of the heating. Varying the frequency and power characteristics to the coil can also aid in melt stirring and homogenization. When thick cross sections of components are heated or when materials are heated in an electrically conductive crucible, a lower frequency in the range of 5–30 Hz is used. An application of 60-Hz three-phase power can effectively stir the contents of the molten crucible. The depth of penetration decreases as the frequency increases. The depth of penetration from 100 to 400 Hz is very shallow in the metal. If the frequencies are 480 Hz or higher, only microscopic parts and skin effects or surface heating occurs [4, 5].

Because the VIM process occurs under a vacuum, it is ideally suited to the processing of reactive metals and metals that have strong affinities for certain gases like oxygen. For this reason, VIM is widely used in the production of specialized and high-performance alloys, including nickel-based superalloys, stainless steels for nuclear applications, cobalt-based alloys for medical applications, high-purity alloys, and magnetic alloys. This is also why the VIM process is particularly well suited for the melting and casting of uranium, where material quality and contamination by impurities is a paramount concern.

The chamber, generally water cooled, is evacuated by a series of vacuum pumps to quickly reduce the pressure inside. In a typical system, there will be pass-through areas where the power for such items as the induction coil, thermocouples, and instrumentation needs to be connected from the outside of the chamber to sensors and equipment inside the furnace. The chamber needs to be sufficiently leak tight so that it can be pumped down to operating pressures, generally on the order of 1×10^{-3} Torr, and so it can pass a leak rate test. The test consists of pulling a vacuum, securing the vacuum pumps, and allowing the chamber to sit for some predetermined period of time with little or no increase in pressure.

An additional consideration is the choice of vacuum pumps, specifically the use of the oilless variety. When an ultimate vacuum for a vacuum pump is reached, a slight back streaming of oil into the chamber will start to occur. It is important that no matter which vacuum system is selected, a leak rate check is part of the precasting inspection. This is crucial because with a sufficiently large vacuum system, any leakage into the system can be overcome. When this happens, the establishment of a good vacuum will still be evident, but the partial pressure of oxygen will be higher than the chamber pressure would indicate. The gauge can appear to be holding and maintaining a good vacuum that would be required for production of high-purity reactive metals. However, the in-leaking atmosphere can still contain oxygen, nitrogen, hydrogen, and other gases in sufficient quantities to contaminate the molten metal.

Inside the casting chamber is an induction coil, which is really the heart of the VIM system. The coil generally consists of several turns of copper tubing, which carries a high-frequency alternating current. The coil is usually placed as close as practical to the crucible and charge or to the material to be heated or melted. An insulating layer may exist between the coil and the crucible to provide thermal and electrical isolation. The alternating current that passes through the coil produces eddy currents in the conductive materials within the crucible, either causing them to heat and melt or directly heating the crucible to melt the metal.

The crucible containing the metal charge is placed inside the coil. If one were to imagine the field lines the coil created, it would be readily apparent that the field lines are most concentrated on the coil's interior. That is not to say that the field is only on the inside, simply that the field intensity is highest on the coil's interior. The part of the field that is outside the coil's interior is commonly referred to as the *stray field*. This stray field can also heat the furnace walls and ancillary equipment if they are too close to the coil. The induction coil is generally water cooled to prevent overheating [6, 7].

Overheating is bad for a number of reasons, not the least of which is that the resistivity of a metal increases as a function of temperature, and this higher resistivity exacerbates the heating of the coil. Designs have been built in which primary and secondary cooling loops with a heat exchanger or in which the system's coil is gas cooled or even solid metal. For melting, the coil is generally a multi-turn design with insulating spacers to maintain the interval between the energized coils. And some type of support structure is included to allow each turn of the coil to be held in relative position within the chamber with respect to the crucible mold and furnace hardware. Many articles and even books have been dedicated to coil design, so there is no need for further elaboration here. The coil needs to be carefully designed by an expert in order to ensure the most efficient transfer of energy from the power supply to the crucible and to the metal [8, 9].

A recent improvement in VIM induction coils is the addition of flux concentrators. These take some of the stray field that would spread around the outside of the induction coil and redistribute it into the interior, where that energy can be used to heat and melt the charge. When a number of flux concentrators are placed on the outside of the induction coil with the open slot directed toward the crucible and charge, practically all of the coil's current flows near the surface, which faces the load. This improves the coil-to-workpiece magnetic coupling and results in dramatically increased coil efficiency. This improvement in coupling can be directly translated into a reduction in the power required to melt the charge and can also result in lower coil temperatures and drastically improved coil life. If flux concentrators are to be added to an existing coil, it is important to note that the impedance of the coil can be much different. It is important to ensure that the coil properly matches the power supply after the flux concentrators have been applied. With multi-turn coils, care needs to be exercised because the voltage across the coil turns can be significant, and the potential for developing a short-circuit path exists if the coil turns are not electrically isolated with insulation (Fig. 2.2).

The ideal crucible and mold material for VIM would be chemically inert to molten uranium; would have good thermal conductivity, a low coefficient of thermal expansion, high-temperature toughness, and thermal shock resistance; and could heat rapidly in an induction field (or be transparent to allow the direct heating of the charge material). These properties allow the crucible and mold to heat and cool rapidly.

Chemical inertness reduces the amount of contaminants in the melt from the crucible and mold. One often overlooked aspect of this is the outer surface of the mold. Although only the inside of the crucible and mold comes in contact with

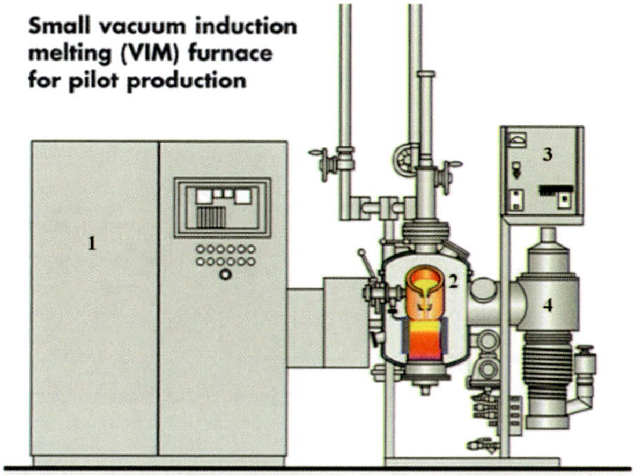

Fig. 2.2 This is a schematic of a small tilt pouring VIM caster for pilot-scale castings showing (1) power supply, (2) chamber with crucible and mold, (3) control panel, and (4) vacuum system

the molten metal, interactions with moisture and thermal breakdown products in the crucible and mold can react with the melt through the gas phase. While coatings are routinely used to protect the inside of the crucible and mold, it is common for the outside of the crucible and molds to be overlooked as a potential source for chemical contamination through the gas phase. Graphite is commonly used as crucible and mold material because graphite has excellent thermal properties, it is inexpensive, it machines easily to tight tolerances, it has high resistivity in an electrical conductor, and it readily accepts coatings. Because graphite reacts rapidly with molten uranium, isolation of the metal from the crucible and mold is critical if graphite is used. It is also important to preheat the graphite before coating to ensure that moisture and volatile organic compounds are reduced or eliminated (Figs. 2.3 and 2.4) [4–8].

It is important to choose a mold coating on the basis of the purity requirements of the finished product and the presence of alloying elements. Typically, metal oxide ceramics in a suspension with a binder are used as mold coatings. Although commercially available mold coatings may work for some applications, quite often, process-specific mold coatings have to be formulated. This is often a case of trial and error. Some candidate materials include yttrium oxide, zirconium oxide, magnesium oxide, erbium oxide, or combinations thereof. No single one is perfect. For example, magnesium oxide does react with molten uranium to a slight extent and results in contamination of the melt and increased oxidation of the metal. Zirconium oxide can present explosion hazards in casting cleanup if pickling of the metal is required (Fig. 2.5) [10–12].

Erbium oxide is prohibitively expensive and does not create a particularly durable mold coating. Yttrium oxide applies well and is relatively durable. However, yttrium oxide does dissolve slightly, and after a number of runs, contaminations by yttrium may build up in a product over time. Combinations like magnesium zirconate perform well when applied by brush but perform best when applied by flame spraying [6].

Fig. 2.3 VIM caster showing (1) power supply, (2) chamber (with crucible and mold inside), (3) control panel, and (4) vacuum system

Fig. 2.4 This is a VIM coil used in a tilt pour system for casting of reactive metals

Because of costs involved with some of these ceramic materials, coating of the noncontact areas (the outside of the mold, for instance) can be accomplished with less expensive coatings such as aluminum oxide, which work well for this purpose but would be unacceptable for use on the inside of the crucible or mold. Another important consideration is the choice of binders because a binder that burns off can often be a source of carbon contamination. It is also important to apply the correct amount of mold wash because an excessive amount of coating can cause it to spall

Fig. 2.5 This is the coil that was shown in Fig. 2.4 installed in a system. The molten metal in the crucible is a surrogate for this demonstration

off and result in contamination; conversely, a coating that is too thin or that has pinholes can cause contamination as well [5].

Casting atmosphere is a key factor to achieving acceptable metal yield and meeting desired chemical specifications. *Vacuum induction melting*, by definition, is performed in a vacuum. Typical evacuation of the furnace consists of using a roughing pump to reduce the pressure to approximately 1.5 mmHg. To then decrease the pressure to 25 μm, a turbomolecular pump, diffusion pump, cryogenic pump, or similar system is employed. While the metal is being melted, these low pressures are maintained. To outgas the metal before casting, the metal should be raised above the melt temperature with an additional 50–200 °C of superheat to allow for efficient outgassing. Key to metal quality is to limit in-leakage and to eliminate contamination of the atmosphere as a result of off-gassing from molds, stack components, and materials of furnace construction. Atmospheric levels of moisture that are present in graphite and insulation can react with stack and mold components, resulting in oxidation of metal. Increased levels of carbon, hydrogen, and other contaminants in the casting can be a result of transport into the melt through the gas phase.

To accurately measure temperature in the metal casting range is difficult. In an ideal situation, the casting process would be tolerant enough of variations in temperature to handle errors of 0.5 % (50 °C per 1,000 °C) without any detrimental effects. In reality, however, many uranium casting processes require temperature control that is much more precise. The use of thermocouples is a good solution if they can be located in areas where the induction field does not affect them. In the range of 1,200–1,600 °C, thermocouples can be accurate to within 2–4 °C. It is important to understand where induction fields (including stray fields) are, and it

Fig. 2.6 Example of mold that has been coated to prevent contamination of the metal by the mold and to facilitate releasing the casting from the mold

still may be necessary to use shielding, twisted pairs, common mode, and differential mode filters to remove interference before the signal goes to the amplifier. Many systems have common mode suppression circuits, but they are usually effective only at lower frequencies. One strategy is to briefly switch off the induction field and allow the system to equilibrate for a few minutes to get a more accurate measurement. There are some convenient points where the accuracy of the measurements can be checked. At 668 °C, 772 °C, and 1,132 °C, there are thermal arrests caused by phase changes from $\alpha \rightarrow \beta$ (668 °C), $\beta \rightarrow \gamma$ (772 °C), and γ to molten (1,132 °C). At these temperatures, there will be a short thermal arrest or plateau at the phase change, where the temperature stops increasing and levels out, and then after, the phase change the temperature continues to increase. Knowing this, the operator can check the temperature measurements to ensure that they are accurate at these very important points. If they are accurate at these points, the operator can have a little more confidence in their accuracy at other temperatures (Fig. 2.6).

2.4 Optical Pyrometry

Noncontact optical pyrometry is an excellent method of measuring temperatures because electromagnetic, microwave, or induction fields do not affect it. Various types of optical temperature measurement systems exist. The two most common are the disappearing filament pyrometer and the photoscreenic wedge.

The *disappearing filament pyrometer* uses a filament that can be heated to known values by the introduction of a precise electrical current, which causes it to heat and glow. When the glowing filament in the viewfinder is placed over the hot object in the viewfinder, it will appear hotter (brighter) or colder (darker) than the object to be measured. The current in the filament is adjusted until it disappears when someone is looking at an object to determine its temperature, and at that point, the temperature is determined by looking at the corresponding current versus voltage scale, which typically just reads out directly as a temperature.

A *photoscreenic wedge pyrometer* allows the user to rotate a photoscreenic wedge ring on the housing of the pyrometer while viewing the target. A comparison of an outer ring and a spot inside the ring is adjusted until both are the same color. It should look like a single large spot as opposed to a ring with a spot inside it. Since the light viewed by the operator is monochromic, the readings are not affected by color sensitivities, so it is very easy to get an accurate measurement.

2.5 Infrared Thermometers

There is also a group of optical temperature measurement systems called infrared (IR) thermometers. To understand how these work, it is best to understand how heat works. As an object heats up, the molecules in the object start to vibrate. As the temperature increases, the molecules vibrate faster and faster. IR radiation is given off whenever an object has a temperature above absolute zero. The heat an object emits can be from three possible sources:

1. Heat is reflected off the surface of another object (such as a glowing body or light).
2. An object can transmit heat from its internal temperature.
3. An object can emit heat from its surface.

IR thermometers are designed to capture and quantify emitted heat energy, and it is important for the sake of accuracy that the reflected and transmitted heat energy is ignored and only the emitted heat energy is measured. Two objects that are at exactly the same temperature can appear to have vastly different temperatures, depending on the type and condition of their surface. To get an accurate reading from an object, an emissivity is determined and entered as a correction factor. The emissivity is a ratio of the radiant energy emitted by the surface of an object to the radiant energy emitted by a blackbody at the same temperature. In a single-color pyrometer, this emissivity factor has to be known and entered in manually, or a value is determined experimentally and entered.

One method of eliminating these errors is to use a pyrometer, which reads two colors or temperatures and compares the ratios of the peaks of these temperatures. Using Planck's law and solving for the emissivity, one can correct for emissivity of the material under test to get an accurate temperature measurement [14, 15].

One more consideration is the preparation of the charge material. For ease of handling, the material is usually broken or cut into pieces that are small enough to fit into the crucible. The broken metal should be cleaned of any existing oils or contaminants; this can be done by acid pickling or other cleaning methods. When clean metal is used at the start, the amount of oxide that will be retained in the crucible is reduced. Sometimes if there are heavily oxidized parts or very small parts, a tenacious oxide shell can trap or retain a portion of the molten uranium and make it unavailable for casting. This decreases the metal yield and reduces casting efficiency. Through adequate cleaning of the charge material, these issues can be avoided.

When the metal is molten in the crucible and centered above a heated mold, the final obstacle is to release the metal and allow it to flow into the mold. Several different methods of pouring are available, and all have their strengths and weaknesses:

- Tilt pour
- Rupture disk
- Friction disk
- Pull rod mechanism
- Ball end in cone
- Ball end in bell
- Dud melt

Tilt pour is quite simply a crucible that is equipped with a pour spout and positioned above the mold with a tilting mechanism so the metal can be poured from the crucible into the mold. The advantage of tilt pour is that it is simple, easily controllable, and predictable. The disadvantage is that it pours from the top, so the first part of the melt that enters the mold includes oxides and dross that are floating on the surface. This can allow contaminants to be trapped in the cast part. This method is rarely used for high-performance precision castings for this reason. One variation of this that does yield high-quality, sound casting is the centrifugal caster. Quality precision castings can be made if the crucible has a spout on the side and is positioned adjacent to a mold that is on its side and on an arm that can be spun quickly and maintain the centrifugal and centripetal forces on the casting while the molten metal solidifies. For most uranium applications, this is impractical. However, it is mentioned as a possibility that could yield sound quality parts.

The *rupture disk* is a small disk of carbon or ceramic that is fitted in the lowest part of the crucible. It is strategically weakened to break when struck with a pour rod. The pour rod generally has grooves to allow the metal to flow past the pour rod. It is also customary to have a small reservoir to trap the fragment to prevent it from becoming trapped in the casting or blocking the pour discharge hole. The advantages of the rupture disk are that it is simple, reliable, and inexpensive. The disadvantages are that the pour rod mechanism has to be lowered into the molten pool and can allow molten metal to freeze on the pour rod mechanism. This is often overcome by allowing the pour rod to be lowered into the molten metal and to soak for a few minutes to allow any frozen metal to remelt. The broken edges are exposed to molten metal, and there is the potential for contamination or entrapment of fragments. (Studies have been conducted, and as a practical matter, the chemistries using this method are statistically the same as those using other pour methods.)

A *friction disk* is similar to the rupture disc; however, the center portion of the disk is held in by friction, and the pour rod pushes it through. The friction disk can have a larger top to prevent it from dropping through. The advantages are that like the rupture disc, it is simple, reliable, and inexpensive. Since this method requires a tight friction fit to prevent a premature pour, when the mechanism is activated, it does tend to dislodge the mold wash and expose carbon and has the potential to dislodge fragments of mold wash, which can become entrapped in the casting [15].

Another method that has proved successful is the *pull rod mechanism*. Pull rods have to be held down in the molten metal because they are lighter than molten uranium. In cases such as this, the potential for a pre-pour exists, which will adversely affect the quality of the casting. If the crucible is equipped with a cone, and the pour rod has a ball end, the pour rod can be conveniently forced into the cone and there is no issue with getting the alignment perfect. Other arrangements such as taper in tapered hole, cone in cone, or tapered plug in round hole have all been tried but proved to require angular alignment, which is so precise that it is impractical. The ball end in cone is simple, reliable, reusable, and inexpensive. In hundreds of castings, it has proved to be a simple and effective method for pouring molten metal.

A slightly different variation on the ball end in cone is the ball end in bell. The bell-shaped hole in the crucible has an added advantage over the ball in cone in that it can be designed to allow for better flowing of the molten metal, a gentler and laminar flow, and less potential for erosion of the mold coating in the crucible. It also allows for shaping the pour spout to allow for more precise pouring. It has all of the advantages of the ball in cone with one exception: It does require designing and computer modeling to optimize the effects.

The last method is to merely allow the metal melt to resolidify in the crucible. This is commonly referred to as a *dud melt*. It can be used for consolidation and/or to create a billet to be further processed. In the strictest case, a dud melt is not a casting because it does not involve pouring of the molten metal. It can, however, be done so that the molten metal drips into a hot mold. This method is called *drip casting* and can be used for consolidation but generally does not yield high-quality casting. The advantage is that it is cheap, easy, and relatively foolproof.

2.6 Safety

When a VIM system is operated, a key aspect is safety. Because induction uses a high-voltage electrical source to induce a low-voltage high current in the charge metal, it is important to have electrical isolation between the charge, the coil, and components. Proper furnace design and casting procedures need to take into account personnel safety, training, and protective equipment and awareness of the hazards of molten metal, pyrophoric metals, and inert atmospheres, which can asphyxiate. Because of the electrical hazards involved, bath grounding as well as ground detection should be incorporated into the furnace design.

When personnel work near the coil, it should be de-energized and preferably locked out. Any metal that gets close to the direct induction field or the stray induction field will heat rapidly. There are cases of items such as steel-toed shoes, zippers, and wedding bands causing significant injuries to people who failed to secure the power to induction coils and inadvertently got too close to the induction fields.

2.7 Vacuum Arc Remelting

Vacuum arc remelting (VAR) is a process to melt metal to produce alloys and to produce high-purity feedstock for recasting. VAR is most often used to quickly melt uranium and other high-melting-temperature reactive metals and alloys. The process produces excellent homogeneity and high chemical purity. VAR is rarely used as a direct casting method but is commonly used to produce the starting alloys and charge materials. The VAR process is typically used for high-value metals because it requires additional processing steps in order to improve casting quality or to produce difficult-to-make specialty alloys.

The VAR process uses a crucible (the anode), which is electrically conductive. A water-cooled copper crucible, for example, is commonly used. In addition, one or more central electrodes (the cathode) are used, which can be lowered into contact or close proximity with the material in the crucible. A vacuum of typically 1–100 mTorr (0.13–13 Pa) is established above the metal to be melted, and several kiloamperes of direct current are used to start an arc between the electrode and the metal in the crucible. From there, the arc is maintained to cause the metal to melt (Fig. 2.7) [4, 5, 11–13].

VAR has several advantages over other melting methods:

- Because the solidification rate of the molten metal can be controlled, the microstructure of the metal can be tightly controlled.
- Many impurities that have high vapor pressures are liberated from the melt, and the concentrations in the melt are lowered. Dissolved gases, which are generally considered deleterious to the metals and alloys, are quickly liberated from the metals.
- Centerline porosity and segregation are eliminated (Fig. 2.8).

2.8 Electrodes

Arc melting of uranium can use several different types of electrodes as the cathode. Traditionally, for uranium where minor contamination of tungsten is not an issue, solid tungsten or thoriated tungsten electrodes are used. These are generally

Fig. 2.7 Example of large VAR unit, which makes high-purity alloys for further processing

considered nonconsumable electrodes and can be used for many melt runs before replacement. A run would typically involve placing a small charge of metal into the crucible. With a strong negative electrical charge on the crucible and a strong positive electrical current on the cathode, the current needed to produce and maintain an arc in a vacuum. It is the arc that, when struck, provides the heat necessary to melt the charge metal. The process is akin to welding; as the metal is melted in the arc, the arc can be moved around to melt the whole surface. In some cases with the consumable electrodes, the arc is struck and the metal in the arc melts and falls to the molten surface of the billet. As the melting continues, the underlying metal solidifies, and the electrode is advanced toward the melt to keep it close enough to the surface to keep the arc going.

The vacuum atmosphere is established, and a current is applied to initiate an arc. The power is increased as the electrode is held above the melt and the electrode is advanced into the melting pool. When the entire amount of metal desired is molten, it is allowed to solidify. In a round-bottom crucible, the lower surface of the charge

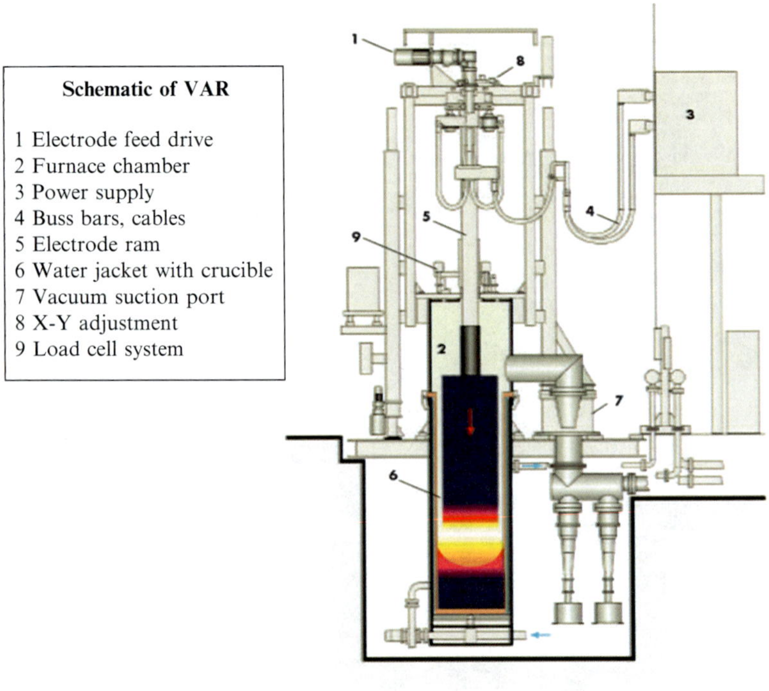

Fig. 2.8 VAR system showing major components

might not stay molten but form a crust or skull. The upper surface can be melted and form a relatively smooth surface. If the surface finish of the bottom of the ingot is unsatisfactory, it is not uncommon to flip the ingot and remelt the other side. This produces a fully melted and uniformly dense ingot that can then be used for feedstock. Small batches of material are generally processed because the surface finish is dependent on maintaining surface tension of the molten pool. Large metal charges can be problematic to keep fully molten, so generally the molten pool is allowed to cool and solidify, and then a new arc is established and additional metal is added to the top of the ingot and melted into it. Uranium and other refractory metals melt at lower temperatures than copper, so it stands to reason that a large quantity of molten uranium pouring into a copper crucible could damage the crucible as well as contaminate the melt. As a practical matter, this is avoided by cooling the crucible, even if it flash solidifies against the crucible after the metal melts, forming a protective coating or skull. The molten uranium can be kept molten and can be consolidated and homogenized without ever risking damage to the lower-melting-temperature copper crucible.

Alloys can also be produced the same way with high-temperature or difficult-to-alloy metals. One distinct advantage to this method is that any volatile contaminants are removed in this process. The arc can be controlled by its position relative to the center of the molten pool. Likewise, a magnetic field can also be used to help direct the arc toward the center of the molten pool. After the arc is struck, the power can be increased to enhance the melting speed. Generally, 200–300 A of current is enough to make a 2-in. ingot, whereas a 7-in. ingot might require 2,000 A of power at 15–30 V.

For high-purity alloys, consumable electrodes are generally used. These consist of making consolidated rods of uranium or alloy and feeding these into the molten pool. These rods can be cast or made up of consolidated metal turnings scrap or powder. This has the advantage of not adding any contaminants to the molten pool and improves purity further by removal of any volatile contaminants. Any of the previously mentioned techniques that use tungsten electrodes can be used with consumable electrodes with excellent results. Another approach is to have the electrode be able to swing in an arc and maintain a short interval from the crucible. This allows more metal to be melted and the metal to be mechanically mixed or stirred. The resulting ingot usually has a very rough bottom surface, which then is generally melted after the ingot is flipped. The consumable ingots can be welded together or may be secured in the cathode holder with a conductive collet. Sometimes it is desirable to cool the exterior of the collet to keep it from melting or forming a weld or eutectic bond with the electrode. Another approach is to make the electrode threaded and have a threaded cathode collet. When the cathode is too short to be used, a new one can be threaded on the end and the cathode pulled out to an appropriate length and fed in as needed.

2.9 The VAR Chamber

The VAR chamber is constructed so that a vacuum can be established, and the top and bottom are generally constructed so they can be electrically isolated and a strong electrical current can be applied. It is important to be able to observe the melt and to be able to control the distance from the crucible bottom and the relative position from the center to the edge. The bottom of the chamber, which functions as the crucible, should be cooled; water or other suitable cooling media, such as a molten salt or similar mixture, can be used. It is important to note that maintenance and inspection are critical with this type of system. If a leak were to develop and water were to be introduced into the molten metal, steam could develop and the potential for overpressurization or explosion could occur. Water may also represent a neutron moderation hazard for enriched uranium (Fig. 2.9).

It is a good idea to have an adequate vacuum system and have the ability to backfill with an inert gas. The use of gases like argon, which has a low electrical breakdown voltage and a high plasma potential, actually improves the performance of the VAR in melting the charge.

Fig. 2.9 VAR system showing major components for operation

2.10 Pyrometry and Temperature Measurement

Thermal measurement is not as critical with VAR as it is with VIM and microwave melting, because with VAR, usually only a portion of the charge is molten at a time. With nonconsumable electrodes, the metal is melted in small quantities and in small batches. With consumable electrodes, the electrode tip melts and drops into the molten top of the billet and forms the ever-growing billet. It is important to use closed-circuit cameras to view the melt or to have vacuum-tight windows to allow the direct viewing of the melting charge and allow manipulation of the electrode (Fig. 2.10).

Considerations for pyrometry are like those of VIM and microwave temperature measurement. Dual-color pyrometry or direct measurement with shielded thermocouples, neither of which will contaminate the melt, should be used. Since the pyrometry process generally has only part of the billet molten at any particular time, it will occur at a temperature very near 1,132 °C for uranium, and the molten portion is only about a hundred degrees above the melt point of higher melting alloys.

Fig. 2.10 This is an example of a process and over temperature pyrometer being used to control temperature. This configuration is for a microwave caster, but the equipment and setup are the same for microwave, VIM, and VAR

2.11 Microwave Melting Overview of Technology

Of the three melting and casting techniques discussed, microwave metal casting is the newest and the most obscure. Bulk metals do not readily couple directly with microwave energy at room temperatures. Since metals are electrically conductive, they readily reflect the incident energy. Microwave metal melting is counterintuitive, given common misconceptions about the behavior of metals in microwave fields. A few fundamental principles that make this possible follow:

In order to heat and melt bulk metals using microwaves, three basic ingredients are required: a multimode microwave cavity, a microwave-absorbing ceramic crucible, and a thermally insulating casket that is microwave transparent. The metal charge is placed in an open ceramic crucible, and the insulating casket is positioned to completely cover the open crucible. The casket and crucible assembly are then placed into a high-power multimode microwave cavity that can uniformly heat the crucible to the desired temperature. When microwave energy is applied to the cavity, the energy is strongly absorbed by the crucible. The metal charge in the crucible is quickly heated through radiation, conduction, and convection in the heated crucible. The thermally insulating casket increases the energy efficiency of the microwave system by trapping the heat generated in the crucible. In this way, metal objects that could not be directly heated by microwave energy can be melted easily and efficiently [14, 15]. Microwaves are preferentially absorbed by whatever absorbs them best. That might seem obvious, but remembering it can save a lot of frustrations when issues arise. If two items that absorb microwaves on their own and heat up readily are placed into the same chamber, whichever item absorbs the best will heat much more than the other. It is also possible that a very small component in the system might get superheated and the balance of the stack could remain cold. If there is arcing or a plasma formation

Fig. 2.11 View of a casting stack, crucible, and insulation. A lid will be added; no mold is used to melt and solidify a billet. This is called a dud melt

Fig. 2.12 This is an image of a microwave cast uranium billet

in the chamber, the arc or plasma may absorb essentially all of the energy, which could damage equipment while essentially no energy gets to the crucible or mold. As mentioned before, if this behavior is recalled when a problem arises, it might well save expensive equipment or hours of frustration (Figs. 2.11 and 2.12).

2.12 Theory

Metals are electrically conductive and as such microwave energy is readily reflected off the surface. There is a very minute depth to which microwaves penetrate, but for the most part, pieces of metal will not heat in a microwave directly to a temperature that would allow for melting and casting. In order to melt metal, a ceramic crucible is used, which couples (absorbs microwave energy) with the microwaves and converts microwaves into heat. To *suscept* also means to absorb microwaves to convert them into heat. A suscepting crucible is used to heat the metal from room temperature to about two-thirds of its melting point. This occurs by the ceramic crucible's dielectric absorption of microwaves. Dielectric constants, despite the name, are not constant with respect to temperature or electromagnetic frequency. As the temperature increases, the ceramic's ability to absorb microwaves enters into one of four possible scenarios. The ceramic:

- Is transparent to microwaves; it does not absorb microwaves and also does not heat up.
- Absorbs microwaves better as the temperature increases, making the ceramic hotter, which in turn increases its capacity to absorb microwaves. This behavior is called "thermal runaway." If these ceramics are used, applied power has to be decreased to control temperature as the temperature increases.
- Can decrease its ability to absorb microwaves as a function of temperature; so as the ceramic gets hot, it becomes increasingly more difficult to heat and will generally establish a plateau or suddenly drop in temperature once a critical temperature is reached.
- Does not start to absorb until a critical temperature is reached and then it starts to absorb increasingly better as the temperature increases.
- Heats in a linear fashion to the upper limit of what the system can handle with no change in absorption as a function of temperature. This behavior of linear heating is extremely rare (Fig. 2.13).

The crucible, which is coated to prevent reaction with the molten metal, is situated above the mold, which is usually made of graphite and is also coated to prevent reaction with the molten metal. The crucible and mold, which make up the casting stack, are enclosed in ceramic high-temperature insulation. This is referred to as the "casket," which is transparent to microwaves but does not allow heat generated inside the casket to escape. With this arrangement, microwaves in the chamber pass through the ceramic insulation and impinge on the crucible. If the microwaves hit the metal or graphite, they bounce off and continue to bounce around until they are absorbed by the crucible. The crucible heats and the metal inside is heated by conduction and convection. Since the heat generated inside the casket cannot escape, the temperature inside the casket rises quickly and heats the mold and pour rod as well as the metal. Since only what is inside the casket is heated, this process is a relatively simple and efficient method of melting metal. As the metal heats up, it starts to heat directly by ohmic heating mechanisms, and at elevated temperatures, many ceramics used for crucibles start to lose their ability to absorb, so an interesting phenomenon happens (Fig. 2.14).

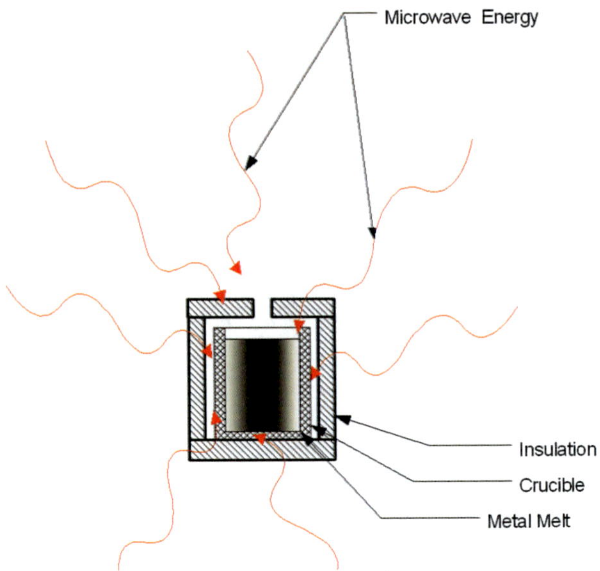

Fig. 2.13 Microwave energy can easily be transferred to metal shapes in a suscepting crucible

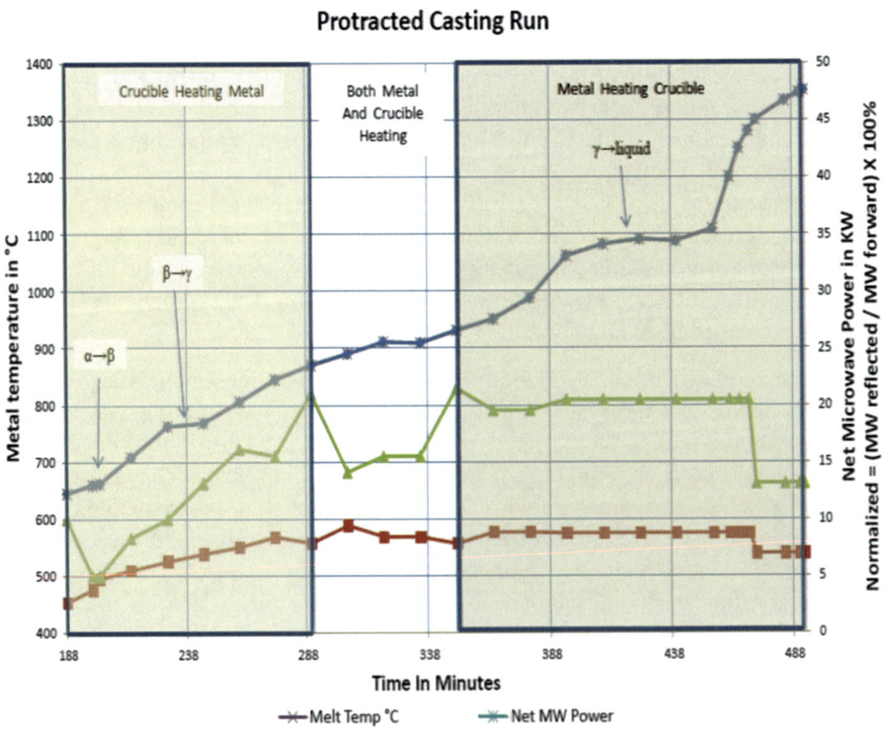

Fig. 2.14 Protracted casting run (extremely long slow casting run) to show temperature, microwave power, and regions where the crucible is the main absorber and where the metal is a better absorber

From room temperature to about two-thirds of the melting point, the crucible heats up and the metal and mold are heated by conduction and convection, so the crucible is the hottest thing in the system. At some point (which is different for every metal), the metal absorbs the microwaves and heats, and at that point, the metal is the hottest thing in the system. This has the advantage of putting the heat where it is needed and allowing for the most efficient application of power to supply the superheat required to cast the metal.

2.13 The Equipment and Setup

A microwave for melting and casting reactive metals in quantities of about 90 lb or less should have a chamber that is large enough to hold the casting stack and leave about a 9-in. space remaining on each side and at the top. The chamber should be microwave tight and have adequate viewing ports (which also prevent microwave leakage) to view the stack and to look down into the melt. It should be fitted with a vacuum roughing pump and be able to be backfilled with an inert gas (Fig. 2.15) [14, 15].

Since the chamber will be run under positive pressure, the vacuum is used to reduce the amount of inert gas required to establish an inert casting atmosphere. The microwave power source should be from 12 to 24 kW, with the understanding that the power is applied at a rate at which the stack absorbs the microwaves. Any extra power is simply wasted as reflected power or can overheat components like windows and waveguides, which could damage the system. The most commonly used frequency is 2.45 gHz.

The microwave generator is made up of some key components. Although these industrial microwave generators are usually purchased as complete units, it is important to know what components are being used in the system. Key components

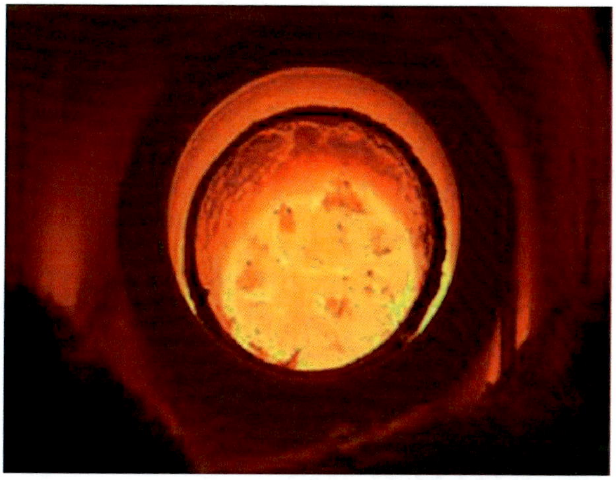

Fig. 2.15 Molten uranium metal; note the insulation casket and suscepting ceramic crucible

in the microwave generator are the power supply, the magnetron, and the isolator. Each of these components can profoundly affect the operation of the system. The complete microwave consists of a number of specialized components, and discussions of each follow.

2.14 Chamber or Applicator

The microwave chamber (also called an applicator) is a metal container that holds the casting stack and charge, and the microwaves are directed into it. Common types of applicators include front loading, drawer, or bottom loading; however, many application-specific chambers could be employed. The metal of construction is important; the losses due to absorption by the walls need to be minimized. Chambers like the one that Fig. 2.11 provides are constructed of aluminum, which has low wall losses and is inexpensive and easy to manufacture. Despite the low melting temperature of aluminum, such a chamber can be used to cast metals up to and including titanium (melting point 1,668 °C) without getting hot. This is because of the excellent insulating properties of the ceramic insulation used in the casket.

It is still a good idea to have thermocouples and thermal interlocks and cutoffs in case of crucible brake or casket damage. However, with proper care, these features will never be needed.

It is important to have some type of pour mechanism to release the molten metal into the mold. Commonly used pour mechanisms are the pull rod and rupture disk. Just as with the VIM, the choice is a matter of practicality and convenience. With the exception of the waveguides, any penetrations into the applicator need to be designed so that microwaves cannot inadvertently leak out. This can be done by incorporating a perforated metal screen like that used in a home microwave or by having a tube narrow enough and thin enough to prevent the microwave energy from escaping. The latter of these concepts is called "waveguide beyond cutoff." It needs to be capable of being evacuated and backfilled with an inert environment and of holding a modest pressure. In addition, it should be smooth walled with sound, smooth welds and be easily cleanable (Figs. 2.16, 2.17, and 2.18) [14, 15].

2.15 Power Supply

One of the key components in the microwave generator is the power supply. The different types of power supplies are described below.

Half-wave voltage doubler (common in home microwave ovens). This type of a power supply is the simplest and takes alternating current as an input and produces twice the voltage in direct current as the output. These are cheap and simple but

Fig. 2.16 Large front loading microwave chamber for casting high-temperature reactive metals

Fig. 2.17 These are uranium chips that are about to be cleaned, pressed, and dried before melting

Fig. 2.18 The solid oxide shells often referred to as *ghosts*, which are left in the crucible when the uranium is melted and poured into the mold

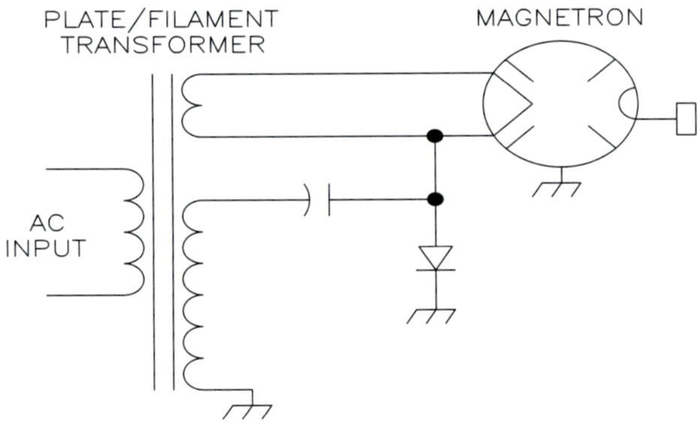

Fig. 2.19 A simplified schematic of a half-wave voltage doubler power supply

have limited duty life and performance. They have a fixed peak output and can be run as a pulsed duty power supply (Fig. 2.19).

Full-wave rectified (linear). This type of power supply takes AC current as its input and converts the whole wave form to an output of constant polarity. The performance of this type of power supply circuit is much better than the half-wave doubler, and it can be operated as a fixed-voltage or variable-voltage power supply. It is more complex and more expensive, but the cost is offset by much better performance and longer service life. These are the types of power supplies in the better industrial-quality microwave generators (Fig. 2.20).

Inverter (switch mode). This type of power supply is the most efficient of the three but suffers from the potential to cause noise problems if not adequately shielded. The advantages are that it can deliver variable output, in pulsed or continuous wave waveform; plus, it has unparalleled performance and runs significantly cooler than the other power supplies. However, it is more complex and more expensive, and it can suffer from shorter service life if it is used for extended periods of time.

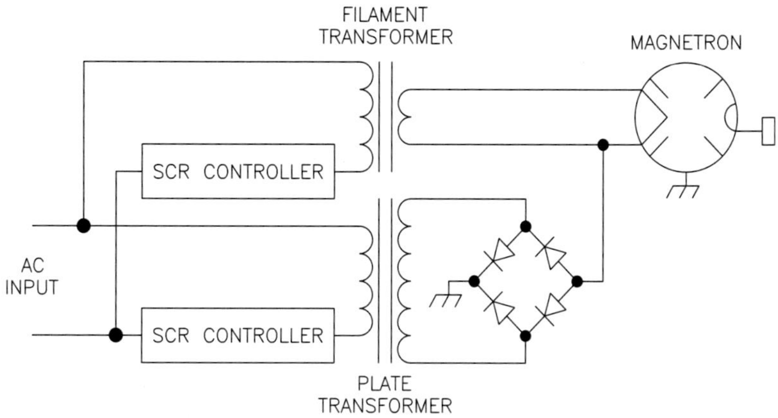

Fig. 2.20 Simplified schematic of a linear or full-wave rectified power supply circuit

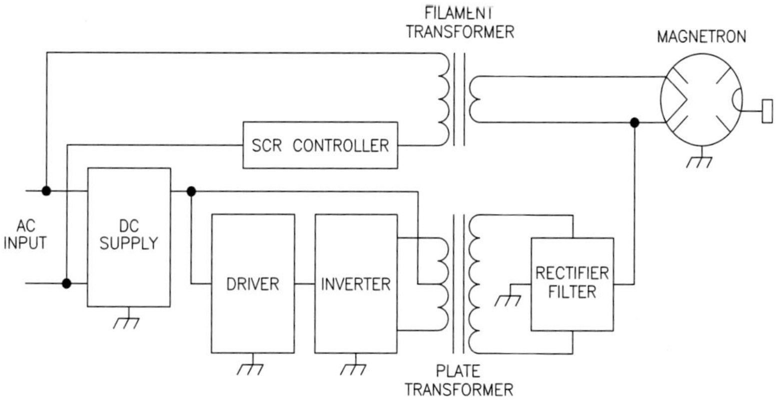

Fig. 2.21 Simplified schematic of a switch mode or inverter type of power supply

For most continuous industrial heating applications, the linear or full-wave rectified power supply is the best option because of its long service life, minimal complexity, and overall value (Fig. 2.21).

2.16 Magnetron

The theory of magnetron operation is based on the motion of electrons under the combined influence of electrical and magnetic fields. For the tube to operate, electrons must flow from the cathode to the anode. Two fundamental laws govern their trajectory:

1. The force exerted by an electrical field on an electron is proportional to the strength of the field. Electrons tend to move from a point of negative potential toward a positive potential. With no magnetic field present, there is a uniform and direct movement of the electrons in an electrical field from the negative cathode to the positive anode.
2. The force exerted on an electron in a magnetic field is at right angles to both the field itself and the path of the electron. The direction of the force is such that the electron proceeds to the anode in a curved rather than a direct path.

Permanent magnets are added above and below the tube structure. The tube is a cavity, which has a number of spaces. Around the rim of the tube is a series of cylindrical resonant cavities on its periphery. These cavities have a slit opening along the interior wall. The electron leaves the cathode and is attracted to the wall, but because of the magnetic bias applied, it cannot travel in a straight line to the wall but approaches the wall in a decaying spiral trajectory. As it does, it sets up a high-frequency resonant electromagnetic field in the cavity. This resonant frequency is determined by the size of the cavities. A small output coupling loop acts as an antenna and allows the electromagnetic waves to exit the tube through a launcher and enter the waveguide.

2.17 Circulator or Isolator

One critical component for high-power microwave operation is the circulator, which consists of an isolator and a dummy load. The isolator is a piece of equipment that prevents microwaves from reentering the waveguide and damaging the magnetron. It is inserted between the magnetron (microwave generator) and the waveguides. It allows microwaves to pass through unimpeded in one direction, but when they enter in the opposite direction, they are redirected into either a solid-state load or a water load. The purpose of the dummy load is to absorb this excess energy and prevent microwaves from making it into the magnetron. Reflected microwave power causes excess heat and, ultimately, catastrophic damage to the magnetron.

The isolator works by establishing a strong magnetic field across the waveguide. Ferrites and magnets on top and bottom function to make sure that microwaves entering can follow only two basic rules: First, they can circle only in one direction, and second, they have to exit at the first possible opportunity. If one were to imagine the isolator as a roundabout with three roads, one road is an entrance, opposite it is an exit, and the third road is a dead end (dummy load). If a car were to enter this roundabout at the entrance, it moves with the direction of traffic and takes the first possible opportunity to leave the roundabout. This is the exit that leads to the applicator. Now, if the car tries to return via the exit, it moves in the direction of traffic and takes the first opportunity to exit, which in this case is a dead end (dummy load) and is absorbed. With this system in place, there is no danger of the magnetron's being damaged by reflected power (Fig. 2.22).

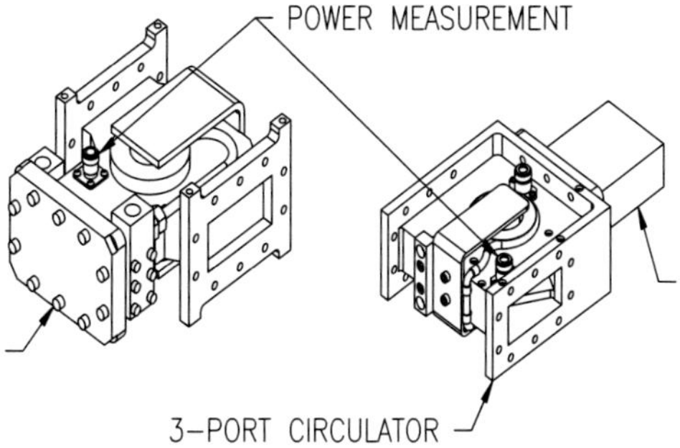

Fig. 2.22 A circulator and isolator showing coupler to give power measurement

2.18 Power Measurement Devices

When high-power microwave generators are used, there is great potential to put a lot of energy into a crucible and casting stack. Conversely, there is great potential to cause quick damage to equipment and components if they inadvertently receive too much microwave energy too quickly. It is the absorbed power that is important:

$$\text{Absorbed power} = \text{forward power} - \text{reflected power}$$

In order to know the absorbed power, it is critical to know the forward and reflected power. These measurements are made using one of several measurement devices. Each of these devices has a purpose, depending on what information is available from other sources.

Directional couplers measure power in one direction. It can be subject to error because it gives information only in one direction; it essentially ignores the power flowing in the opposite direction. A pair of directional couplers, which has one in each direction, is referred to as a dual coupler. This will allow a measurement in both the forward and reflected (rev) directions. The dual coupler reduces this error by showing power flow in both directions. Many times these will be used, even if there is forward or reflected power indications somewhere else in the system, because they can be placed strategically to give information at the point where it is desired.

The nondirectional couplers are inexpensive, but in order to be useful, they have to be placed where power flows in one direction only.

Finally, reflectometers cost more but can provide a lot more information than other types of power measurement devices. These are used to make complex impedance measurements. For most applications, reflectometers are overkill, and the information gathered is not required (Fig. 2.23).

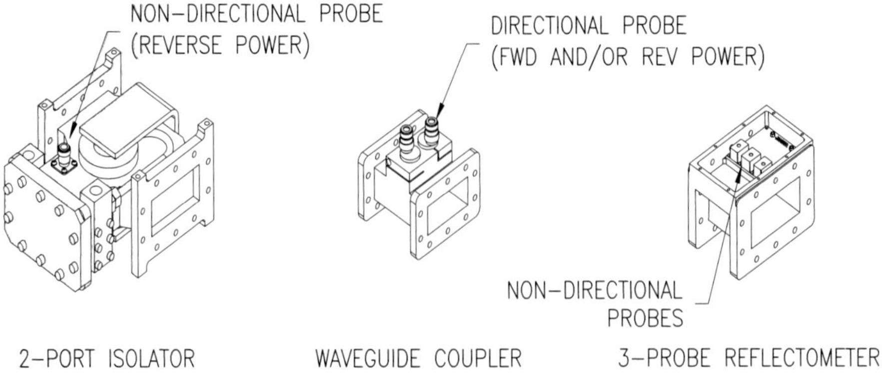

Fig. 2.23 Three different power measurement devices

2.19 Pyrometry

Pyrometry for microwave systems is essentially the same as was discussed for VIM. For microwave applications, the following items are preferred:

- Dual-color or multicolor.
- Small spot size with viewing port (or camera).
- Temperature range that spans the range of interest many times; several will have to be used for different parts of the run.
- Calibration of the pyrometer in accordance with the conditions under which it will be read. If shooting will be done through two windows—one that is a few inches away and at room temperature and the other that is a foot away at 500 °C—it is a good idea to duplicate this setup in the calibration.

Finally, one needs to pay attention to the temperature profile. If uranium is heated at a constant energy input, consistent heating that follows a smooth curve will more or less be achieved. With a few notable exceptions, thermal arrests will occur at 668, 772, and 1,132 °C because of phase changes. These will show up as small plateaus in the otherwise smooth curve. These can also be used to check the relative accuracy of the pyrometric measurements because these thermal arrests or plateaus have to occur, and always occur, at specific temperatures. Knowing this, one can correct for any systematic errors to gauge the temperature at any point with more precision and accuracy than would be possible by simply relying solely on the dual-color pyrometer.

2.20 Impedance-Matching Devices or Tuners

Impedance matching was originally developed for electrical engineering but is particularly applicable to high-power microwave applications. Simply stated, impedance matching maximizes the power transfer and minimizes the reflection from the load.

Several approaches can be used to accomplish this goal, including the following:

- Manual three-stub tuners
- Automatic stub tuners
- Fixed iris
- Internal wave matching structures

Manual three-stub tuners are a cost-effective way to maximize power application while minimizing reflection. The tuner has three stubs, which are effectively quarter-wave chokes. They move into or out of the waveguide and can eliminate part of the signal, which is likely to be reflected back into the waveguide. Typically the stubs will be turned to extend them into the waveguide while monitoring the change in reflected power. If the reflected power decreases, the stub is turned further into the waveguide. If the reflected power increases, it is withdrawn. This is performed in a systematic method, going from one tuner to the next, until the minimum reflected power is attained. Generally, this approach is adequate. However, it is possible the lowest reflected power that can be obtained by tuning the stubs in one combination might represent a local minima. But using a different combination may result in a lower overall reflected power. If it is critical to obtain the best possible impedance match, different combinations of tunings need to be tried, the positions noted (most tuners have a micrometer or a digital position indicator), and the absolute minima noted. These are inexpensive and effective methods of impedance matching.

Automatic stub tuners operate like the manual tuners but have a computerized program and built-in reflectometer, so multiple combinations of tunings are attempted, and the absolute minima is achieved in a matter of seconds rather than after laborious minutes of manual tuning. The system is constantly checking for standing wave measurements and adjusts accordingly. This is an expensive system but has the advantage of being able to adjust to constantly changing loads. It also relieves the operators of the responsibility of constantly adjusting for operating peak efficiency.

For a steady-state system where little systematic changes are observed, a cheaper alternative is to place an impedance-matching feature in the waveguide or to fit the window with an iris. These wave matching features do have to be determined by trial and error but are very inexpensive to employ. Typical shapes that are used interior to the waveguide are blocks of copper or aluminum that obstruct a portion of the waveguide from top to bottom, blocks of copper or aluminum that obstruct a portion of the waveguide from side to side, or pieces that block one or more corners. Another approach is to place a conductive metal sphere inside the waveguide on the bottom or top center wall. The position of this is important so it can be put in place, moved around until it is in the right spot, and then permanently installed. Finally, an iris or partial opening between the waveguide and the process chamber can be used. The iris generally blocks the top and/or bottom, left and/or right side, or one or more corners. If an iris or waveguide matching structure is used, no other impedance-matching feature should be used (e.g., manual or automatic stub tuners) (Fig. 2.24).

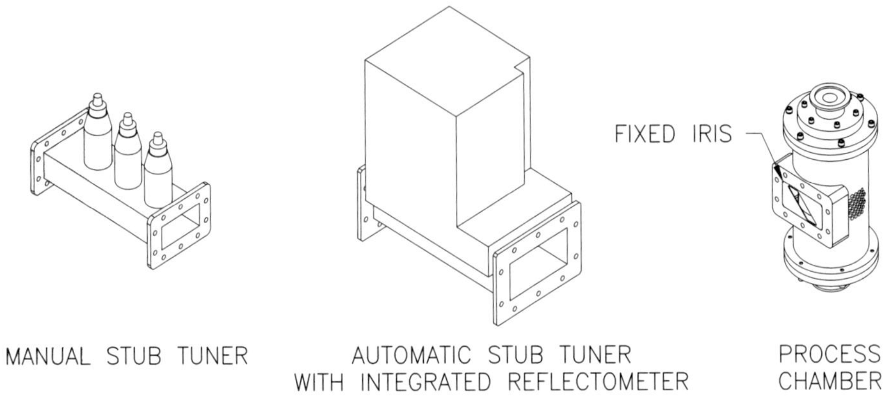

Fig. 2.24 Examples of impedance-matching features

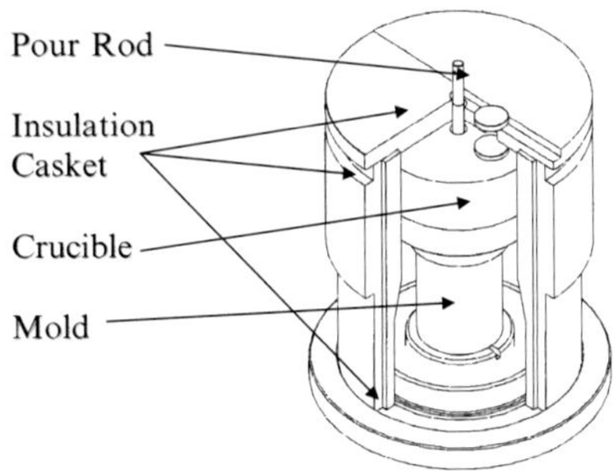

Fig. 2.25 Cutaway of the casting stack showing the crucible, mold, and insulation casket

2.21 Casket

The *casket* is the term used to refer to the thermally insulating structure around the casting stack. It is typically a rigid fiberboard of 80 % Al_2O_3 and 20 % SiO_2. It is important that it be transparent to microwaves at the frequency of interest and that it be able to withstand high temperatures. A good density range is nominally 30 lb/ft^3 and 85 % open porosity. These are general guidelines to be used as a starting point. The insulating casket could be made with completely different materials, providing it does two things: allows the microwaves to enter the stack unimpeded and does not allow heat to escape. This creates a superheated interior, which allows the metal to heat and melt (Fig. 2.25).

2.22 Casting Stack

The casting stack consists of the crucible, mold, and ancillary items such as the pour rod and windows. As discussed previously, the crucible is made of a ceramic, which couples with the microwaves. The mold is typically made of coated graphite, which can contain the molten metal without reacting with its chemistry. The pour rod is also typically coated graphite.

Of course, if the goal is simply melting and consolidation, a closed-bottom crucible with a slight draft angle on the side can be used. A crucible that is going to be used to pour molten metal should also be designed with a central hole for pouring (alternatively, a spout could be used for tilt pouring), a shallow taper on the sides, and a slight draft angle sloping from the sides of the crucible to the pour hole. With this general design, the metal can melt; when poured, the metal will flow out completely, and if no pour is made and the metal freezes in the crucible, it can be removed without sacrificing the crucible. If this were to occur, it is generally a good idea to break, shear, or otherwise reduce the metal in size before remelting. It could also be placed in a slightly larger crucible or even placed on its side before remelting. If the frozen metal is simply reheated in the same crucible without one of these things being done, as the metal heats, the expansion could crack the crucible and result in an unsuccessful pour or damage to the equipment. Of course, all of these components must be of criticality safe dimensions, and to reduce the potential for a criticality accident, the entire system must be evaluated if the metal to be cast is fissile, fertile, or fissionable.

2.23 Vacuum and Inert Gas Backfill System

Unlike VIM or VAR, the microwave is operated at slightly positive pressure with a small flow of gas, which is vented to remove any moisture and any gases that are given off by the insulation, crucible, or metal as they are heated. To achieve this, it is a good idea to pull a rough vacuum and then backfill with an inert gas. Positive pressure is key because when microwave power is applied to a gas at reduced pressure, the potential for creating a plasma is greatly increased. Some gases are less likely to produce a plasma, and as the pressure increases, the plasma potential decreases.

It is helpful to look at the Paschen discharge curves for the atmosphere desired for the casting operation in order to select a gas that is less likely to be a problem. It is also possible to bleed in a less reactive "quench" gas to reduce the plasma potential or to stop one when it forms. If a plasma forms, it is advisable to turn off the power completely for a few seconds and then turn it back on slowly. If a quench gas is used, it may alleviate any further problems.

If a plasma reoccurs, the addition of a higher plasma potential gas or a quenching blends like Ar 90 %, H_2 10 %, or N_2 or any other compatible high plasma potential gas. If the application of the quench gas is unsuccessful, simply turning

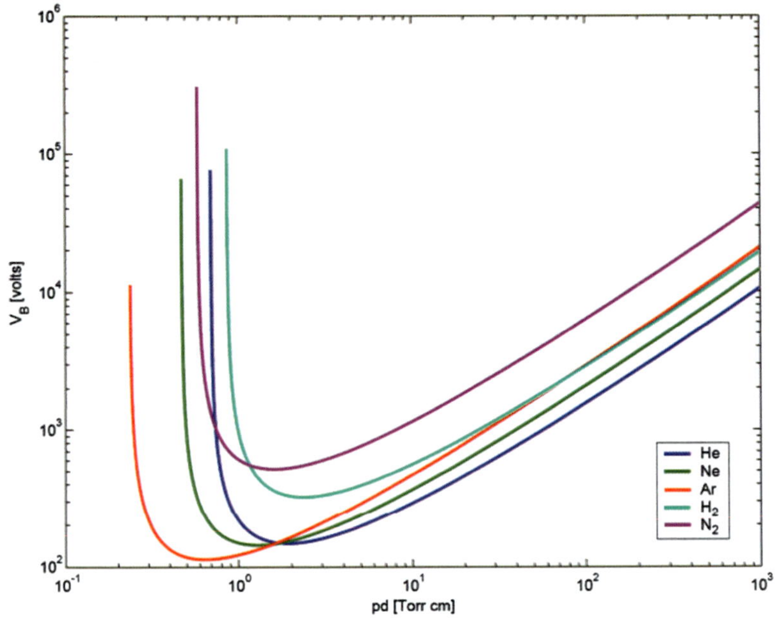

Fig. 2.26 Paschen discharge curve of several gases

the system off and back on might take care of the plasma problem as well. If plasma formation occurs, it is important to pay attention to the reflected power levels because plasma formation is often a symptom of excess power in the chamber. Only absorbed power is heating the crucible and metal, so reflected power is wasted and can lead to issues later (Fig. 2.26).

The Paschen discharge curve is an indicator of the relative plasma potential only. These curves were derived under a different set of conditions, so no direct mathematical correlation of microwave energy to a specific casting stack and the likelihood of plasma formation can be inferred. However, the relative behavior and pressure trends hold true in microwave applications.

2.24 Summary and Conclusion

When a method of heating and casting uranium or other reactive metals is being considered, a number of factors must be taken into account:

- Ease of operation
- Ease and cost of maintenance
- Ergonomics
- Contamination control
- Ultimate chemical requirements of final product
- Mechanical properties of cast material

- Size, shape and complexity of part, etc.
- Capital equipment cost
- Cost to operate
- Casting capacity and size of part
- Service life of equipment
- Number of castings per week

There is no off-the-shelf solution to the problems routinely encountered when casting uranium. Every alloy, charge size, part type, etc., will present unique challenges that will need to be addressed before entering the foundry. This information was intended to be a brief overview to make the reader aware of the current state of the various technologies and help them determine which options might best suit their needs.

Acknowledgements

Figure credits
Photos for Figs. 2.1, 2.7, 2.12, and 2.15 were taken by Brett Pate.
Photos for Fig. 2.6 were taken by Matt Marsicek.
Photos for Figs. 2.2, 2.3, 2.4, 2.5, 2.8, and 2.9 were provided by ALD Vacuum Technologies.
Figures 2.19, 2.20, 2.21, 2.22, 2.23, and 2.24 were provided by John Gerling.
Photos for Figs. 2.11, 2.14 2.16, 2.17, and 2.18 were provided by MS Technologies Inc.
Figures 2.13, 2.14, 2.25, and 2.26 were provided by the author.
Photo for Fig. 2.10 was taken by Hugo Huey.

References

1. Dahl A, Cleaves H (1949) The freezing point of uranium. J Res Natl Bur Stand 43:513–517
2. Baumrucker C, Warren B (1953) Unpublished information
3. Sadiku OMN (2007) Elements of electromagnetics, 4th edn. Oxford University Press, New York
4. Harrington C, Ruehle A (1959) Uranium production technology. D. Van Nostrand Co. Inc., Princeton, NJ
5. Patton F, Googan J, Griffith W (1963) Enriched uranium processing. The Macmillan Co., A Pergamon Press Book, New York
6. Wilkinson W (1962) Uranium metallurgy, vol I. Wiley, New York
7. Wilkinson W (1962) Uranium metallurgy, vol II. Wiley, New York
8. Wilhelm H (1942) The casting of uranium rods at Iowa State College. Internal publication (A-4045): Iowa
9. Holden A (1942) Physial metallurgy of uranium. Addison-Wesley, Reading, MA
10. Blumenthal B (1955) Melting of high purity uranium. University of Michigan Library Jan, 1952
11. Googan J, VanDermeer R, Koger J, Donnelly (1981) Metallurgical technology of uranium and uranium alloys vol. I. ASM American Society for Metals Press, OH
12. Googan J, VanDermeer R, Koger J, Donnelly (1981) Metallurgical technology of uranium and uranium alloys vol. II. ASM American Society for Metals Press, OH
13. Googan J, VanDermeer R, Koger J, Donnelly (1981) Metallurgical technology of uranium and uranium alloys vol III. ASM American Society for Metals Press, OH
14. Gupta M, Leong W, Eugene W (2007) Microwaves and metals. Wiley, Asia
15. Brooks K, Industrial-Scale Melting of Uranium Using Microwave Energy. Industrial Heating, February 8, 2011

Chapter 3
Machining of Uranium and Uranium Alloys with Coated Cutting Tools

Mark J. Jackson, Grant M. Robinson, Michael D. Whitfield, Rodney G. Handy, and Jonathan S. Morrell

Abstract The machining of depleted uranium and its alloys is discussed in this chapter. Traditionally, these materials have been machined using uncoated cutting tools with limited success. New developments in titanium-based coatings such as cation-substituted $Ti_{l-x-y-z}Al_xCr_yY_2N$ alloys, with $y = 0.03$ and $z = 0.02$, have been shown to offer enhanced high-temperature oxidation resistance compared to presently used TiN and $Ti_{l-x}Al_xN$ films that are deposited to cutting tool surfaces. Layers (3 μm thickness) deposited by unbalanced magnetron sputter deposition to small grain WC–Co cutting tool inserts which had been ion etched in situ using a steered Cr-metal-ion cathodic arc discharge at an Ar pressure of 6×10^{-4} mbar (0.45 mTorr) have been tested and found to be beneficial in terms of tool life. The metal-ion etching promoted initial local epitaxy on individual substrate grains, while the overall film texture evolved through competitive growth to (111) in $Ti_{0.44}Al_{0.53}Cr_{0.03}N$ alloys and (200) in $Ti_{0.43}Al_{0.52}Cr_{0.03}Y_{0.02}N$ alloys. Although $Ti_{0.04}Al_{0.53}Cr_{0.03}N$ layers exhibited a columnar microstructure similar to that previously observed in $Ti_{l-x}Al_xN$ alloys, the addition of 2 mol% YN resulted in significant grain refinement giving rise to an equiaxed structure. The Knoop microhardness of $Ti_{0.43}Al_{0.52}Cr_{0.03}Y_{0.02}N$ alloys was $HK_{0.025} = 2{,}700$ kg mm^{-2} compared to 2,400 kg mm^{-2} for $Ti_{0.44}Al_{0.53}Cr_{0.03}N$. Rapid oxidation of coatings, as determined from thermogravimetric measurements, ranged from ≈600 °C for TiN, 870 °C for $Ti_{0.466}Al_{0.54}N$, and 920 °C for $Ti_{0.44}Al_{0.53}Cr_{0.03}N$ to 950 °C for

M.J. Jackson (✉) • G.M. Robinson • M.D. Whitfield
Center for Advanced Manufacturing, Purdue University, West Lafayette, IN, USA
e-mail: jacksonmj04@yahoo.com; grantmrob@gmail.com; michaeldarrenwhitfield@gmail.com

R.G. Handy
Department of Engineering Technology & Construction Management, University of North Carolina-Charlotte, 9201 University City Blvd., Charlotte, NC 28223, USA
e-mail: rhandy@uncc.edu

J.S. Morrell
Y-12 National Security Complex, 1 Bear Creek Road, MS #8097, Oak Ridge, TN 37831, USA
e-mail: morrelljs@y12.doe.gov

J.S. Morrell and M.J. Jackson (eds.), *Uranium Processing and Properties*,
DOI 10.1007/978-1-4614-7591-0_3,

$Ti_{0.43}Al_{0.522}Cr_{0.03}Y_{0.02}N$. Machining experiments indicated that cutting tool life is improved significantly using yttrium-doped titanium-based coatings when machining uranium alloys.

Keywords Depleted uranium • Machining • Coated cutting tools • Yttrium

3.1 Introduction

Uranium and uranium alloys are considered to be easy and difficult to machine depending upon the alloying content. Depleted uranium is machined at high rates of production, and problems with machining are associated with galling, abrasiveness, work hardening, pyrophoricity, reactivity with coolants, reactivity with tool materials, and toxicity. Figure 3.1 shows the reactivity of uranium when it is being machined. The image shows a steady stream of sparks generated even during the simplest of machining operations.

There are three classes of uranium alloys. These alloys are described as:

- Dilute alloys containing less than 0.4 wt% of alloying elements
- Intermediate alloys containing between 0.4 and 4 wt% of alloying elements
- Stainless alloys containing more than 4 wt% of alloying elements

Uranium consists of an orthorhombic α-uranium matrix with carbides and oxides as principal inclusions. Alloying elements used include titanium that makes uranium hard and niobium that makes uranium corrosion resistant. Depleted uranium and uranium alloys are used in nuclear weapons, calorimeter plates,

Fig. 3.1 Totally enclosed machining of uranium showing the effects of pyrophoricity during machining

ballistic penetrators, radiation shielding, counterweights, and gyroscope rotors. The machinability of uranium is affected by the following characteristics:

- Shape memory
- Non-isotropic coefficient of thermal expansion
- Pyrophoric, toxic, and radioactive properties

Shape memory is the property of a material to return to its original shape after a change in temperature and is prevalent in alloyed uranium than in unalloyed uranium. The non-isotropic coefficient of thermal expansion causes the material to distort under a thermal influence. This means that the material expands in two crystallographic directions rather than in three when subjected to a thermal influence. Unalloyed uranium expands in two principal directions and contracts in the third. During machining, the cutting tool must be kept sharp and the temperature of the material must be kept as low as possible [1]. Uranium and its alloys can generally be machined to tolerances within the range of a standard lathe. Tolerances of the order of 8 μm are not uncommon. However, cutting tools must remain sharp to produce accurate tolerances [2]. The machining conditions required for cutting uranium and its alloys (drilling, tapping, turning, facing, and milling) are shown in reference [2] and will not be repeated here. A bibliography is shown at the end of this chapter that directs the reader to a variety of published works on the machining of uranium and uranium alloys.

3.2 Wear of Cutting Tools

Cutting tool wear is the product of a combination of load factors on the cutting edge of the tool. Wear is the result of interactions between tool, workpiece material, and machining conditions. The load factors are mechanical, thermal, chemical, and abrasive.

Apart from the static components of mechanical loads, there are various dynamic components arising from chip formation itself. Metal cutting generates a lot of heat on the face of the chip and on the flank of the insert. Thermal loads are considerable on the tool material, and during interrupted machining operations, the thermal load pulsates, thus leading to thermal fatigue of the tool material. The typical wear zones on the cutting edge are shown in Figs. 3.1 and 3.2. The formation of a chip means that a fresh metal interface is continuously produced and forced at very high pressure and temperature along the tool material. The zones produced make it an attractive environment for diffusion and chemical reactions to occur (Fig. 3.3).

As a result of load factors exerted on the cutting tool edge, a few basic mechanisms dominate metal cutting. These mechanisms are:

1. Diffusion wear—affected by chemical loading on the tool and is controlled by the metallurgical composition of the tool and coating material.
2. Abrasive wear—affected by the hardness of the tool material and is controlled by the carbide content of the cutting tool material.

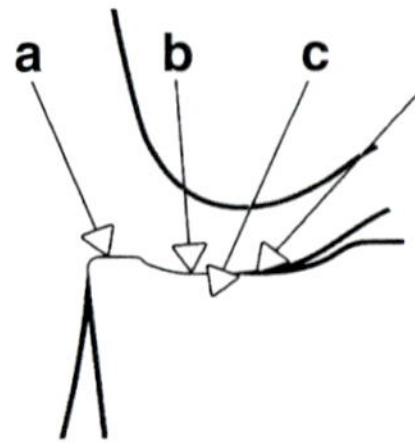

Fig. 3.2 Wear zones on the cutting tool caused by chip formation [1]. The main load factors are (**a**) mechanical, (**b**) thermal, (**c**) chemical, and (**d**) abrasive. *Reproduced courtesy of Sandvik and reprinted with permission of Springer publishers*

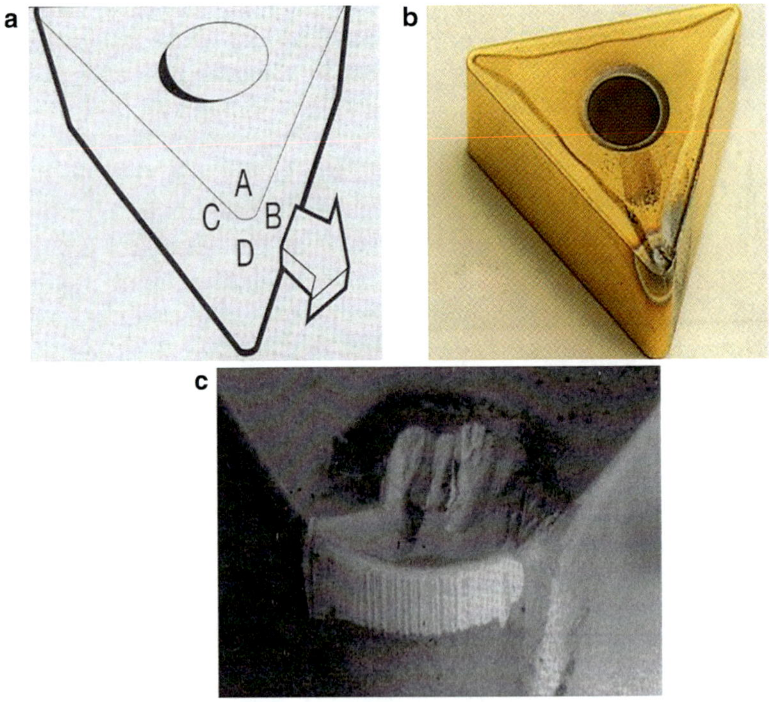

Fig. 3.3 (**a**) Areas of wear on the cutting tool insert shown in Fig. 3.2, (**b**) physical cutting tool insert showing worn edges described in (**a**), and (**c**) well-developed flank and crater wear at the nose of the insert [1]. *Reproduced courtesy of Sandvik and reprinted with permission of Springer publishers*

3. Oxidation wear—causes gaps to occur in coated films and results in a loss of the coating at elevated temperatures.
4. Fatigue wear (static or dynamic)—this is a thermomechanical effect and leads to the edges of the cutting tool to break down.
5. Adhesion wear—occurs at low machining temperatures on the chip face of the tool and leads to the formation of a built-up edge and the continual breakdown of the built-up edge and the tool edge itself.

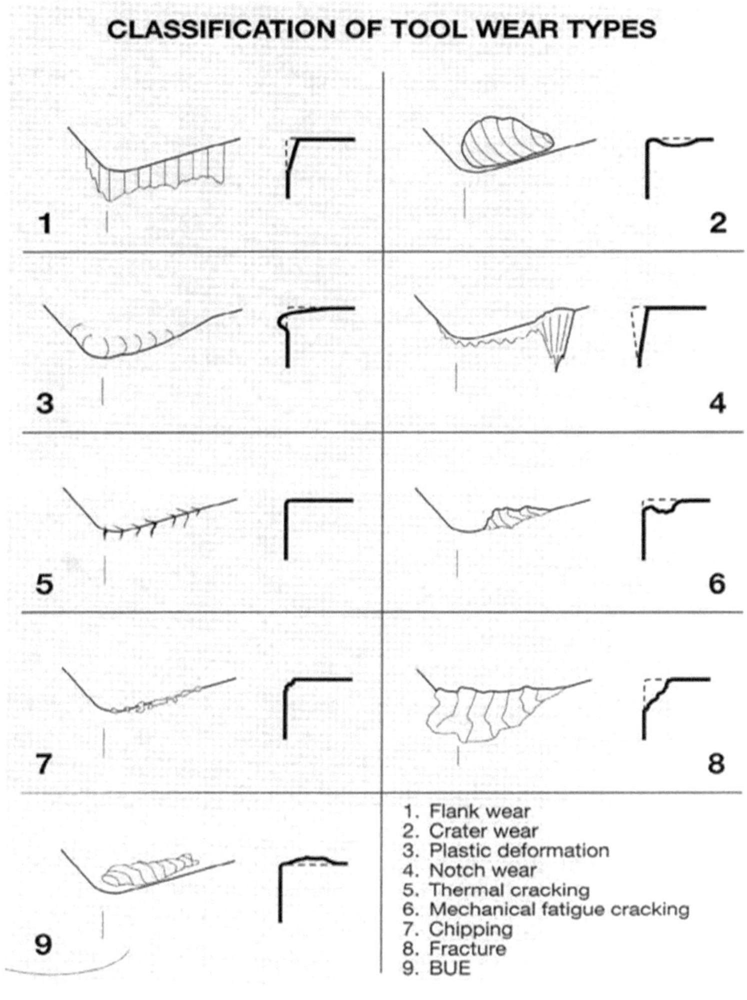

Fig. 3.4 Classification of cutting tool wear caused by a combination of different tool wear mechanisms [1]. *Reproduced courtesy of Sandvik and reprinted with permission of Springer publishers*

The classification of cutting tool wear is shown in Fig. 3.4. The criterion of machining, i.e., removal rate, surface roughness, tool life, and chip control, depends on the development of tool wear.

Figure 3.5 shows the first three types of tool wear known as flank wear, crater wear, and plastic deformation. Flank wear is caused by the abrasive wear mechanism and is caused by the chip moving against the face of the cutting tool. This type of wear is normal and is progressive. Crater wear is caused by a combination of abrasive and diffusion wear mechanisms. High tool and high hot hardness of the tool material minimizes the tendency for crater wear to occur. Crater wear should

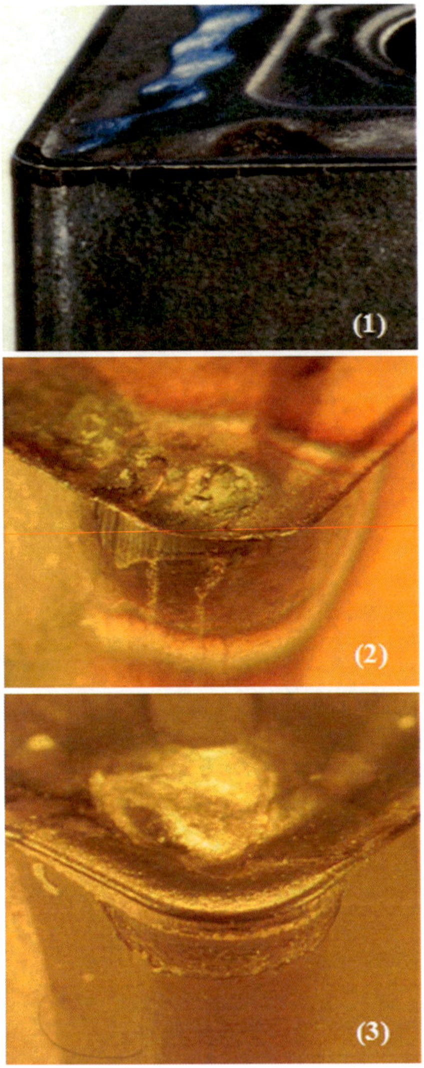

Fig. 3.5 Cutting tool wear types: (1) flank wear, (2) crater wear, and (3) plastic deformation [1]. *Reproduced courtesy of Sandvik and reprinted with permission of Springer publishers*

be avoided as it changes the geometry of the tool and weakens the cutting edge. Plastic deformation takes place as a function of high pressures and temperatures on the cutting tool. Figure 3.6 shows notch wear, thermal cracking, and mechanical fatigue cracking of the cutting tool.

Notch wear on the trailing edge is a type of adhesion wear but can be caused by oxidation. The wear is localized at the end of the cut where air can penetrate the cutting zone. Notch wear on the leading edge is mechanical and weakens the cutting edge. Thermal cracking is a form of fatigue and is due to thermal cycling, cracks form in the tool material, and the edge of the tool is lost through machining. Varying the chip thickness affects the cutting temperature at the chip/tool interface

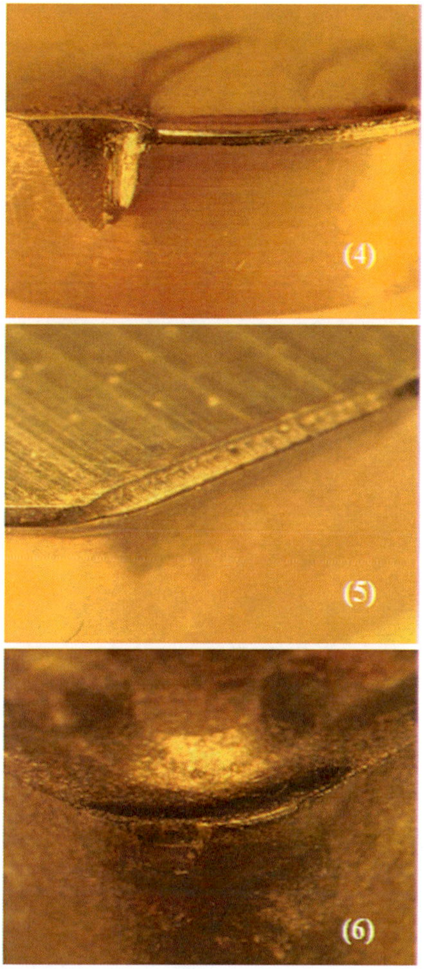

Fig. 3.6 Cutting tool wear types: (4) notch wear, (5) thermal cracking, and (6) mechanical fatigue cracking [1]. *Reproduced courtesy of Sandvik and reprinted with permission of Springer publishers*

and can lead to the rapid breakdown of the tool material. Mechanical fatigue cracking takes place when cutting force shocks are excessive.

Figure 3.7 shows the wear of the cutting edge caused by chipping, fracture, and the built-up edge. Chipping of the cutting edge is cyclic and is caused by the fatigue of the cutting tool due to intermittent cutting. Fracture is the catastrophic breakdown of the tool material and is caused by heavy cuts or very high cutting forces. Finally, the built-up edge changes the geometry of the cutting tool and leads to an increase in the level of cutting force. When the built-up edge breaks off, the cutting tool force reduces which leads to a cycling of the load. The tool edge eventually breaks off.

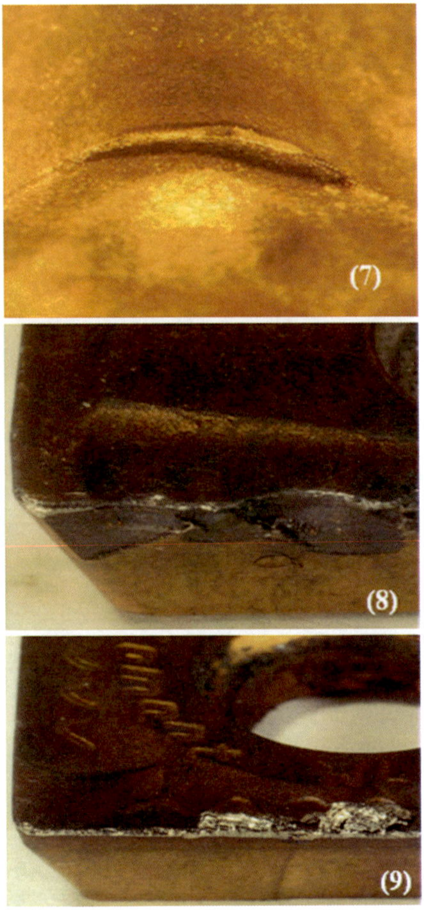

Fig. 3.7 Cutting tool wear types: (7) chipping, (8) fracture, and (9) built-up edge [1]. *Reproduced courtesy of Sandvik and reprinted with permission of Springer publishers*

3.3 Machining of Uranium Alloys

Milling, threading, drilling, tapping, and turning and facing of uranium and its alloys are reviewed in the publication by Aris [2]. Uranium and its alloys are tough and soft and tend to form a built-up edge. Carbide inclusions tend to make them abrasive as well. Flank wear is the normal type of wear encountered during the machining of uranium alloys when using the correct machining parameters [2]. The cutting edge appears to be mechanically worn, not chipped or notched. Traditionally, cutting tools that provide the best results for turning and facing uranium alloys

Table 3.1 Typical machining parameters when turning and facing uranium alloys [2]

	Roughing	Finishing
Depth of cut (mm)	1.27–2.54	0.0025–012
Cutting speed (m/min)	30–107	122
Feed (mm/rev)	0.15–0.2	0.05–0.1

Reprinted with permission of Springer publishers

and unalloyed uranium are uncoated grades of high tantalum carbide content having the nominal compositions [2]:

1. 74 wt%WC-20 wt%TaC-6 wt%Co
2. 50 wt%TaC-44 wt%WC-6 wt%Co

The tools have positive rake angles between 5° and 12°. For harder alloys, tungsten carbide inserts work better where abrasion, rather than heat at the tool tip, is the primary wear mechanism. In the past, machining with coated cutting tools has generally failed due to cracking of the coating and excessive wear of the coating. High tantalum carbide content tool inserts are preferred when machining unalloyed uranium because they can tolerate high levels of heat generated at the tool tip. However, higher tantalum carbide contents lower the hardness of the tools that leads to excessive wear of the tool edge. Typical machining parameters when turning and facing uranium alloys are shown in Table 3.1 [2].

Initial investigations into the deposition and microstructural properties of metastable $Ti_{l-x}Al_xN$ coatings were first published over a decade ago [3–6]. The principal reason for their increased use over TiN coatings in dry cutting and milling operations stems from enhanced high-temperature oxidation resistance. TiN oxidizes rapidly at temperatures above ≈600 °C to form a layer that is predominantly rutile TiO_2, which has a much larger molar volume than TiN and results in discontinuous spallation. In contrast, McIntyre et al. [7] showed that annealing $Ti_{0.5}Al_{0.5}N$ in oxygen gives rise to the formation of a stable passive oxide double layer, the upper one being Al-rich and the lower one being Ti-rich, with no measurable nitrogen in either. Inert marker experiments demonstrated that oxidation of $Ti_{0.5}Al_{0.5}N$ proceeds by the simultaneous outward diffusion of Al to the oxide/vapor interface and inward diffusion of O to the oxide–nitride interface.

A potential approach to further improve high-temperature performance is the addition of small concentrations of alloying components that are known to provide even higher oxidation resistance [8]. In this article, we present the first results on the microstructure and oxidation resistance of $Ti_{l-x-y-z}Al_xCr_yY_2N$ films deposited using a combined cathodic arc/unbalanced magnetron sputter (CA/UBM) deposition process [9, 10]. The paper also describes the improvement in cutting tool life during dry machining operations using a series of coated tool inserts machining uranium alloys, which suggest that yttrium-doped coatings may have a beneficial effect on cutting tool life compared to TiN and TiAlN coatings.

3.3.1 Experimental Procedures

3.3.1.1 Deposition of Coatings to Cutting Tools

All films were deposited in a Hauzer Techno Coating Europe B.V. HTC 1000-4 combined CA/UBM system [9]. The growth chamber contains four 60 cm × 19 cm × 1.2 cm vertically mounted targets (two opposing pairs, each separated by 100 cm) which can function in either CA or UBM mode. The base pressure of the system is less than 10^{-5} mbar (7×10^{-6} Torr). Three radiant heater arrays allow independent closed-loop control of the growth temperature (450 °C) using two chromel–alumel thermocouples attached to the surface of dummy substrates. For this study, the following targets (>99.8 % pure) were used: (a) a hot isostatic pressed Cr plate, (b) two cast $Ti_{0.5}Al_{0.5}$ plates, and (c) a forged $Ti_{0.48}Al_{0.48}Y_{0.04}$ plate. Sputtering was carried out in mixed N_2 (99.999 %) and Ar (99.999 %) discharges with the Ar flow maintained constant at 200 sccm, while the N_2 flow was regulated to maintain the total pressure at 3.4×10^{-3} mbar (2.5 mTorr).

The inserts used in these experiments were fine-grain Co-bonded tungsten carbide inserts polished to an average peak-to-valley roughness $R_a = 10$ nm and were cleaned as described by Petrov et al. [10] and mounted in a substrate holder with a 3-axis planetary rotation table. The average target-to-substrate distance was 25 cm. Immediately prior to deposition, the substrates were subjected to −1.2-kV Cr-steered-arc metal-ion etching for 20 min at a total pressure of 6×10^{-4} mbar (0.45 mTorr). Cr was chosen as an ion etchant in order to suppress droplet deposition on the substrate [11]. A 0.2-μm-thick $Ti_{l-x-y}Al_xCr_yN$ base layer was then deposited by UBM in which the power to the two $Ti_{0.5}Al_{0.5}$ targets P_{TiAl} was 8 kW each, while $P_{Cr} = 0.5$ kW and $P_{TiAlY} = 0$. Power to the $Ti_{0.48}Al_{0.48}Y_{0.04}$ was then set at 8 kW to deposit 3-μm-thick $Ti_{l-x-y-z}Al_xCr_yY_zN$ films. The properties of $Ti_{l-x-y-z}Al_xCr_yY_zN$ were compared with equivalent thickness $Ti_{l-x-y}Al_xCr_yN$ layers deposited under the same conditions as the base layer and $Ti_{l-x}Al_xN$ ($P_{Cr} = 0$) layers deposited using three $Ti_{0.5}Al_{0.5}$ targets operated at $P_{TiAl} = 8$ kW. In all cases, the substrate bias during film growth was maintained constant at −75 V.

Surface morphologies of as-deposited and annealed samples were examined using scanning electron microscopy (SEM), while microstructures and phase compositions of films and film–substrate interfaces were determined using a combination of X-ray diffraction (XRD) and cross-sectional transmission electron microscopy (XTEM). XTEM samples were prepared by mechanical thinning and ion milling as described by Petrov et al. [10]. Area-averaged elemental depth profiles were measured by 2-MeV He Rutherford backscattering spectroscopy (RBS) and secondary neutral mass spectrometry (SNMS) using a 16-keV Ga^+ primary beam. Local compositional profiles were investigated using energy-dispersive X-ray (EDX) analyses of XTEM samples in a Vacuum Generators HB5 scanning TEM (STEM) equipped with a field-emission source and operated at 100 kV. During STEM analyses, samples were traversed with respect to a stationary electron beam focused to a diameter of ≈1 nm, and X-ray spectra were collected at a take-off angle of 40°. Corrections for atomic number were carried out using the MAGIC-V computer code [12]. Oxidation rates in air were assessed by thermogravimetric analyses (TGA) in isothermal (at 800–950 °C for up to 5 h) and linear-temperature-ramp (400–1,000 °C at 1 °C min^{-1}) modes. Knoop microhardness values were determined using a load of 25 g.

3.3.1.2 Machining Experiments

Machining experiments were performed to assess the life of newly developed titanium-based coated tools. The assessment of machinability used in these experiments pertains to the development of Taylor's tool life equations for uncoated and coated cutting tools. Taylor's tool life equation is stated as

$$V \cdot T^n = C \tag{3.1}$$

where V is the cutting speed in meters per minute (m/min), T is the tool life in minutes, n is an exponent dependent upon the machining conditions and the tool and workpiece compositions, and C is a constant. For each tool life equation generated, a sample of the metal is turned at a specific cutting speed, and the time it takes to wear 0.3 mm of the flank face of the insert away from the cutting tool is noted and used in the calculation. For the machining of uranium alloys in this study, a computer numerically controlled lathe was used to vary the cutting speed. An Emco Maier CNC lathe was used to turn bars of uranium alloy that were 1-in round bars with a measured surface roughness of 25 μm. The type of cut used was a roughing cut using the parameters shown in Table 3.1. A maximum cut was 2.54 mm in order to produce a thick chip. Water coolant was applied to the chips as they collected in the chip tray. This was performed so as not to ignite the chips after they were deposited into the chip tray. This was conducted in order to machine the metal "dry" in order to confirm or refute the effectiveness of the coated tool inserts. The uncoated cutting tool inserts were composed of 94 wt% tungsten carbide bonded with a 6 wt% cobalt binder. The coated cutting tools were supplied with a superlattice coating composed of 1,000 deposited layers to form a coating thickness of 3 μm. The alloy used for machining experiments was an intermediate uranium alloy with 0.75 wt% titanium content usually used for armor-piercing applications. The maximum feed rate used in these experiments was set at 0.2 mm per revolution of the spindle. The tools were initially sharp and were not previously used in any other application. The tool inserts were coated at the Bodycote-SHU facility housed within the Materials Research Institute at Sheffield Hallam University. The initial cutting speed used for the experiments was set at 30 m/min. The tool life was measured by inspecting the cutting tool until 0.3 mm of the flank had worn away. Further increments of cutting speed were made until a maximum cutting speed of 70 m/min had been achieved.

3.3.2 *Experimental Results and Discussion*

3.3.2.1 Microstructure, Composition, and Properties of Deposited Layers

Elemental analysis by RBS and STEM–EDX revealed that the $Ti_{1-x}Al_xN$, $Ti_{1-x-y}Al_xCr_yN$, and $Ti_{1-x-y-z}Al_xCr_yY_zN$ films were stoichiometric with N/metal ratios 1 ± 0.03 and compositions of $Ti_{0.46}Al_{0.54}N$, $Ti_{0.44}Al_{0.53}Cr_{0.03}N$, and $Ti_{0.43}Al_{0.52}Cr_{0.03}Y_{0.02}N$, respectively.

XRD θ-2θ scans showed that as-deposited $Ti_{0.44}Al_{0.53}Cr_{0.03}N$ and $Ti_{0.43}Al_{0.52}Cr_{0.03}Y_{0.02}N$ layers were single phase in the B1-NaCl structure with lattice parameters a_o of 0.4170 nm and 0.4220 nm, respectively. Comparison of integrated peak intensities showed that $Ti_{0.44}Al_{0.53}Cr_{0.03}N$ alloys exhibit a strong (111) preferred orientation with an intensity ratio $I_{111}/I_{\Sigma} = 0.84$; however, the incorporation of 2 mol% YN resulted in a (200) texture with $I_{002}/I_{\Sigma} = 0.67$. For comparison, the (111) and (200) intensity ratio values for randomly oriented TiN [13] are 0.28 and 0.37, respectively. Peak widths were found to increase significantly with the addition of YN. For example, the full-width-at-half-maximum intensity of the $Ti_{0.43}Al_{0.52}Cr_{0.03}Y_{0.02}N$ (111) peak was larger by a factor of 5 (1° vs. 0.2°) than that of $Ti_{0.44}Al_{0.53}Cr_{0.03}N$. From XTEM analyses discussed below, this is primarily due to the strong effect of Y on grain refinement.

Figure 3.8a, b shows bright-field XTEM images and selected-area electron diffraction (SAED) patterns from as-deposited $Ti_{0.44}Al_{0.53}Cr_{0.03}N$ and $Ti_{0.43}Al_{0.52}Cr_{0.03}Y_{0.02}N$ films on stainless steel. Examination of the film–substrate interface reveals local epitaxy on underlying substrate grains that evolves into competitive columnar growth at a thickness greater than 100–150 nm. The formation of a similar interfacial microstructure was described in detail by Petrov et al. [10] for the case of $Ti_{1-x-y}Al_xNb_yN$ alloys deposited on steel substrates following $Ti_{0.50}Al_{0.50}$ CA-ion etching. In $Ti_{0.44}Al_{0.53}Cr_{0.03}N$ films, the individual columns are typically single grains giving rise to uniform XTEM contrast over extended column lengths. The SAED image in Fig. 3.8a, obtained with a 0.5-μm-diameter aperture focused at the middle of the film, shows a B1-NaCl structure pattern with rings of spots consistent with a relatively large grain size.

$Ti_{0.43}Al_{0.52}Cr_{0.03}Y_{0.02}N$ samples exhibit a transition, above the base layer, to a much less-pronounced columnar growth with nearly equiaxed grains. The addition of YN resulted in disruption, through continuous re-nucleation, of local epitaxy on individual columns. Similar microstructures have been observed to form when re-nucleation was generated in TiN [14] films by excessive ion-irradiation damage and in $Ti_{1-x}Al_xN$ [15] metastable alloys by wurtzite–AlN phase precipitation during growth. In the present experiments, we attribute the re-nucleation to Y surface segregation during deposition. Consistent with the small grain size, the SAED pattern in Fig. 3.8b exhibits broad continuous rings of approximately uniform intensity.

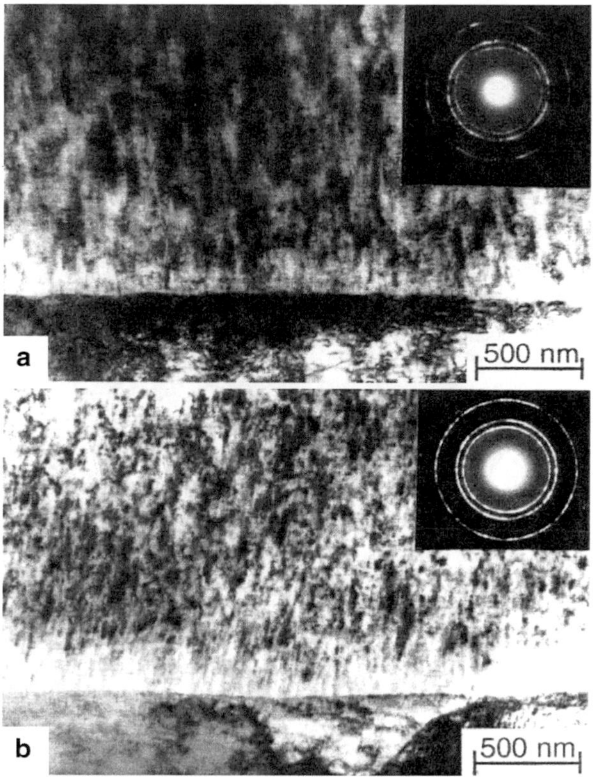

Fig. 3.8 XTEM bright-field images and SAED patterns from as-deposited (**a**) $Ti_{0.44}Al_{0.53}Cr_{0.03}N$ and (**b**) $Ti_{0.43}Al_{0.52}Cr_{0.03}Y_{0.02}N$ alloy layers on steel substrate (Courtesy of Prof. Munz). *Reprinted with permission from Springer publishers*

The Knoop microhardness for $Ti_{0.46}Al_{0.54}N$ and $Ti_{0.44}Al_{0.53}Cr_{0.03}N$ was $HK_{0.025} = 2{,}400$ kg mm^2, consistent with previous reports [6] for $Ti_{0.5}Al_{0.5}N$. The incorporation of YN increased $HK_{0.025}$ to 2,700 kg mm^2. The surface roughness, measured using a stylus profilometer, was $R_a = 0.038–0.040$ μm for all films. VDI Rockwell-C indentation adhesion evaluation revealed an indent quality [16, 17] of 1 indicating excellent adhesion.

3.3.2.2 Oxidation Behavior of Coatings

Figure 3.9a, b are plots of TGA oxidation rate measurements for TiN, $Ti_{0.46}Al_{0.54}N$, $Ti_{0.44}Al_{0.53}Cr_{0.03}N$, and $Ti_{0.43}Al_{0.52}Cr_{0.03}Y_{0.02}N$ layers. The onset for rapid oxidation of TiN during 1 °C min^{-1} thermal ramping (Fig. 3.9a) occurs at $T_{ox} \approx 600$ °C, while T_{ox} for $Ti_{0.46}Al_{0.54}N$, $Ti_{0.44}Al_{0.53}Cr_{0.03}N$, and $Ti_{0.43}Al_{0.52}Cr_{0.03}Y_{0.02}N$ alloys increases to ≈870, 920, and 950 °C, respectively.

Isothermal measurements at 900 °C (Fig. 3.9b) also show that alloying dramatically enhances the oxidation resistance. The data not only confirms the superior oxidation resistance of $Ti_{l-x}Al_xN$ over TiN [5, 7] but also demonstrates that the

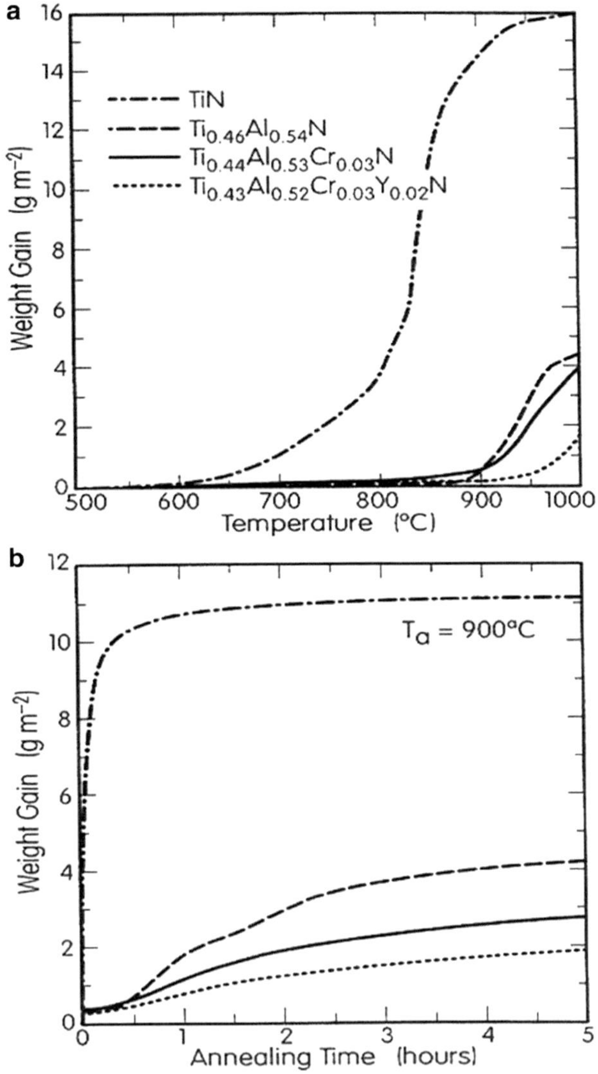

Fig. 3.9 Thermogravimetric oxidation rate measurements in air during (**a**) a linear-temperature ramp at 1 °C min^{-1} and (**b**) an isothermal anneal at 900 °C (Courtesy of Prof. Munz). *Reprinted with permission from Springer publishers*

successive addition of small amounts of CrN and YN to $Ti_{l-x}Al_xN$ provides further significant improvements. Oxidation reaction paths were determined using a combination of SNMS, SEM, XTEM, and STEM–EDX analyses.

Figure 3.10a, b are SNMS profiles from $Ti_{0.44}Al_{0.53}Cr_{0.03}N$ and $Ti_{0.43}Al_{0.52}Cr_{0.03}Y_{0.02}N$ alloys annealed for 1 h at 900 °C. For clarity of presentation, only the major signals (N, O, Ti, Al, and Fe) are plotted. Both samples exhibit a two-layer, Al-rich followed by Ti-rich (i.e., Al-depleted) oxidation zone similar to that reported [7] for the initial stages of $T_{0.5}Al_{0.5}N$ oxidation.

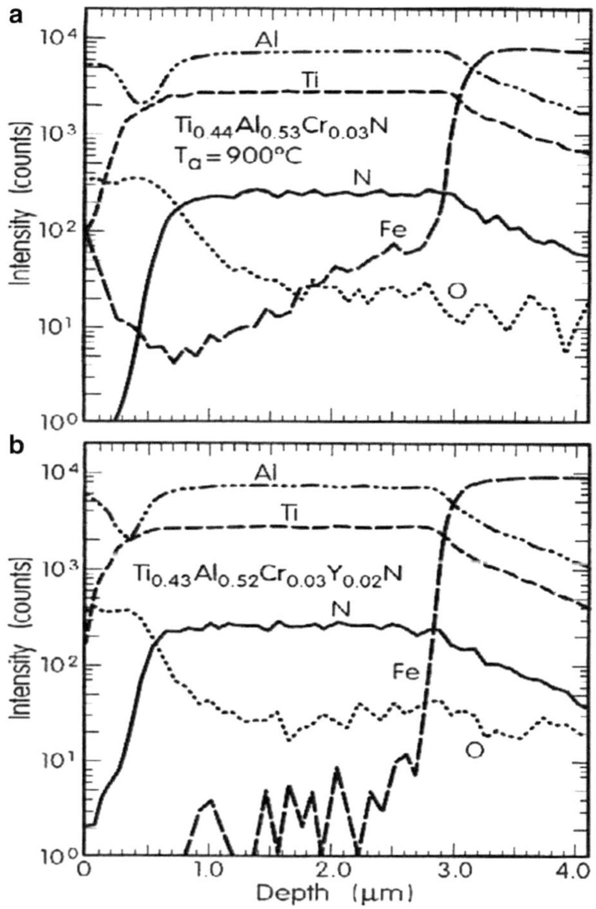

Fig. 3.10 SNMS profiles from (**a**) $Ti_{0.44}Al_{0.53}Cr_{0.03}N$ and (**b**) $Ti_{0.43}Al_{0.52}Cr_{0.03}Y_{0.02}N$ alloys annealed for 1 h at 900 °C (Courtesy of Prof. Munz). *Reprinted with permission from Springer publishers*

Figure 3.10 shows that the oxide bilayer thickness decreases from 0.8 to 0.6 μm with the addition of YN. In the case of $Ti_{0.44}Al_{0.53}Cr_{0.03}N$, the Fe substrate profile is markedly broadened penetrating the entire nitride film and accumulating at the free surface. In contrast, no measurable Fe penetration is observed in $Ti_{0.43}Al_{0.52}Cr_{0.03}Y_{0.02}N$ layers. Figure 3.11 is a collage of SEM and XTEM images from a $Ti_{0.44}Al_{0.53}Cr_{0.03}N$ layer annealed for 1 h in air at a higher temperature of 950 °C. The large-scale SEM micrograph of the XTEM sample in Fig. 3.11a reveals massive oxidation together with the formation of micron-size voids in the substrate. The film region near the interface, marked sub-layer 1 in Fig. 3.11a and shown at higher resolution in the XTEM image in Fig. 3.11b, exhibits a columnar structure similar to as-deposited nitride layers. The SAED pattern shows that this region remains NaCl structure after annealing. The darker-contrast sub-layer 2, magnified in Fig. 3.11c, consisted of columnar nitride grains that are partially oxidized. The column boundaries are eroded and open as a result of cation out-diffusion. Above region 2, several oxide sub-layers (sub-layers 3–6) are visible with

Fig. 3.11 A collage of SEM and XTEM images from a $Ti_{0.44}Al_{0.53}Cr_{0.03}N$ layer annealed for 1 h in air at 950 °C. (**a**) A large-scale SEM micrograph of the XTEM sample. (**b–d**) Higher-resolution XTEM images of the indicated regions, SAED pattern corresponding to sub-layer 1 (Courtesy of Prof. Munz). *Reprinted with permission from Springer publishers*

increasing grain size toward the free surface. Figure 3.12 is a typical STEM–EDX compositional depth profile of the sample presented in Fig. 3.11, revealing a complex sequence of compositional modulations. Based upon the XTEM, SAED, and STEM results, together with previous detailed studies of $Ti_{0.5}Al_{0.5}N$

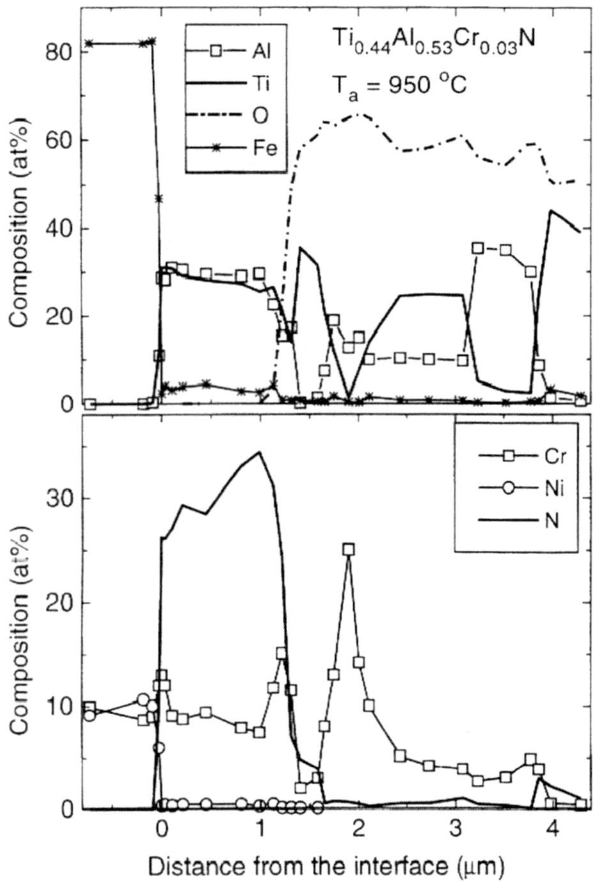

Fig. 3.12 STEM–EDX compositional depth profile through the cross-sectional sample shown in Fig. 3.11 (Courtesy of Prof. Munz). *Reprinted with permission from Springer publishers*

oxidation [7], the profiles in Fig. 3.12 are interpreted as follows. The top three sub-layers correspond to those observed in $Ti_{0.5}Al_{0.5}N$ after prolonged annealing at $T_{ox} > 900\ ^\circ C$ [7]. Large facetted TiO_2 crystallites with a rutile structure (see upper part of Fig. 3.11a, d) have developed on top of the initially formed Al-rich/Ti-rich oxide bilayer through massive out-diffusion of Ti. The oxide bilayer, which initially is fine-grained and continuous, has coarsened through recrystallization and becomes porous as seen in Fig. 3.11. Thus oxygen can, at this stage, more rapidly access the oxidation front where STEM–EDX profiles show the formation of a second Al-rich/Ti-rich oxide bilayer. The new feature at this second stage of oxidation is that Cr accumulates to ≈15 at.% at the lower interface and 25 at.% at the upper interface of the second Ti-rich oxide sub-layer. This is another indication of the out-diffusion of substrate species during annealing. Indeed, the content of Cr in the substrate near the film–substrate interface is significantly depleted, <10 at.% compared to the initial ≈18 at.%. The out-diffusion of Cr is directly related to the formation of voids in the substrate. The loss of Cr and Fe from the steel is the cause of the voids observed in the substrate near to the interface

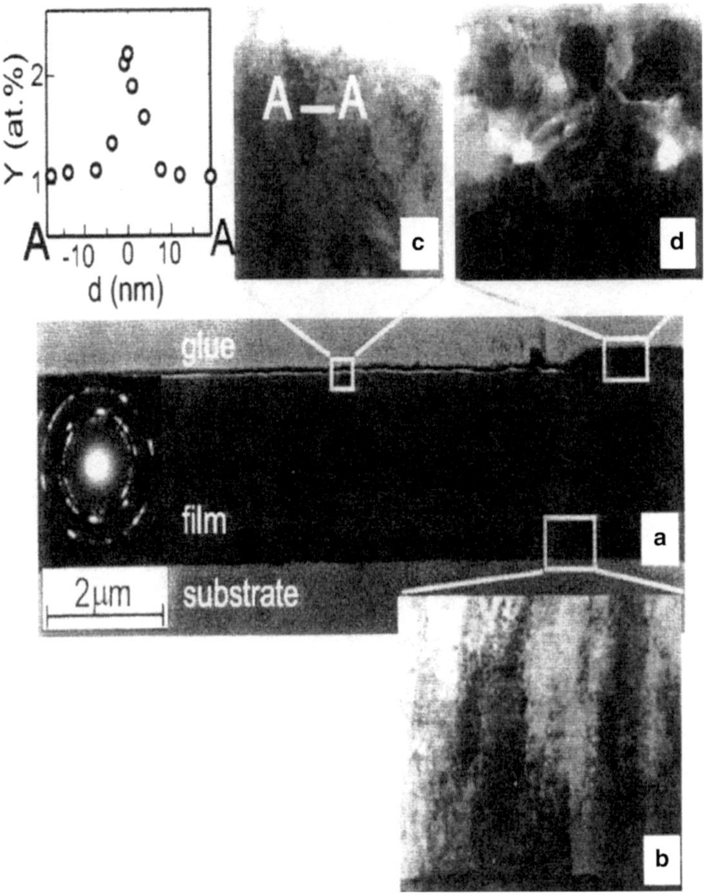

Fig. 3.13 A collage of SEM and XTEM images from a $Ti_{0.43}Al_{0.52}Cr_{0.03}Y_{0.02}N$ layer annealed for 1 h in air at 950 °C. (**a**) A large-scale SEM micrograph of the XTEM sample. (**b–d**) Higher-resolution XTEM images of the indicated regions; the SAED pattern corresponds to the nitride film. The inset shows the Y concentration from a STEM–EDX profile along the line A–A in c (Courtesy of Prof. Munz). *Reprinted with permission from Springer publishers*

(see Fig. 3.11a). Similar analyses were carried out on $Ti_{0.43}Al_{0.52-}Cr_{0.03}Y_{0.02}N$ films annealed under identical conditions. The SEM micrograph in Fig. 3.13a shows that, compared to $Ti_{0.44}Al_{0.53}Cr_{0.03}N$, the film is still dense with relatively little evidence of oxidation. XTEM images of the interfacial (Fig. 3.13b) and upper (Fig. 3.13c) regions of the annealed sample reveal that the film–substrate interface remains abrupt and the $Ti_{0.43}Al_{0.52}Cr_{0.03}Y_{0.02}N$ microstructure is virtually unaffected by the anneal. SAED patterns from the base layer (not shown) and the bulk film are essentially identical to those from as-deposited layers. The microstructure of the oxide layer, shown in the XTEM micrograph in Fig. 3.13d, exhibits a dense, partially crystalline surface overlayer with a thickness of between 100 and 150 nm

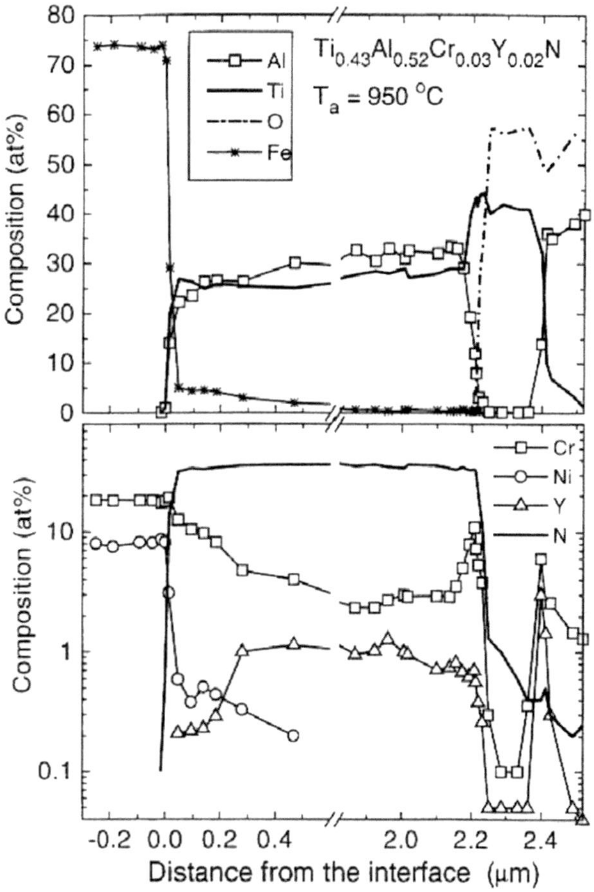

Fig. 3.14 STEM–EDX compositional depth profile through the interfacial and oxidized regions of the cross-sectional sample shown in Fig. 3.13 (Courtesy of Prof. Munz). *Reprinted with permission from Springer publishers*

and an under-dense, although coarser (grain size ≈100 nm), polycrystalline underlayer. The total thickness of the oxide layer on $Ti_{0.43}Al_{0.52}Cr_{0.03}Y_{0.02}N$ is ≈0.4 μm compared to greater than 3 μm on $Ti_{0.44}Al_{0.53}Cr_{0.03}N$ after 1 h at 950 °C. Most of the oxide layer delaminated during XTEM sample preparation and is only visible in the upper right-hand corner of Fig. 3.13a.

Figure 3.14 is a typical STEM–EDX compositional depth profile through the film–substrate interfacial region and the oxidized region of the $Ti_{0.43}Al_{0.52}Cr_{0.03}Y_{0.02}N$ sample shown in Fig. 3.13. The interface width (the distance over which the Fe, Al, and Ti signals change from 15 to 85% of their saturation values) remains the same as for as-deposited samples, ≈15 nm.

The concentrations of Cr and Fe in the base layer near the film–substrate interface are ≈12 and 5 at.%, respectively, and decrease rapidly in the bulk $Ti_{0.43}Al_{0.52}Cr_{0.03}Y_{0.02}N$ film. The profile through the oxidized region again shows an Al-rich/Ti-rich bilayer with CrO_x accumulation on both sides of the Ti-rich oxide layer. In this case, however, not only is the oxide bilayer much thinner but Cr

Table 3.2 Cutting tool life experimental results

Cutting tool	Spindle speed (m/min)	Cutting tool life
Uncoated tool (WC–Co)	30	1 min
	50	9 s
	70	3 s
TiN-coated WC–Co	30	1.5 min
	50	16.5 s
	70	5 s
$Ti_{0.46}Al_{0.54}N$-coated WC–Co	30	2 min
	50	27 s
	70	10.5 s
$Ti_{0.44}Al_{0.53}Cr_{0.03}N$-coated WC–Co	30	2.5 min
	50	37 s
	70	15 s
$Ti_{0.43}Al_{0.52}Cr_{0.03}Y_{0.02}N$-coated WC–Co	30	3.9 min
	50	63 s
	70	27 s

Reprinted with permission of Springer publishers

concentration is substantially reduced (<10 at.% Cr). YO_x also accumulates at the interface between the Ti-rich and Al-rich oxide sub-layers but not at the lower Ti-rich oxide–nitride interface. STEM–EDX compositional profiles across grain boundaries in annealed $Ti_{0.44}Al_{0.50}Cr_{0.04}Y_{0.02}N$ films show a significant segregation of yttrium, Y. An example, corresponding to the line labelled A–A in Fig. 3.13c, is reproduced in the upper left-hand corner of Fig. 3.13. The concentration profile of Y at the boundary is enhanced more than a factor of two. The presence of excess Y and YO_x at film grain boundaries may play a decisive role in reducing bulk film oxidation rates through inhibition of both cation out-diffusion and O in-diffusion.

3.3.2.3 Cutting Tool Life and Wear Behavior

The life of the uncoated and coated cutting tools was compared after measuring the progression of flank wear of the cutting tool inserts after machining at different cutting speeds under the conditions stated in Sect. 3.2. The results of the tool life experiments are presented in Table 3.2. Under the machining conditions stated in Sect. 3.2, the dominant wear behavior observed during the machining experiments was flank wear. The data provided above allows us to calculate the Taylor tool life equation for each cutting tool type for the experimental conditions stated. The following equations relate to the cutting tool types described:

1. Uncoated WC–Co: $V{\cdot}T^{0.28} = 30$
2. TiN-coated WC–Co: $V{\cdot}T^{0.30} = 34$
3. $Ti_{0.46}Al_{0.54}N$-coated WC–Co: $V{\cdot}T^{0.35} = 38$
4. $Ti_{0.44}Al_{0.53}Cr_{0.03}N$-coated WC–Co: $V{\cdot}T^{0.37} = 42$
5. $Ti_{0.43}Al_{0.52}Cr_{0.03}Y_{0.02}N$-coated WC–Co: $V{\cdot}T^{0.39} = 51$

The results presented show that the tool life is increased as the coating is improved with small additions of elemental chromium and yttrium. This effect coincides with the observations made in Sect. 3.3.2 regarding the oxidation behavior of the coatings treated with various elemental concentrations of chromium and yttrium. Compared to other materials, the tool life equations presented appear to show that machining uranium alloys results in an extremely low tool life compared to more commonly used engineering steels with microstructures that tend to favor good machinability. The machinability of uranium alloys appears to be similar to that of austenitic stainless steels, i.e., very low tool life even at moderately high cutting speeds. The chief characteristic property of uranium is its dependence on temperature for changing its ductility and its tendency to work harden which makes it difficult to machine. In this respect, it is noted that machining uranium resembles machining austenitic steels with a manganese content of 13 wt% [18]. The observations of wear behavior tend to show that the normal mode of wear during machining is progressive flank wear.

However, it must be noted that cutting tool life experiments were based on roughing machining operating conditions and did not include experimental determination of tool life during finish machining operations.

3.4 Conclusions

The incorporation of 3 mol% CrN in $Ti_{1-x}Al_xN$ alloys did not significantly change film hardness or microstructure. The latter remained columnar with individual columns consisting of single grains over extended vertical distances. Adding an additional 2 mol% YN increased the film hardness by $HK_{0.025} \approx 300$ kg mm^{-2}, while Y segregation during growth promoted continuous re-nucleation which resulted in a considerable grain refinement and a more equiaxed structure.

Thermogravimetric analysis in oxidizing ambient atmospheres showed that the onset of rapid oxidation was increased from $\approx$ 600 °C for TiN to 950 °C for $Ti_{0.43}Al_{0.52}Cr_{0.03}Y_{0.02}N$, compared to 870 °C for $Ti_{0.46}Al_{0.54}N$ and 920 °C for $Ti_{0.44}Al_{0.53}Cr_{0.03}N$. The initial oxidation reaction path was found to be similar for the three alloys with the formation of an Al-rich surface oxide and Ti-rich oxide underlayer. Annealing $Ti_{0.44}Al_{0.53}Cr_{0.03}N$ layers on steel substrates for 1 h at 950 °C results in massive oxidation with cation out-diffusion, giving rise to void formation and under-dense column boundaries extending nearly to the film–substrate interface. Voids are also observed on the substrate side of the film–substrate interface due to rapid out-diffusion of Cr, which is rejected by Ti-rich sub-layers in the oxidized film and accumulates at adjacent boundaries, and Fe. In stark contrast, the addition of only 2 mol% YN to form $Ti_{0.43}Al_{0.52}Cr_{0.03}Y_{0.02}N$ reduces the oxide thickness from >3 to $\approx$0.4 μm while significantly inhibiting out-diffusion of substrate species. STEM–EDX profiles, obtained after annealing, show that Y segregates to the nitride grain boundaries. This may explain the enhanced high-temperature oxidation resistance of $Ti_{0.43}Al_{0.52}Cr_{0.03}Y_{0.02}N$

alloy as Y and YO_x inhibit grain-boundary diffusion of both cation species toward the free surface and oxygen penetration into the film.

The machining characteristics of coated cutting tools did have some benefit to extending the life of the cutting tool when machining uranium alloys. The benefits associated with using titanium-based coated tools were further enhanced by using a superlattice coated cutting tool. This ensured that thermal fatigue cracking of the coating did not interfere with the normal operation of the cutting insert when machining uranium alloys. The use of coated tools appeared to have a similar tool life compared to austenitic stainless steels. However, further experiments are required to determine the life of the tools when finish machining operations are involved.

Acknowledgements The authors acknowledge permission from Springer publishers to reproduce the article (license # 2922220397554—June 4, 2012): M. J. Jackson and G. M. Robinson, *Machining Depleted Uranium Using Coated Cutting Tools*, Journal of Materials Engineering and Performance, **15**, 2006, 161–171, that has been updated to reflect advances in the machining of depleted uranium and alloys. Copyright is retained by Springer publishers and is published "With kind permission from Springer Science+Business Media: M. J. Jackson and G. M. Robinson, *Machining Depleted Uranium Using Coated Cutting Tools*, Journal of Materials Engineering and Performance, **15**, 2006, 161–171, and any original (first) copyright notice displayed with material."

References

1. Metal Cutting Technology—Technical Guide, published by Sandvik-Coromant (Sweden), 2010, Section A (Turning), A1–A153
2. Aris J (1989) Metals handbook—machining. In: Aris J (ed) Machining of uranium and uranium alloys, vol 16, 9th edn. ASM International, Materials Park, OH, pp 874–878, ISBN0-87170-007-7
3. Knotek O, Böhmer M, Leyendecker T (1986) Structure and properties of Ti and Al hard compound films. J Vac Sci Technol A 4:2695–2710
4. Jehn H, Hofmann S, Rückborn V-E, Münz W-D (1986) Morphology and properties of sputter (Ti, Al) N layers on high-speed steel as a function of deposition temperature and sputtering atmosphere. J Vac Sci Technol A 4:2701–2705
5. Münz W-D (1986) TiAlN films: an alternative to TiN coatings. J Vac Sci Technol A 4:2717–2725
6. Håkansson G, Sundgren J-E, McIntyre D, Greene JE, Münz W-D (1987) Microstructure and physical properties of polycrystalline metastable Ti0.5Al0.5N alloys grown by d.c. magnetron sputter deposition. Thin Solid Films 153:55–62
7. McIntyre D, Greene JE, Håkansson G, Sundgren J-E, Münz W-D (1990) Oxidation of metastable single phase polycrystalline Ti0.5Al0.5N Films. J Appl Phys 67:1542–1553
8. Münz W-D (1990) Oxidation resistance of hard wear resistant Ti0.5Al0.5N coatings grown by magnetron sputtering deposition. Werkstoffe Korros 41:753–754
9. Münz W-D, Schulze D, Hauzer FJM (1992) A new method for hard coatings—arc bond sputtering. Surf Coat Technol 50:169–178
10. Petrov I, Losbichler P, Bergstrom D, Greene JE, Münz W-D, Hurkmans T, Trinh T (1997) Large scale fabrication of hard superlattice films by combined stress arc evaporated unbalanced magnetron sputtering. Thin Solid Films 302:179–192

11. Münz W-D, Smith IJ, Lewis DB, Creasey S (1997) Droplet formation on steel substrates during cathodic arc steered metal ion etching. Vacuum 48:473–481
12. Colby JW (1980) Quantex-ray instruction manual. Kevex, Foster City, CA
13. JCPDS International Center for Powder Diffraction Data (1989) Powder diffraction file for tin [6-642]. Swarthmore, PA
14. Petrov I, Hultman L, Helmersson U, Sundgren J-E, Greene JE (1989) A newly developed ion implanter for industrial applications. Thin Solid Films 169:299
15. Adibi F, Petrov I, Hultman L, Wahlström U, Shimizu T, McIntyre D, Greene JE, Sundgren J-E (1991) Low energy ion irradiation during growth of TiN. J Appl Phys 69:6437
16. VDI-Richtlinien 3198 (1992) Beschichten von Werkzeugen der Kkaltmassivumformung. Beuth Verlag, Berlin
17. Jackson MJ, Robinson GM (2006) Machining depleted uranium using coated cutting tools. J Mater Eng Perform 15:161–171
18. Yemel'Yanov VS, Yesstyyukhin AI (1969) The metallurgy of nuclear fuel—properties and principles of the technology of uranium. Thorium and Plutonium, Pergamon Press, Oxford

Bibliography

Aris J (1989) Metals handbook—machining. In: Aris J (ed) Machining of uranium and uranium alloys, vol 16, 9th edn. ASM International, Materials Park, OH, pp 874–878, *ISBN0-87170-007-7*

Boland JF, Sandstrom DJ (1974) Mechanical fabrication, heat treatment and machining of uranium alloys. Publication LA-UR-74-113, Los Alamos Scientific Laboratory, Los Alamos, Mexico, On behalf of the US Atomic Energy Commission, Contract # W-7405-Eng, 36, 1974

Conboy J, Shevchik P (1983) DU chip recovery program—phase I: a machining study for the production of contaminant-free chips. Report # AD-A131 389/9, South Creek Industries, Inc., July 1983

Denst A, Ross HV (1955) How to machine uranium. Am Machinist 99(16):95–97

Fuller JE, Lynch JK (1971) Machining study of a uranium-4 wt.% niobium alloy. Report # RFP-1743, Rocky Flats Plant, Rockwell International, November 1971

Hurst JS, Read GM (1972) Machining depleted uranium. Publication Y-SC-39, Union Carbide Corporation, Nuclear Division, Y-12 Plant, Oak Ridge, Tennessee, April 24 1972

Latham-Brown CE, Porter F. Atmosphere assisted machining of depleted uranium (DU) penetrators. Report # AD-A182 138/8/XAB, US Army Armament Research, Development and Engineering Center

Machining and Grinding of Ultra High Strength Steels and Stainless Steels. AEC/NASA Handbook SP-5084, for US Army Missile Command, Authored by Battelle's Columbus Laboratories, Battelle Memorial Institute, Columbus, OH

Morris TO (1981) Machining of uranium and uranium alloys. Paper presented at the Uranium Technology Seminar, American Society of Metals, Gatlinburg, TN, 1–31 May 1981

Olofson CT, Meyer GE, Hoffmanner AL (1976) Processing and applications of depleted uranium alloy products. Report # MCIC-76-28, pp. 60–83, Metals and Ceramics Information Center, Battelle Columbus Laboratory, Columbus, OH, September 1976

Stephens WE (1967) Uranium machining. Technical Paper MR 67-222, American Society of Manufacturing Engineers, Dearborn, Michigan, May 1967

Wright WJ (1967) The machining of nuclear materials. Production Engineer 46(11):638–650

Chapter 4
Grinding of Uranium and Uranium Alloys

Mark J. Jackson, Micheal D. Whitfield, Grant M. Robinson, Rodney G. Handy, and Jonathan S. Morrell

Abstract The grinding of uranium and its alloys is not very well understood, and the existing information that relates grinding process parameters and grinding wheel structure to the grinding performance is outdated at best. Since the initial papers published on grinding uranium were published more than 30 years ago, advances in grinding and the development of abrasive grains and bonding systems have accelerated to such an extent that the old published data is not very reliable. Therefore, this chapter explains the current advances in abrasive grain technology and also covers the recommendations made by previously generated data and how the data could be improved using the new types of grains and bonding systems.

Keywords Uranium • Uranium alloys • Grinding • Abrasive grains • Grinding wheels

M.J. Jackson (✉) • M.D. Whitfield • G.M. Robinson
Center for Advanced Manufacturing, Purdue University, West Lafayette, IN, USA
e-mail: jacksonmj04@yahoo.com; michaeldarrenwhitfield@gmail.com; grantmrob@gmail.com

R.G. Handy
Department of Engineering Technology & Construction Management, University of North Carolina-Charlotte, 9201 University City Blvd., Charlotte, NC 28223, USA
e-mail: rhandy@uncc.edu

J.S. Morrell
Y-12 National Security Complex, 1 Bear Creek Road, MS #8097, Oak Ridge, TN 37831, USA
e-mail: morrelljs@y12.doe.gov

J.S. Morrell and M.J. Jackson (eds.), *Uranium Processing and Properties*,
DOI 10.1007/978-1-4614-7591-0_4,

4.1 Introduction

4.1.1 Silicon Carbide

Silicon carbide (SiC) was the first of the synthetic abrasives that ushered in twentieth-century manufacturing. It was first synthesized in commercial quantities around 1891 by Dr. Edward G Acheson [1] who gave it the trade name "Carborundum" and was initially produced in small quantities and sold as a substitute for diamond powder for lapping precious stones at $880/lb (at 1891 dollar value!). With process optimization, the price fell precipitously $0.10/lb in 1938. Today (2013), the price is about $0.80/lb. The heart of the process is the Acheson resistance heating furnace, an adaption of the Cowles electric batch smelting furnace patented just a few years previously in 1885, in which quartz silica sand and petroleum coke are reacted at a temperature of around 2,400 °C [2]. The overall reaction is described by the carbothermic reduction equation:

$$SiO_2 + 3C \rightarrow SiC + 2CO$$

The furnace is prepared by placing a large carbon resistor rod on a horizontal bed or trough of raw materials to which a heavy current is applied. The raw material also includes sawdust to add porosity to help release the CO and salt to remove iron impurities. The whole process takes from 36 h to 10 days and yields typically 10–50 t of product. From the time it is formed, SiC remains a solid as no melting occurs (SiC sublimates at 2,700 °C). Pure SiC is colorless. Two grades of SiC are produced for abrasive applications—"green" and "black." Green SiC is the purer produced from a virgin mix of sand and coke; black SiC is produced from recycled feed including amorphous SiC from previous furnacing cycles; the black coloration comes from iron impurities. Green and black products are also sorted in terms of distance from the carbon rod, with some green material being obtained closer to the electrode even with recycled feed.

The Acheson process has remained essentially unchanged for many decades. As such the primary driver for the location of manufacturing has been cheap, easily available electric power—most commonly hydroelectric power. The original Acheson furnaces were driven by power from Niagara Falls although North American production is now severely challenged by operating costs; global manufacturing is now dominated by China with almost half the market. Other countries with significant grain production include Brazil, Russia, and Vietnam. More recently Czech, Spain, and Bhutan have come on line although much of this product is for other applications than abrasives. There has been a large upsurge in interest in SiC for applications such as tank and body armor, heat resistant bodies for kiln ware, high-temperature electronic devices aeroengine components, and wire saw applications for electronics spurring research into both manufacturing processes and SiC material properties. Silicon carbide is the hardest of the conventional abrasives with a Knoop hardness of 2,500 and a surface Knoop microhardness of

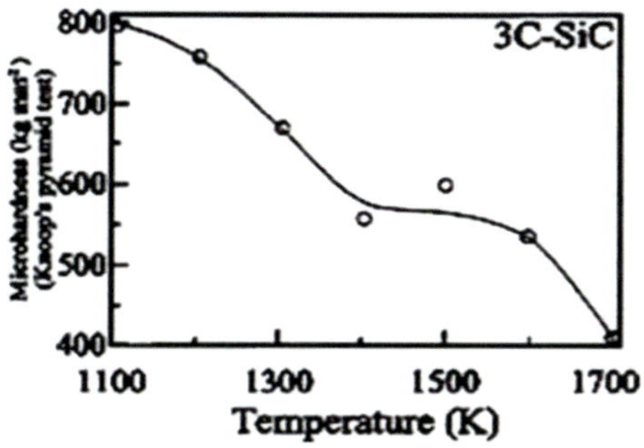

Fig. 4.1 Hardness variation with temperature for SiC. Reprinted "With kind permission from Springer Science+Business Media: 'Machining with Abrasives', Edited by Mark J. Jackson and J. Paulo Davim, Springer-Verlag, New York, November 2010, pp 1–423. ISBN 978-1-4419-7301-6"

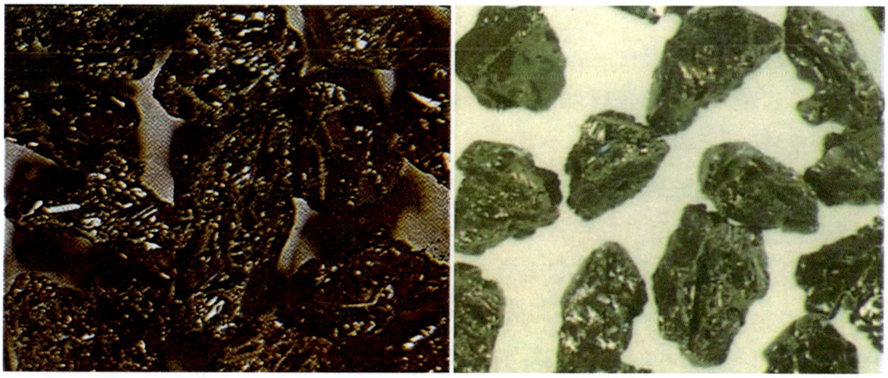

Fig. 4.2 Examples of black and green SiC. Reprinted "With kind permission from Springer Science+Business Media: 'Machining with Abrasives', Edited by Mark J. Jackson and J. Paulo Davim, Springer-Verlag, New York, November 2010, pp 1–423. ISBN 978-1-4419-7301-6"

2,900–3,100 kg mm^2 at room temperature. Microhardness decreases with temperature as shown in Fig. 4.1 [3, 4].

Green silicon carbide is the higher-purity silicon carbide manufactured with typically >98.5 % of SiC. The crystal type is alpha phase silicon carbide in the form of hexagonally shaped platelets. Black silicon carbide is of a lower purity (95–98 %) and consists of the alpha phase with both hexagonal and rhombohedral forms. The green is slightly harder but more friable and angular. For this reason, green silicon carbide is used for grinding hard metals such as chilled cast iron rolls, titanium, and metal and ceramic cutting materials. Black silicon carbide is used more for grinding of soft nonferrous metals and nonmetallics like rubber, wood, ceramics, and glass.

Both SiC grades are more friable than fused alumina grain (Fig. 4.2). SiC does show reactivity or solubility with iron, limiting its use to grinding ferrous materials.

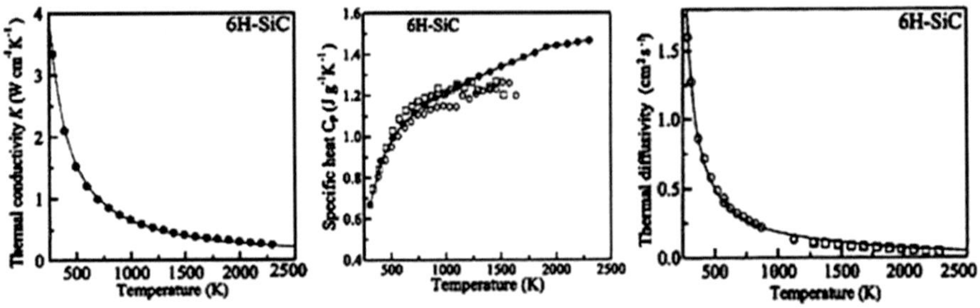

Fig. 4.3 Thermal Properties—κ, C_p, and $(\kappa C_p)^{1/2}$ of SiC. Reprinted "With kind permission from Springer Science+Business Media: 'Machining with Abrasives', Edited by Mark J. Jackson and J. Paulo Davim, Springer-Verlag, New York, November 2010, pp 1–423. ISBN 978-1-4419-7301-6"

It is also susceptible to oxidation at higher temperatures. Thermal properties are shown in Fig. 4.3 [5]. The demand for SiC is such that the price of abrasive grain has risen steeply in the last few years on a trend that is expected to continue due to energy costs and increased demand for nonabrasive applications. This is presenting a major challenge to the abrasive industry to find alternative grinding solutions such as agglomerated aluminum oxide grains.

4.1.2 Fused Alumina

The second great innovation in abrasive technology at the turn of the twentieth century, after the Acheson process for SiC synthesis, was the invention of the Higgins electric arc furnace for the production of electro-fused alumina (or "Alundum") by Aldus C. Higgins of the Norton Company in 1904 [6]. Prior to this, wheel makers had used naturally occurring aluminum oxide in the forms of the minerals emery and corundum, but the variability in chemical and mechanical properties made controlling wheel formulations difficult. Emery abrasive use now is limited to coated paper.

Raw bauxite is the starting material for all fused aluminum oxide grain (Fig. 4.4). It consists of up to 60 % alumina in the form of the minerals gibbsite ($Al(OH)_3$), boehmite (γ-AlO (OH)), and diaspore (α-AlO (OH)), together with the iron oxides goethite and hematite, the clay mineral kaolinite, and small amounts of anatase and titania, TiO_2. Australia is the largest producer with almost a third of world production, followed by China, Brazil, Guinea, and Jamaica [7]. Bauxite along with coke and iron is the direct feed material for fusion to create the brown fused alumina (BFA), a family of abrasives containing controlled amount of up to 4 % titania. Bauxite can also be purified prior to fusion by the Bayer process that was invented in 1887 by Karl Bayer in Russia. In this case, bauxite is heated in a pressure vessel with sodium hydroxide solution at 150–200 °C. After separation of

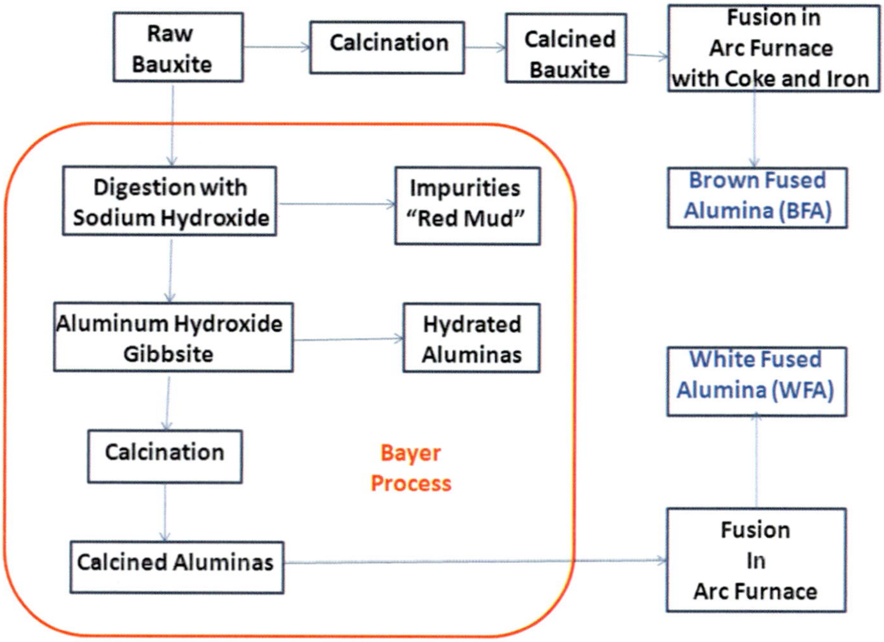

Fig. 4.4 Processing routes for fused alumina. Reprinted "With kind permission from Springer Science+Business Media: 'Machining with Abrasives', Edited by Mark J. Jackson and J. Paulo Davim, Springer-Verlag, New York, November 2010, pp 1–423. ISBN 978-1-4419-7301-6"

the iron-based residue (red mud) by filtering, pure gibbsite is precipitated by cooling the liquid and seeding with fine-grained aluminum hydroxide. Gibbsite is then converted into aluminum oxide by calcining. The Bayer process removes almost all of the natural impurities present in the raw bauxite but leaves behind 0.1 % to 0.4 % soda (Na_2O) in the purified calcined alumina. This is the feed for the production of white fused alumina (WFA) and its family of abrasives. The Bayer process increases the cost of WFA feed material by about a factor of five compared with bauxite feed for BFA. A Higgins electric arc furnace consists of a thin steel or aluminum shell on a heavy metal hearth. A wall of water running over the outside of the shell cools it sufficiently to maintain the shell integrity in combination with a thin layer of aluminum oxide that forms on the inside due to its extremely poor thermal conductivity. Steel was historically the normal shell material, as it has a relatively high melting point, but aluminum is now preferred especially for WFA fusions to prevent discoloration from iron rust contamination. Feed material is poured into the bottom of the furnace and a carbon starter rod laid on it. Two or three large vertical carbon rods are then brought down to touch and a heavy current applied. The starter rod is rapidly consumed, but the heat generated melts the bauxite, which then becomes an electrolyte. Feed material is added continually over the next several hours to build up the volume of melt to as much as 20 t. Current flow is controlled by adjusting the height of the electrodes that are eventually consumed in the process. The reaction conditions of a BFA fusion are a result of the added coke which reacts with oxygen in the impurities to produce carbon monoxide, reducing the silica to silicon and iron oxide to iron which combine along with the added iron to form a heavy, highly fluid, ferrosilicon phase. Silica is also lost as fume due to the high temperatures. In addition, titania levels may be adjusted by reduction to titanium that precipitates out with the ferrosilicon. A typical fusion of 30 t may take around 20 h to completely fill and melt the contents of the furnace pot; cooling time may be up to 4 days. The cooling is directional; the insulating outer layer of the melt is quenched next to the pot and highly microcrystalline. There is then a large crystalline, dendritic, growth region in a radial direction with solidification in toward the center of the pot as heat flows from the center out. The pot has a high profile with an aspect ratio of about 1:1. Impurities will concentrate in the liquid phase in the center and toward the bottom of the forming ingot. After cooling, the ingot must be broken up and hand sorted to remove the primary concentration of impurities. Additional iron and ferrosilicon are subsequently removed with magnetic separators during crushing.

The reaction conditions for a WFA fusion are in general not considered reducing in that the only carbon present is from the electric arc and starter rods. The biggest concern is the conversion of the residual soda from the Bayer process feed into sodium β-alumina, which crystallizes as soft hexagonal plates in the alumina. As sodium β-alumina has a lower melting point than alumina, it will again concentrate in the in portions of the ingot that solidify last. The Higgins furnace has evolved from an original design with a small capacity of 1–5 t to furnaces today of up to 40 t with pot diameters of up to 3.5 m (12 ft) and a power supply of up to 4MVA. It requires 2.2MVA hour to produce 1 t of BFA and 1.5MVA hr to produce 1 t of WFA. However, with increasing capacity and efficiency demands, there has been a move toward even larger tilting furnaces up to 6 m in diameter that can pour the molten alumina into pots with water-cooled hearths. These furnaces use a power supply of as much as 10MVA or even greater and can pour up to 24 t every 4 h while maintaining a more consistent batch-to-batch chemistry. Pour pot design for use with tilt furnaces can have a major impact on grain structure and chemistry (Fig. 4.5). For example, for a WFA fusion pour into a high-profile pot, the cooling process and output are similar to a Higgins furnace, i.e., a large alumina crystallite size with dendritic growth and a very low sodium β-alumina content after sorting.

However, pouring into a low profile pot (aspect ratio $<<1$) onto a cold hearth results in a much faster cooling rate, a fine crystallite alumina structure, and a much more evenly dispersed sodium β-alumina content. The high thermal gradient when cooling a WFA ingot in a deep profile pot causes crystallization of α-alumina in a dendritic habit made up of intergrown rhombohedra extending along the thermal gradient. This type of crystal is the result of the edges of the rhombohedron growing much faster than the faces (Fig. 4.6). Abrasive grain made from this will tend to fracture in relatively large fragments along defined planes but be very self-sharpening.

Crystallization in low profile pots will show structures with less directional growth and that are finer in crystal size resulting in smaller fragments during grain

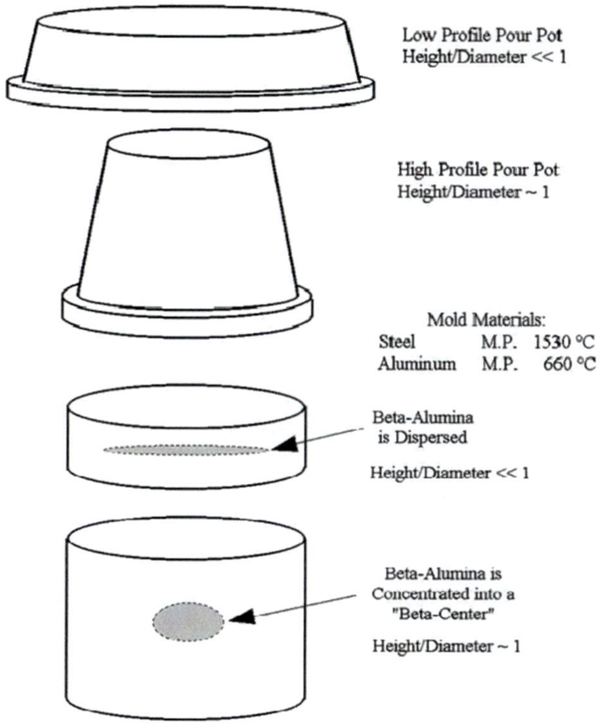

Fig. 4.5 Pour pot designs and resulting distribution of sodium β-alumina content. Reprinted "With kind permission from Springer Science+Business Media: 'Machining with Abrasives', Edited by Mark J. Jackson and J. Paulo Davim, Springer-Verlag, New York, November 2010, pp 1–423. ISBN 978-1-4419-7301-6"

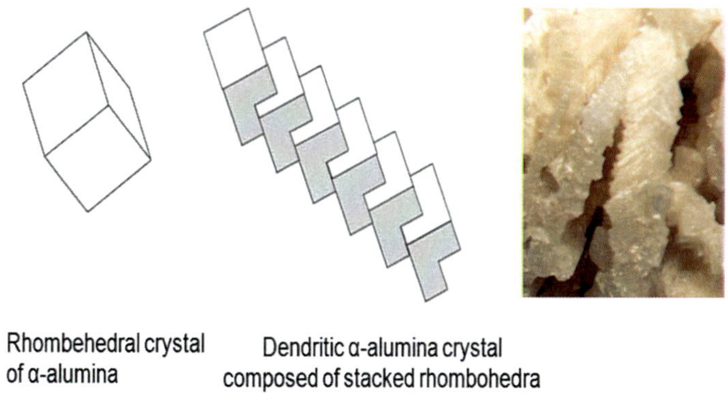

Fig. 4.6 Structural characteristic of WFA grain [8]. Reprinted "With kind permission from Springer Science+Business Media: 'Machining with Abrasives', Edited by Mark J. Jackson and J. Paulo Davim, Springer-Verlag, New York, November 2010, pp 1–423. ISBN 978-1-4419-7301-6"

fracture; the material is also about 10 % softer from higher sodium β-alumina contamination. For BFA fusions in tilt pour furnaces, about 25 % of the furnace content is poured each time while most of the ferrosilicon collects at the bottom where it can accumulate over many regular pours until it is removed in a "deep pour" into a rail cart with a bed of sand. The quality and consistency from large pour furnaces tend to be better because it is less expensive to monitor per tonne taking regular dip samples, in particular avoiding over-reduction of titania. Pour molds tend to be low profile and the resulting grain a mixture of dendritic with finer equiaxial α-alumina together with some ferrosilicon inclusions. For further discussion on grain fusion and furnacing, see Wolfe et al. [8], Lunghofer et al. [9], and Whiting Equipment of Canada [10]. The fusion of "Alundum" began over a hundred years ago on the shores of the Great Lakes using cheap hydroelectric power from Niagara Falls. Today, 2013, only three plants remain in North America with just 5 % of global capacity although fused alumina has been substituted to a significant degree with more recent ceramic technology and manufacture as discussed below. In the last 12 years, China has increased its capacity for fused alumina, especially BFA, to over 60 % USGS [11, 12] with integrated manufacturing close to the bauxite mines and the largest capacity tilt pour furnaces. Eastern Europe, India, South Korea, and South America also continue to increase their prominence in world markets. Low-cost electrical power availability, furnace capacity, quality control, and cost of raw material sourcing should be the dominant factors influencing the relative competitive stance of each.

4.1.2.1 Grain Types

Properties of the grain depend both on the fusion process and chemistry but also on the subsequent comminution process. The ingot is initially split and sorted and then pre-crushed in steel jawed Barmac and beater crushers. All these processes are high impact and will create major fractures resulting in a grain that is sharp edged, flawed, and anisotropic, typically like a sliver in shape. Subsequent processing in steel or rubber lined ball mills will have a tendency to reduce grain size by rounding of the grain edge. In this manner, it is possible to control shape to a degree to have either angular or blocky forms from the same material. The grain availability can be divided into BFA- and WFA-based families:

Brown fused aluminum oxide: contains 2–4 % titania which enhances toughness. This is still the most widely used abrasive in wheels to grind high-tensile-strength materials and for rough grinding, deburring, and snagging, as well as to cut low-alloy, ferrous materials, and is generally viewed as the "workhorse" of the industry. Brown fused alumina is a tough, sharp, but blocky abrasive. Depending on the processing regimes, the grain is typically about 50 % single crystal and can be provided in high, medium, and low density based on shape packing characteristics. The grain may also be calcined after sizing to toughen it by annealing cracks generated in the crushing processes. The material is sometimes termed blue-fired

Fig. 4.7 Examples of white, pink, and red fused alumina grain. Reprinted "With kind permission from Springer Science+Business Media: 'Machining with Abrasives', Edited by Mark J. Jackson and J. Paulo Davim, Springer-Verlag, New York, November 2010, pp 1–423. ISBN 978-1-4419-7301-6"

BFA as the grain changes color due to surface oxidation of impurities. Specialty coating such as silane (for resin bonded wheels to resist coolant interactions) or red iron oxide (for resin and rubber bonded wheels to increase surface area) may also be applied.

Low titania ("light" or "semi-friable") brown fused aluminum oxide: has 1–2 % TiO_2 content and is used in bonded or coated applications that require an abrasive that is slightly tougher than white aluminum oxide. Reducing the titania content reduces the abrasive's toughness, but increases its friability. Light BFA is commonly used in depressed center wheels, in cutoff wheels, and for surface and cylindrical grinding of heat-sensitive metals, alloys, etc., where cool but fast cutting is required. The grain can be supplied with similar post and surface treatments as regular BFA.

White fused alumina: is standard multicrystalline WFA with sodium β-alumina contamination and is the most friable grain in the fused alumina family. It is considerably harder than BFA. Commonest applications include the grinding of tool, high-speed, and stainless steels (Fig. 4.7).

Fig. 4.8 Example of sintered extruded brown alumina grain (Courtesy Saint-Gobain Abrasives). Reprinted "With kind permission from Springer Science+Business Media: 'Machining with Abrasives', Edited by Mark J. Jackson and J. Paulo Davim, Springer-Verlag, New York, November 2010, pp 1–423. ISBN 978-1-4419-7301-6"

Single crystal white fused alumina: is single crystal grain that has been produced in deep pour fusion pots and separated from any sodium β-alumina contamination. This is the hardest and most brittle of the alumina family of grains used most commonly for grinding tool and very high-alloy steels that are very sensitive to heat.

Pink alumina: is WFA to which <0.5 % chromium oxide has been added in the fusion process to produce a grain that is slightly tougher than regular WFA used for grinding unhardened high-alloy steels (Fig. 4.7).

Ruby alumina: is WFA to which 3 % chromium oxide has been added to provide additional toughness over pink alumina (Fig. 4.7).

It can be inferred there is a steady increase in toughness but reduction in hardness in the following order:

Single crystal WFA → WFA → pink WFA → ruby WFA → light BFA → BFA → blue-fired BFA

In general, the wheel maker will blend various grain types and sizes to combine the properties of each. In addition to chrome, other metal oxide additions have been investigated including vanadium and beryllium but not found to be commercially viable.

Sintered alumina: is a family of grains developed in the 1950s produced from *unfused* alumina. Several processes exist based on both raw bauxite and Bayer processed aluminas. The most common is to use a feed material of raw bauxite milled to <5 μm. The mix with binder is first extruded to produce rods that are cut into short cylinders or cones in the green state. They are then fired in rotary kilns at 1,350–1,500 °C using natural impurities in the bauxite as sintering agents (Fig. 4.8) [13]. The resulting grain is extremely tough especially at the relatively large

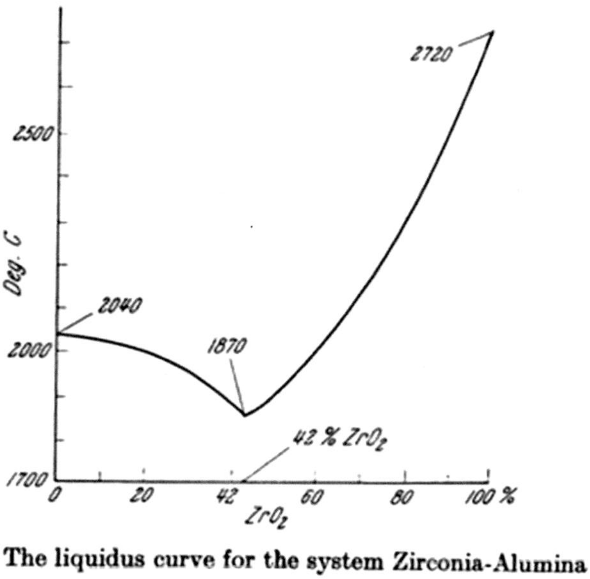

Fig. 4.9 Phase diagram for the zirconia–alumina system. Reprinted "With kind permission from Springer Science+Business Media: 'Machining with Abrasives', Edited by Mark J. Jackson and J. Paulo Davim, Springer-Verlag, New York, November 2010, pp 1–423. ISBN 978-1-4419-7301-6"

sizes the technology allowed to be produced (8#–20#), and the material found great success, until the advent of alumina–zirconia grain, in billet conditioning and other rough grinding operations. It is still used as a blend component with alumina–zirconias especially in the grinding of stainless steels.

4.1.3 Alumina–Zirconia

Zirconia is a very high-temperature refractory material, tougher but softer than alumina. It also has a higher melting point than alumina making fusion based on the Higgins furnace more demanding in terms of containment and control. Fortunately, from a manufacturing viewpoint (if not a grinding perspective), the thermal conductivity of zirconia like alumina is very low, making the Higgins fusion route still possible. However, as can be seen the liquidus curve for the zirconia–alumina system (Fig. 4.9) [14], compositions of these two materials combined, with the zirconia contents kept under about 65 %, have comparable or lower melting points to alumina alone. This allows for relative ease of control of a combined alumina–zirconia fusion process including pouring from a tilt furnace.

Interest in zirconia as a potential abrasive grain or component of grain has been evident from the literature since at least the mid-1950s, e.g., [15, 16], in part because a relatively pure form known as baddeleyite began being produced in quantity at that time as a by-product of heavy metal mining especially of uranium ores in countries such as Russia, Brazil, and South Africa. Another much commoner source of zirconia is as zircon (zirconium silicate $ZrSiO_4$) found as sands in the USA, Australia, and Brazil. Zircon sand can be refined by fusing it with coke, iron, and lime until the silica is reduced and separates out as a denser, relatively low-viscosity ferrosilicon liquid. Alumina–zirconias can be produced in a similar way by addition of Bayer process alumina to the fusion. Zirconia has the interesting property that it is metastable in the tetragonal state at certain crystal sizes when held under constraint. For pure zirconia, the grain size upper limit is about 0.1–0.3 μm. With additions of small amounts of the alkali oxides CaO or MgO or rare earth oxides such as Y_2O_3 or CeO_2, this limit can be raised into the micron range. Without constraint, the tetragonal crystal converts to the monoclinic phase with a significant increase in volume of about 6 %. If a crack from an active fracture intersects with a tetragonal crystal, it releases the constraint, but in the process, the volume expansion to the monoclinic phase dissipates the ability of the crack tip to propagate. The result is an increase in the K_{1C} fracture toughness of the grain of an order of magnitude.

The technical superiority of a fused grain of zirconia over one of alumina especially at very coarse sizes for rough grinding was recognized by the mid-1960s but was cost prohibitive, although alumina–zirconia blends showed advantage [17]. It was also recognized that an alumina–zirconia eutectic produced a strong structure due to a uniform dispersion of fine zirconia crystals in an alumina matrix. However, any excess alumina or zirconia from the eutectic would grow to a considerable crystal size depending on cooling rates from the melt. Rapid quench was therefore identified as a necessary prerequisite of a processing route [18]. This required quench rates of 100 °C/s, two orders of magnitude faster than previously. Numerous attempts were made during the late 1960s and 1970s to develop a viable process [19–27], using various inert cooling media or hearth plates, but it was a process developed by Scott [28, 29] of the Norton Company that was to prove commercially and technically effective. The process from the original patent is illustrated in Fig. 4.10. Molten alumina–zirconia from an electric arc tilt furnace is poured into the relatively thin spaces between a plurality of relatively thick heat sink plates of graphite or iron as they pass underneath, before being emptied at a discharge station by the plates separating. The result is a very fine structure of α-alumina with high tetragonal zirconia content. The zirconia is in the form of rods (or platelets), which on average are less than 0.3 μm in diameter. The solidified melt is made up of cells or colonies typically 40 μm or less across their width. Groups of cells having identical orientation of microstructure form grains that typically include from 2 to 100 or more cells or colonies [30]. Figure 4.11 shows a TEM micrograph that illustrates the fine, rodlike zirconia structures within the larger cell [31].

After solidification, the material is comminuted by standard methods of crushing, milling, and sizing to product grain. The processing will lead to some conversion of the tetragonal to the monoclinic phase depending on the processing energy and especially on the final grain size. The smallest grain will lose much of its tetragonal toughening favoring this type of grain for use in coarse-sized, roughing operations.

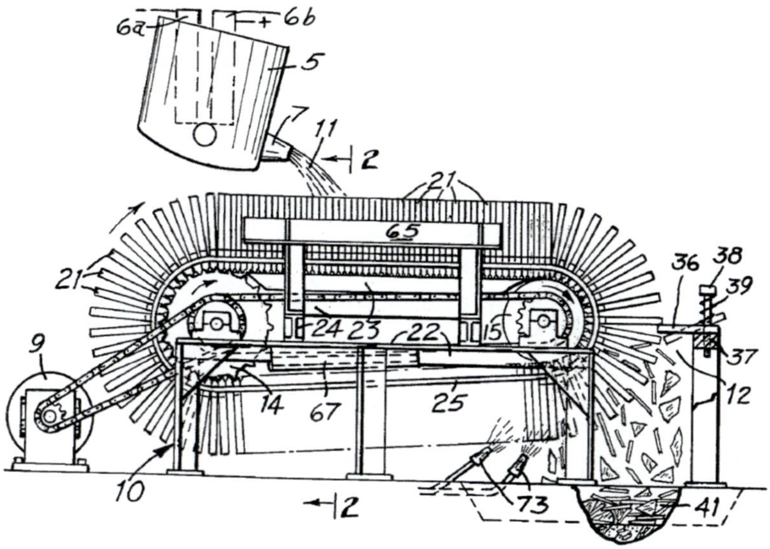

Fig. 4.10 Scott's patent detail for producing rapid quench alumina–zirconia. Reprinted "With kind permission from Springer Science+Business Media: 'Machining with Abrasives', Edited by Mark J. Jackson and J. Paulo Davim, Springer-Verlag, New York, November 2010, pp 1–423. ISBN 978-1-4419-7301-6"

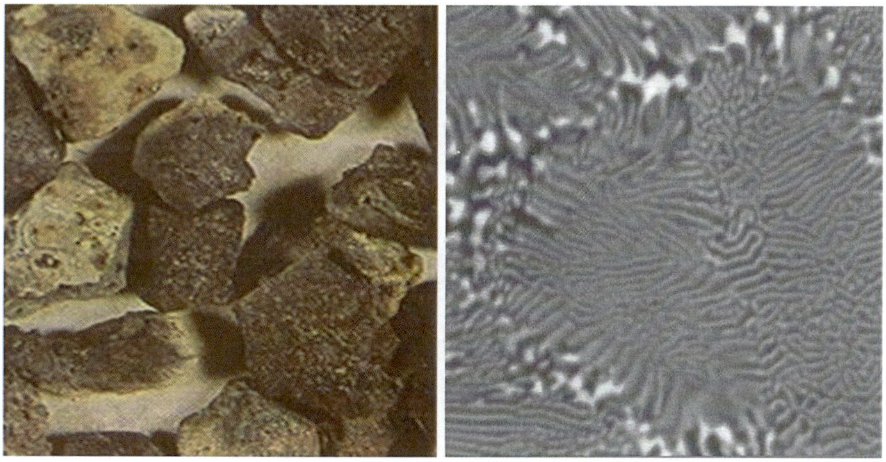

Fig. 4.11 Fused alumina–zirconia grain and a TEM micrograph of its rodlike zirconia structures [31]. Reprinted "With kind permission from Springer Science+Business Media: 'Machining with Abrasives', Edited by Mark J. Jackson and J. Paulo Davim, Springer-Verlag, New York, November 2010, pp 1–423. ISBN 978-1-4419-7301-6"

The vast majority of alumina–zirconia grain for grinding wheels contained 25 % zirconia; sold under the trade names of ZF or ZS Alundum, depending on the comminution method; and used in hot pressed resin bonds for rough steel, titanium, and nickel alloy billet conditioning or for foundry snagging. Grain size can be as coarse as 4# (0.26″ or 6.8 mm) and either used as a single grain type or blended to include extruded sintered brown alumina (for finish), SiC (grinding titanium), or regular alumina. Billet conditioning is very aggressive form of grinding. The operation is run dry with wheel speeds up to 80 m/s (16,500 sfpm) and spindle power as great as 500Hp on the most modern equipment. The workpiece is still often red hot from the furnace. Metal removal rates are extraordinary and can exceed 2,500 lb/h ($Q' \approx 40 in^3/in/min$ or 400 $mm^3/mm/s$) on steel or 400 lb/h ($Q' \approx 12 in^3/in/min$ or 120 $mm^3/mm/s$) on titanium, far exceeding most other metal removal processes. Other applications include the regrinding of rail track in situ to remove fatigue cracks using special trains travelling at speeds of up to 6 mph (10 km/h). Grinding dry, in all these processes the grains self-sharpen from cracks generated by thermal shock. What is perhaps most striking about these processes is that the wheels are heavily reinforced with binder and grain and is processed at less than 200 °C, yet the interface temperatures in grinding can far exceed 1,000 °C to the point of melting the metal. It is only the extremely poor thermal conductivity of the grain that prevents the rapid degradation of the bond around the grain base, while the high temperatures erode the bond between the grains at the grinding interface to give chip clearance. Eutectic alumina–zirconia grain containing 40 % zirconia is also produced and sold under brand names of NZ Plus, NZ®, NZP®, and NorZon®. It is used primarily for coated applications and as such requires a different balance of hardness (−) and toughness (+) properties and shape, compared to grain for grinding wheels.

4.1.4 Engineered Alumina Abrasives

Engineered abrasive have microstructures that have been produced with controlled crystallite sizes from the submicron to micron level by processes other than simple fusion and comminution. These include sol–gel/sintering and agglomeration techniques. The result is a family of grain types that micro-fracture at controlled micron or submicron levels and have the ability to be micro-trued, enhancing both the wheel life and process control compared with fused aluminum oxide grain.

4.1.4.1 "Ceramic" Sol–Gel Abrasives

The development and commercial success of first the sintered extruded alumina family of grains and then the rapid chilled fused alumina–zirconia grain had a major impact on the research programs of abrasive manufacturer in regard to the importance of control of grain crystal size. Furthermore, for alumina grain, it was known

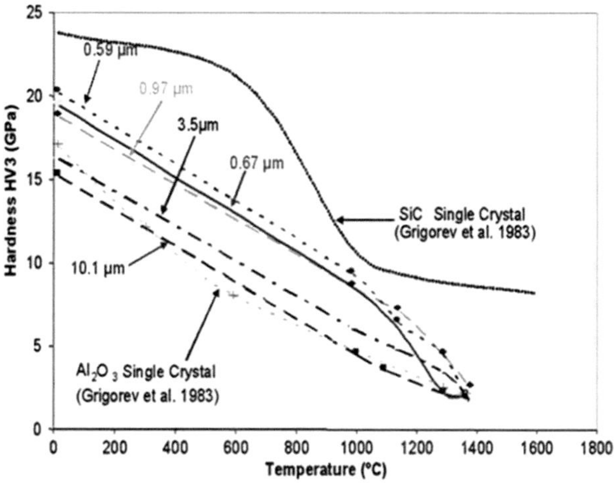

Fig. 4.12 Effect of crystal size on alumina grain hardness. Reprinted "With kind permission from Springer Science+Business Media: 'Machining with Abrasives', Edited by Mark J. Jackson and J. Paulo Davim, Springer-Verlag, New York, November 2010, pp 1–423. ISBN 978-1-4419-7301-6"

that reducing the crystal size from the macroscale equivalent to a single crystal per abrasive grain, common in fused material, to micron or ideally <0.5 µm crystalline structures significantly enhanced grain properties such as hardness (Fig. 4.12) [32].

The response was, rather than using traditional fusing or sintering processes with their general limitations on cooling and crystallization rates, to consolidate microstructures from finer building blocks by sintering well-dispersed submicron precursors by the so-called "sol–gel" route. This allowed the consolidation of an α-alumina-based submicron, highly homogeneous, and fully densified grain structure. The starting point of this new process is the manufacture of boehmite, γ-aluminum oxide hydroxide γ-AlO (OH), from a modified version of the Ziegler process originally developed for the production of linear alcohols [33]. The material is produced as a submicron, narrowly sized powder which when mixed with water and a suitable acid dispersant forms an agglomerate-free sol–gel of aluminum hydrate ($Al_2O_3{\cdot}H_2O$) with a dispersant size of about 100 nm. The sol–gel is then dehydrated/shaped and sintered (Fig. 4.13).

The biggest hurdle to overcome was in the sintering process to maintain a uniform submicron crystal size and full densification. Firing sol–gel from a standard commercial boehmite at 1,400–1,500 °C produces a large amount of porosity and relatively large grains >1 µm in size. This is believed due to high activation energy to convert from a transitional τ- to α-alumina phase resulting in infrequent nucleation with rapid uncontrollable growth rates. Attempting to control growth rates with lower temperatures, e.g., 1,200 °C, merely leads to larger crystals with higher porosity. There are two routes that have been developed to reduce the activation energy and control crystal size and densification. The first is the creation of a bi- or multi-composite structure through the use of modifying agents, and the second is the controlled creation of a single α-alumina structure through the use of seeding agents (Fig. 4.14a). The earlier patents report the use of magnesia [34] which upon sintering forms a bi-composite structure of α-alumina plus a spinel

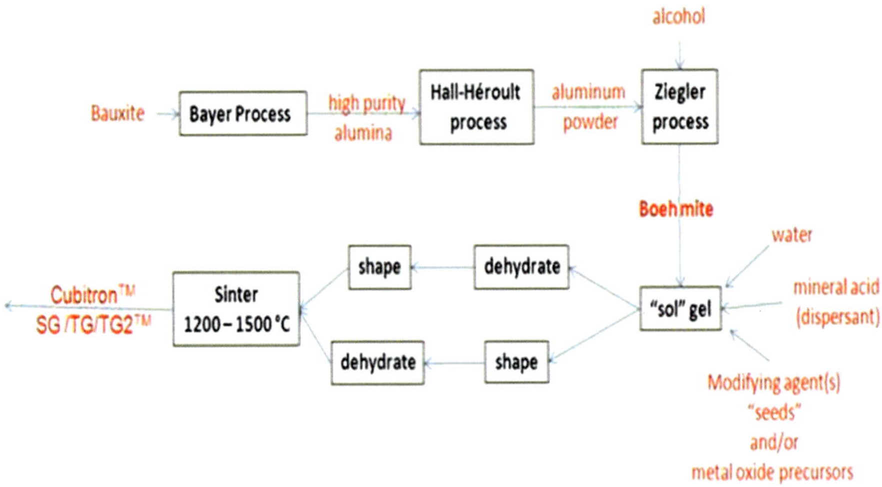

Fig. 4.13 Manufacturing route for production of "ceramic" alumina grain. Reprinted "With kind permission from Springer Science+Business Media: 'Machining with Abrasives', Edited by Mark J. Jackson and J. Paulo Davim, Springer-Verlag, New York, November 2010, pp 1–423. ISBN 978-1-4419-7301-6"

structure of magnesium aluminate at about 25 % by volume as shown in Fig. 4.14b. Note the fine needlelike spinel structure and the still relatively coarse α-alumina phase. This particular grain was used primarily for low-force coated abrasive applications. Later, numerous patents report various multiphase systems using various modifying agents including zirconia, manganese oxide, chromia, nickel oxide, and numerous rare earth oxides. One particularly effective material contains magnesia together with yttria and other rare earth oxides such as lanthana and neodymia to produce a dense and hard (19GPa) grain. In Fig. 4.14c, the microstructure shows a finer α-alumina phase (although still relatively coarse compared to the feed material) but with a submicron "magnetoplumbite"-type structure of needles/plates formed from the modifiers [35].

The structures created by the modifiers are believed to provide strength akin to rebar in reinforced concrete. This particular grain is the 3M™ Cubitron™ 321 grain [36]. The alternate route to controlling crystallization rates is by "seeding" the sol–gel with nano-sized (<100 nm) α-alumina or other materials with a crystallographic match to α-alumina such as α-ferric oxide or various titanates. Additions of 1–5 % of seeding agent creates a heterogeneous nucleation condition by increasing the number of nucleation sites from $10^{11}/cm^3$ to $10^{14}/cm^3$ and an average crystal size of about 400 nm (Fig. 4.14d) [37, 38]. This type of grain is sold commercially under the name Norton SG™. One limitation of such fine crystal size is surface reactivity with standard vitrified bonds for fabricating grinding wheels. Bonds had to be developed to be fired at <1,000 °C rather than the 1,200 °C of older bonds used for fused alumina abrasives [39].

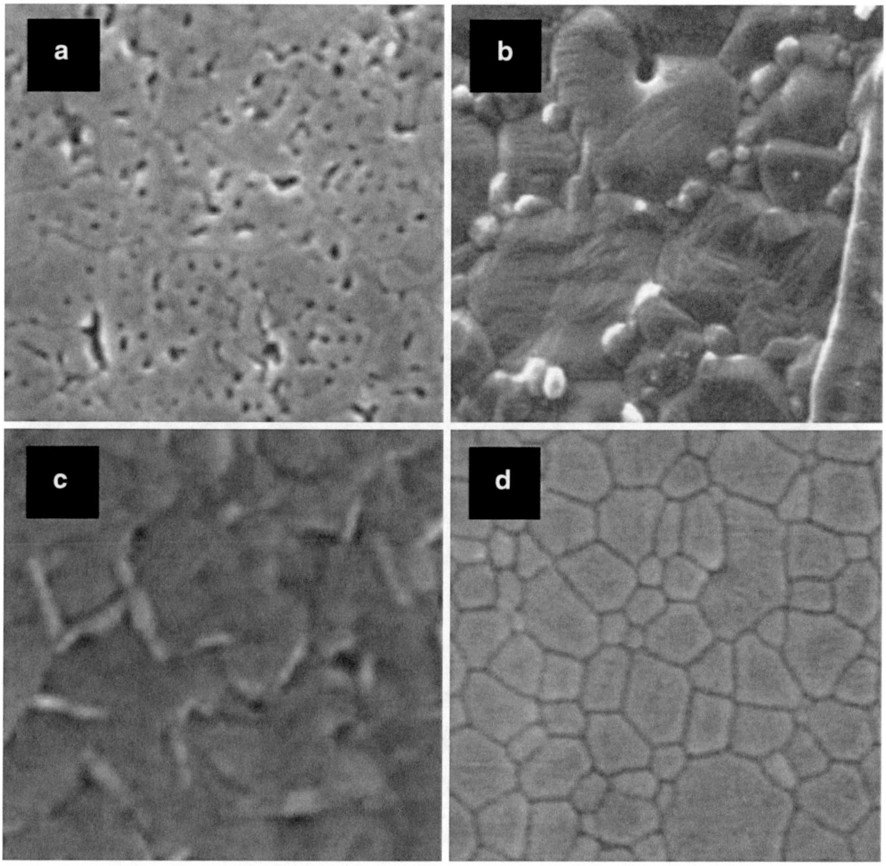

Fig. 4.14 (**a**) Sintered alumina microstructure from boehmite with no modifying agent (image size 3 × 3 μm); (**b**) sintered alumina microstructure from boehmite and magnesia modifying agent (image size 3 × 3 μm); (**c**) sintered alumina microstructure from boehmite, magnesia, yttria, lanthana, and neodymia modifying agents (image size 1.5 × 1.5 μm); and (**d**) sintered alumina microstructure from boehmite with seeding agent (image size 1.5 × 1.5 μm). Reprinted "With kind permission from Springer Science+Business Media: 'Machining with Abrasives', Edited by Mark J. Jackson and J. Paulo Davim, Springer-Verlag, New York, November 2010, pp 1–423. ISBN 978-1-4419-7301-6"

A comparison of Fig. 4.15(a–d) indicates that the single-phase seeded microstructure is finer than the multiphase microstructure and would be expected to be slightly harder and tougher. It would also be expected to give a longer life but require higher force to micro-fracture when used as abrasive grain or should be used at a lower concentration in a blend. The multiphase grain would be slightly freer cutting and also less reactive with high-temperature vitrified bonds. These differences however are relatively minor compared to the difference in overall performance of this family of abrasives relative to fused alumina.

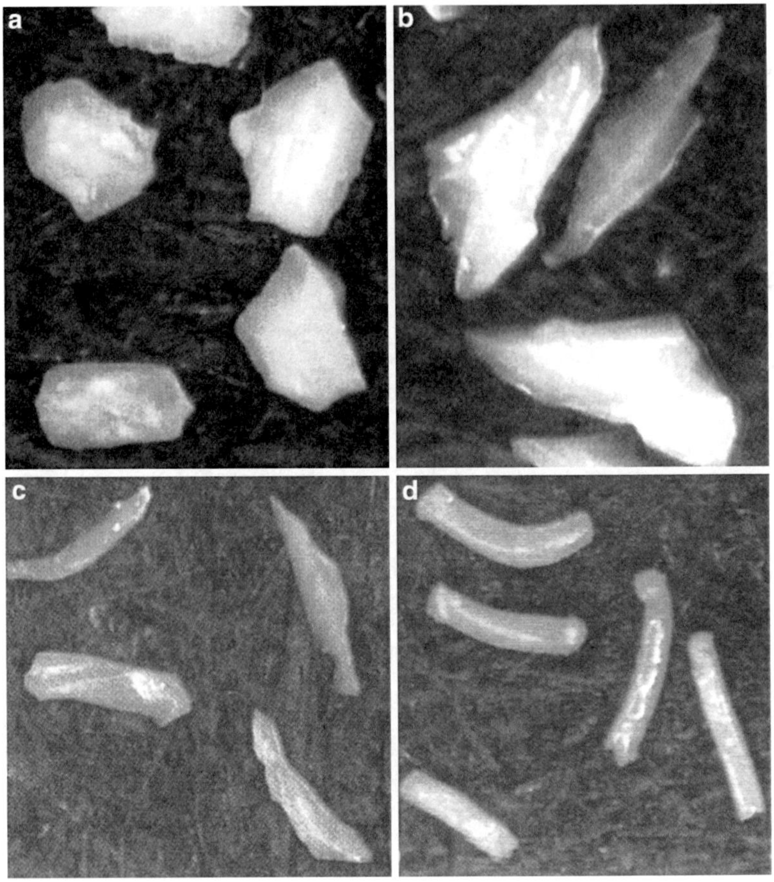

Fig. 4.15 (**a**) Tough blocky ceramic grain produced by milling, (**b**) friable angular grain produced by crushing, (**c**) weak extreme angular grain produced by crushing in green state, and (**d**) extruded TG2™ ceramic grain (Courtesy Saint-Gobain Abrasives). Reprinted "With kind permission from Springer Science+Business Media: 'Machining with Abrasives', Edited by Mark J. Jackson and J. Paulo Davim, Springer-Verlag, New York, November 2010, pp 1–423. ISBN 978-1-4419-7301-6"

Furthermore, further performance optimization can readily be obtained by wheel formulation and especially grain shape. Sol–gel manufacturing allows a much greater manipulation and control of grain shape. Standard crushing and milling methods can produce the typical strong blocky or weak angular shapes. The angularity can be further increased by careful processing of soft, dried pre-sintered material (Fig. 4.16). As expected, these grains are also relatively weak but extremely successful if orientated on, for example, a coated application with relatively low grinding forces.

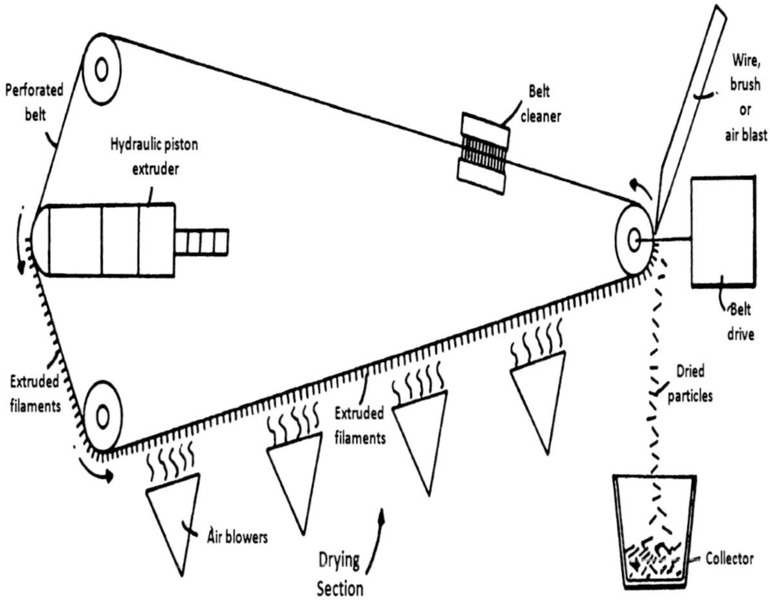

Fig. 4.16 Manufacturing process for producing extruded ceramic grain [62]. Reprinted "With kind permission from Springer Science+Business Media: 'Machining with Abrasives', Edited by Mark J. Jackson and J. Paulo Davim, Springer-Verlag, New York, November 2010, pp 1–423. ISBN 978-1-4419-7301-6"

More interesting, however, are novel techniques [40] that have been developed to extrude as rectangular prisms with extraordinary aspect ratios and having the appearance of smooth, surface defect-free "worms" (Fig. 4.14d).

Norton uses TG™ grain with an aspect ratio of 5 and TG2™ with an aspect ratio of 8 [41]. Not only do these grains maintain a high toughness but they also have a very low packing density. Typical blocky grain may pack to about 50 % by volume; an extruded grain with an aspect of eight has a packing density closer to 30 %. This provides for a very high level of permeability and excellent coolant access in the final fabricated wheel. Owing to the toughness, shape, and ability to provide coolant, the stock removal capabilities on tough superalloys such as Inconel or Rene exceed that of cBN grain, for example, by an order of magnitude.

The most recent variant on the SG-type abrasive is a grain called Quantum™ which maintains the submicron crystallite size and associated hardness of the SG abrasive family of grains but has controlled levels of inclusions to promote micro-fracture that lower force levels (Fig 4.17). This also allows the grain to be micro-trued with dress depths in the 5–15 μm range to generate sharp, fractured but durable cutting edges.

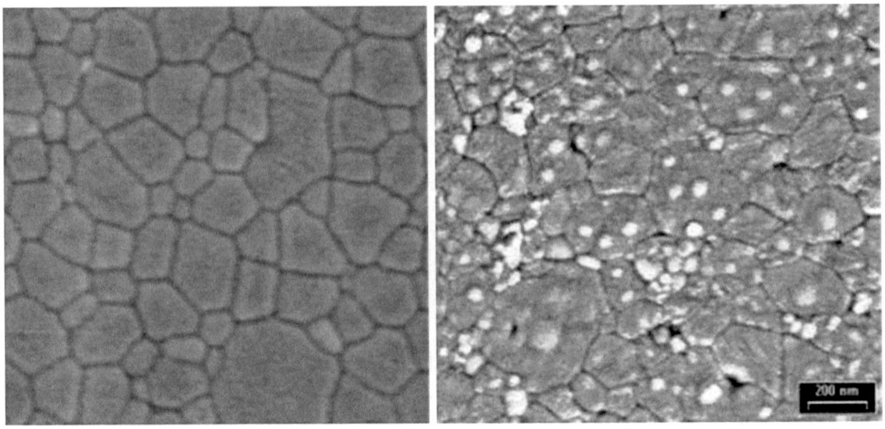

Norton SG Grain microstructure Norton NQ Grain microstructure

Fig. 4.17 Comparison of Norton SG and Quantum grain microstructure. Reprinted "With kind permission from Springer Science+Business Media: 'Machining with Abrasives', Edited by Mark J. Jackson and J. Paulo Davim, Springer-Verlag, New York, November 2010, pp 1–423. ISBN 978-1-4419-7301-6"

4.1.4.2 Agglomerated Grain

So far this discussion has covered grain produced by a fusion route followed by comminution and by chemical precipitation, sintering, and comminution. The former process creates grains with crystallites comparable in size to the grain, i.e., 50–200 μm, while the latter creates crystallite sizes in 0.2–5 μm range. The latest edition to the engineered grain family is to produce "agglomerated" grain made by a fusion, comminution, agglomeration, sintering, and recomminution. The resulting grain has a controlled crystallite size that bridges the gap between SG and fused. Since the size, shape, and chemistry of the crystallites are controlled by the initial comminution process, the possible variations in resulting grain grinding properties are enormous. Furthermore, the options for blending of SG, NQ, and Vortex grains in the same wheel offer an extraordinary range of complimentary grain properties that are only just beginning to be optimized.

As an example, it has been found that agglomerated grains pack to give a naturally high level of porosity in the resulting wheel structure making them very attractive for creep feed grinding (Fig. 4.18). In addition, the creation of very sharp crushed crystallites in the initial comminution process combined with controlled strength in the agglomeration binder allows controlled crystallite breakout to limit wear flat formation, resulting in a very cool grind on heat-sensitive materials. This makes agglomerated grain-based wheel structures a very promising replacement for SiC.

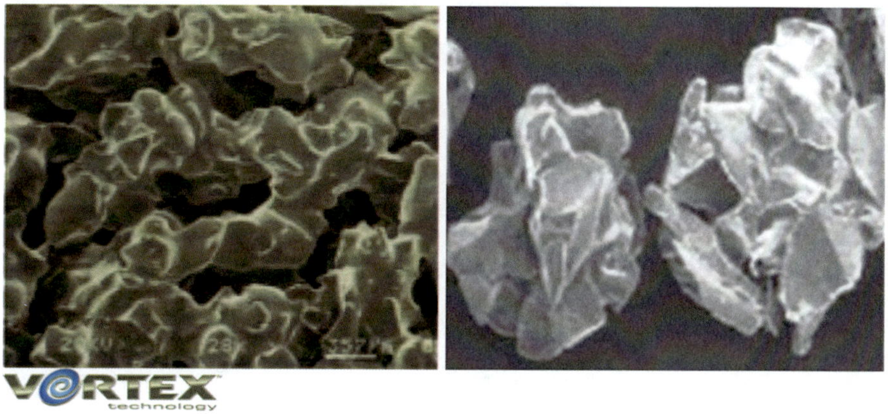

Fig. 4.18 Norton Vortex™ agglomerated aluminum oxide grain. Reprinted "With kind permission from Springer Science+Business Media: 'Machining with Abrasives', Edited by Mark J. Jackson and J. Paulo Davim, Springer-Verlag, New York, November 2010, pp 1–423. ISBN 978-1-4419-7301-6"

4.2 Grinding Practice

It has been reported in the literature (Morris (1981), bibliography) that quick-stop experiments have been conducted on the machining of uranium and its alloys. Although the work falls short of a full scientific characterization of machining uranium, from a qualitative viewpoint, it is interesting to note that under the conditions of machining with a high depth of cut, the chips produced are segmented but form a continuous length of chip as shown in Fig. 4.19.

It is noted that a full and characteristic scientific study of chip formation machining uranium and its alloys will yield a much better understanding of chip formation and flow but also would attempt to predict thermal profiles during machining and grinding that would show the optimum depth of cut that would prevent cut chips from igniting in the surrounding atmosphere. Tool characteristics could be predicted using analytical techniques that have not been applied to the problems associated with machining and grinding of uranium.

Morris (1981, bibliography) provided grinding parameters for grinding an alloyed uranium product. Morris stated that vitrified wheels were best suited for precision grinding of uranium alloys using a medium grit size, soft, and open structure wheel, while centerless grinding operations require a medium grade wheel with an open structure. Both wheels used silicon carbide grains (green type) as the abrasive. Morris's data is shown in Table 4.1.

In a another study on the effects of grinding uranium alloys conducted by Hoffmanner and Meyer [42], the grinding ratio and power consumed were monitored when grinding solution-treated and solution-aged uranium alloys using a

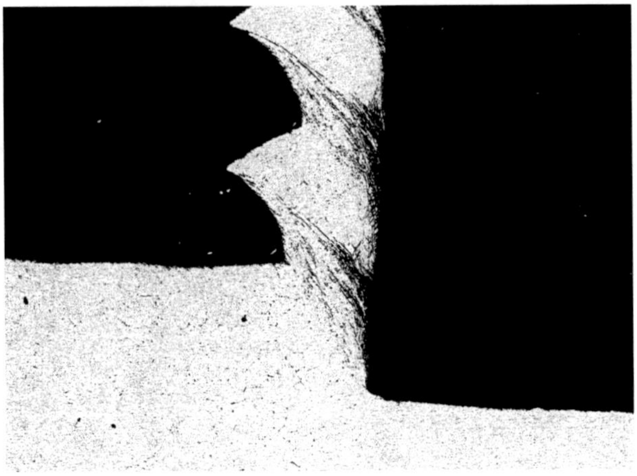

Fig. 4.19 Segmented continuous chip during a uranium machining operation that has been arrested using a quick-stop arrangement (after Morris 1981, see bibliography)

Table 4.1 Grinding parameters for both precision surface and centerless grinding of a uranium alloy (after Morris 1981, see bibliography)

Grinding parameters	Precision grinding		Centerless grinding	
Abrasive types	Regular silicon carbides, green		Regular silicon carbides, green	
Grit size	Medium (46–80)		Medium (46)	
Wheel grade	Soft (G or H)		Medium (I or J)	
Structure	Open (11–13)		Open (10)	
Bond	Vitrified (V)		Resinoid (B)	
Operation	*Roughing*	*Finishing*	*Roughing*	*Finishing*
Feed				
Down feed, in per pass	0.005	0.002 stock removed on diameter inch	0.005[a]	0.002[a]
Cross feed, inch	[b]	[b]		
Through feed of work, ipm	–	–	75–125	75–125
Speeds				
Table, ipm	[b]	[b]		
Wheel, sfpm	3,000–5,000	3,000–5,000	5,500–6,000	5,500–6,000
Grinding fluids	2–5 % grinding fluid inhibitor and water		Soluble oil (30:1), excessive flood	

[a]Stock removed on diameter, inch
[b]Cross feed, table speed, or rotating speed must be kept low enough to prevent burning of work

variety of wheel types and grinding fluids. Figure 4.20 shows that grinding ratio decreases as a function of infeed rate and surface condition depending upon grinding wheel type. Figure 4.21 (from their study [42]) shows that the power required to grind the material increases as infeed rate increases depending upon the grade of grinding wheel and grinding fluid. The reader of this chapter is encouraged

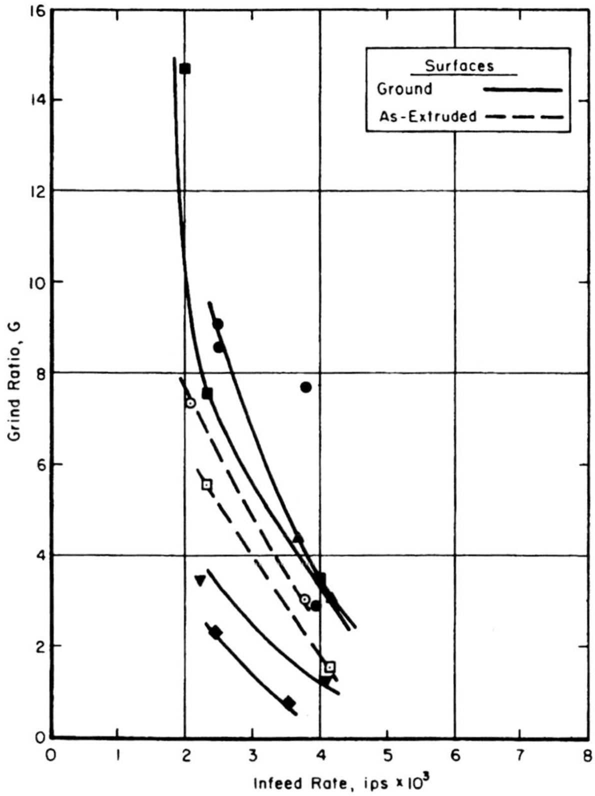

Fig. 4.20 Grinding ratio dependence on infeed rate for solution-treated and solution-aged specimens with a particular grinding fluid [42]

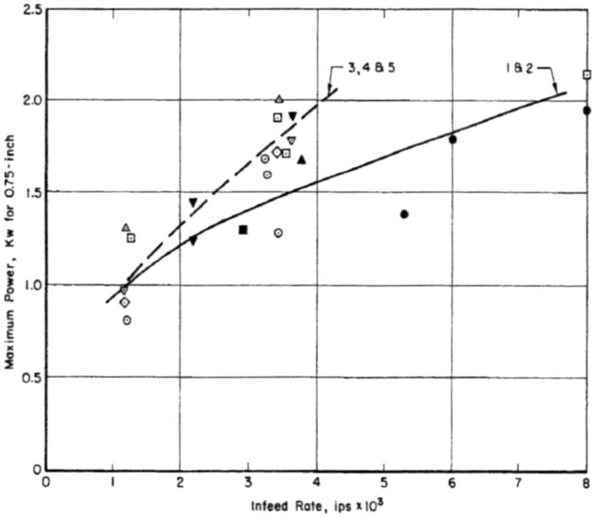

Fig. 4.21 Power dependence on infeed rate for solution-treated and solution-aged specimens with a particular grinding fluid [42]

to read the paper [42] as the authors present a very thorough study of grinding uranium alloys in the plunge form centerless grinding mode.

In addition to presenting the effects of grinding parameters on process performance, Hoffmanner and Meyer [42] present economic data that shows how well each combination of wheel type and grinding fluid contributes to reducing the cost per component ground. Their studies showed that:

1. Good lubricating properties are essential for grinding uranium alloys.
2. Sulfur and chlorine additives improve the performance of aluminum oxide wheels.
3. Phosphorus and boron additives to the fluid enhance grinding of uranium alloys.
4. Single-pass grinding was more efficient than double-pass grinding when grinding uranium alloys.
5. Minimum dressing time and high infeed rates were critical for reducing grinding costs.
6. Vitrified wheels made with aluminum oxide grits could be used efficiently in plunge form centerless grinding operations.

4.3 Discussion and Conclusions

This chapter has focused on developments in abrasive grains and historical grinding practice associated with uranium and uranium alloys. Since most published studies on grinding uranium occurred during the later 1970s and early 1980s, it appears that new developments in abrasive grits and bonding systems may well provide the opportunity to improve the opportunity to grind those materials. The recent developments in the production of agglomerated grains provide the possibility of grinding without significant heat generation. In combination with high-strength glass ceramic bonding systems, these developments should be applied to the further development of grinding uranium and uranium alloys.

Acknowledgments The authors also thank Springer for allowing the authors to reproduce their own material that appeared in "Machining with Abrasives" published by Springer as a chapter on abrasive tools and bonding systems by M. J. Jackson and M. P. Hitchiner (license numbers: 2972251046327, August 18, 2012). This material is published "With kind permission from Springer Science+Business Media: 'Machining with Abrasives,' Edited by Mark J. Jackson and J. Paulo Davim, Springer-Verlag, New York, November 2010, pp 1–423. *ISBN 978-1-4419-7301-6*".

References

1. Acheson EG (1893) Production of artificial crystalline carbonaceous material. US Patent 492,767, (1893)
2. Cowles AH, Cowles EH (1885) US Patent 319945, (1885)

3. Kern EL, Hamill DW, Deem HW, Sheets HD (1969) Mater Res Bull, (Special Issue 4) S25–S32
4. Kern EL, DW, Hamil HW, Deam HD. Sheets, Proceedings of the International Conference on Silicon Carbide, University Park, Pennsylvania, PA, USA, 20–23 October 1968
5. Nilsson O, Mehling H, Horn R, Fricke J, Hofmann R, Muller SG, Eckstein R, Hofmann D (1997) Determination of the thermal diffusivity and conductivity of monocrystalline silicon carbide. High Temperatures-High Pressures 29:73–79
6. Tymeson MM (1953) The Norton Story. Publ. Norton Co., (1953).
7. Wikipedia (2012) Bayer Process. Online encyclopedia reading. Accessed 20 Aug 2012
8. Wolfe LA, Lunghofer EP, (1998) New fused alumina production in South Korea and Australia. 13th Industrial Minerals International Congress, Kuala Lumpur, Malaysia, 26–29 April 1998
9. Lunghofer EP, Wolfe LA (2000) Fused brown alumina production in China. Posted: August 6, 2000 http://www.ceramicindustry.com/copyright/77d58fabca9c7010VgnVCM100000f932a8c0_... Accessed 24 Sep 2009
10. Whiting Equipment Canada Inc. (2009) Metallurgical equipment. Commercial brochure
11. U.S.G.S. Geological Survey (2009) Mineral commodity summaries. Reston, Virginia
12. U.S.G.S. Geological Survey (2008) Minerals Yearbook Abrasives. Manufactured Abrasives, Reston, Virginia
13. Wellborn WW (1991) Synthetic minerals—the foundation stone of modern abrasive tools. AES Magazine 31(1):6–13
14. Coes L (1975) Abrasives. Springer, New York, p 65
15. Polch (1956) US Patent 2769699 11, 1956
16. Robie NP (1959) Abrasive material and method of making same. US Patent 2877104, 10 Mar 1959
17. Foot DG (1965) Mixture of fused alumina and fused granules in bonded abrasive articles. US Patent 3,175,8943/30, 1965
18. Marshall DW (1965) Fused alumina-zirconia abrasives. US Patent 3181939, 4 May 1965
19. Cichy P (1973) Apparatus for producing oxide refractory material having fine crystal structure. US Patent 3,726,6214/10, 1973
20. Richmond WQ, Cichy P (1975) Apparatus for producing oxide refractory material having fine crystal structure. US Patent 3,861,849, 21 Jan 1975
21. Richmond WQ, Cichy P (1975) Semi-continuous process for producing oxide refractory material having fine crystal structure. US Patent 3,928,515, 23 Dec 1975
22. Sekigawa H (1976) Process for manufacturing high strength Al2O3-ZrO2 alloy grains. US Patent 3,977,132, 31 Sep 1976
23. Ilmaier B, Zeiringer H (1977) Method for producing alumina and alumina-zirconia abrasive material. US Patent 4,059,417, 22 Nov 1977
24. Cichy P (1977) Continuous process for producing oxide refractory material. US Patent 4,061,699,6 Dec 1977
25. Ueltz HFG (1980) Fused alumina-zirconia abrasive material formed by an immersion method. US Patent 4,194,887, 25 Mar 1980
26. Richmond WQ (1983) Process for making oxide refractory material having fine crystal structure. US Patent 4,415,510, 15 Nov 1983
27. Richmond WQ (1984) Oxide refractory material having fine crystal structure and process and apparatus for making same. US Patent 4,439,895, 3 Apr 1984
28. Scott JJ (1976) Progressively or continuously cycle mold for forming and discharging a fine crystalline material. US Patent 3,993,119, 23 Nov 1976
29. Scott JJ (1978) Method of producing abrasive grits. US Patent 4,070,796, 31 Jan 1978
30. Rowse RA, Watson GR (1975) Zirconia-alumina abrasive grain and grinding tools. US Patent 3,891,408, 24 June 1975
31. Bange D, Wood B, Erickson D (2001) Development and growth of sol gel abrasives grains. Abrasives Magazine June/July (2001), pp 24–30

32. Webster JA, Tricard M (2004) Innovations in abrasive products for precision grinding. CIRP Innovations in Abrasive Products for Precision Grinding Keynote STC G, 23 August (2004)
33. Bauer R (1987) Process for production of alpha alumina bodies by sintering seeded boehmite made from alumina hydrates. US Statutory Invention Disclosure H000189, June 1 1987
34. Leitheiser MA, Sowman HG (1982) Non-fused aluminum oxide-based abrasive mineral. US Patent 4,314,827, 9 Feb 1982
35. Wood WP, Monroe LD, Conwell SL (1989) Abrasive grits formed of ceramic containing oxides of aluminum and rare earth metal, method of making and products made therewith. US Patent 4,881,951, 21 Nov 1989
36. Bange D, Wood B, Erickson D (2001) Development and growth of sol gel abrasives grains. Abrasives Magazine, June/July (2001), pp 24–30
37. Cottringer TE, van de Merwe RH, Bauer R (1986) Abrasive material and method for preparing the same. US Patent 4,623,364, 18 Nov 1986
38. Schwebel MG (1988) Process for durable sol–gel produced alumina-based ceramics, abrasive grain and abrasive products. US Patent 4,744,802, 17 May (1988)
39. Rue CV (1985) Vitrified bonded grinding wheels containing sintered gel aluminous abrasive grits. US Patent 4,543,107, 24 Sep 1985
40. Pellow SW (1992) Process for the manufacture of filamentary abrasive particles. US Patent 5,090,968, 25 Feb 1992
41. DiCorletto J (2001) Innovations in abrasive products for precision grinding. Conference Precision grinding and finishing in the global economy—2001, Oak Brook, IL, 1–3 October 2001
42. Hoffmanner A, Meyer G (1979) Optimization of centerless plunge form grinding of a uranium alloy. ASM Conference Volume on "Machinability Testing and Utilization of Machining Data", ASM International Materials and Metalworking Series, Oak Brook, IL, 297–324

Bibliography

Aris J (1989) Metals handbook—machining. In: Aris J (ed) Machining of uranium and uranium alloys, vol 16, 9th edn. ASM International, Materials Park, OH, pp 874–878, *ISBN0-87170-007-7*

Boland JF, Sandstrom DJ (1974) Mechanical fabrication, heat treatment and machining of uranium alloys. Publication LA-UR-74-113, Los Alamos Scientific Laboratory, Los Alamos, Mexico, On behalf of the US Atomic Energy Commission, Contract # W-7405-Eng, 36

Conboy J, Shevchik P (1983) DU chip recovery program—phase I: a machining study for the production of contaminant-free chips. Report # AD-A131 389/9, South Creek Industries, Inc., July 1983

Denst A, Ross HV (1955) How to machine uranium. Am Machinist 99(16):95–97

Fuller JE, Lynch JK (1971) Machining study of a uranium-4 wt.% niobium alloy. Report # RFP-1743, Rocky Flats Plant, Rockwell International, November 1971

Hurst JS, Read GM. (1972) Machining depleted uranium. Publication Y-SC-39, Union Carbide Corporation, Nuclear Division, Y-12 Plant, Oak Ridge, Tennessee, 24 April 1972

Latham-Brown CE, Porter F. Atmosphere assisted machining of depleted uranium (DU) penetrators. Report # AD-A182 138/8/XAB, US Army Armament Research, Development and Engineering Center

Machining and Grinding of Ultra High Strength Steels and Stainless Steels. AEC/NASA Handbook SP-5084, for US Army Missile Command, Authored by Battelle's Columbus Laboratories, Battelle Memorial Institute, Columbus, Ohio

Morris TO (1981) Machining of uranium and uranium alloys. Paper presented at the Uranium Technology Seminar, American Society of Metals, Gatlinburg, TN, 1–31 May 1981

Olofson CT, Meyer GE, Hoffmanner AL (1976) Processing and applications of depleted uranium alloy products. Report # MCIC-76-28, pp 60–83, Metals and Ceramics Information Center, Battelle Columbus Laboratory, Ohio, September 1976
Stephens WE (1967) Uranium machining. Technical Paper MR 67-222, American Society of Manufacturing Engineers, Dearborn, Michigan, May 1967
Wright WJ (1967) The machining of nuclear materials. Prod Eng 46(11):638–650

Chapter 5
Uranium Processing

Brajendra Mishra, Nathan R. Gubel, and Rahul Bhola

Abstract Due to typically low concentrations of uranium, chemical complexity, and the varying nature of many uranium ores, the economic recovery of uranium often poses difficulty for the extractive industry. Physical concentration techniques (flotation, gravitational, electromagnetic, etc.) have been met with only limited success for uranium. The methods used for the recovery of uranium have been designed to economically treat large ore volumes. Uranium is a very electropositive metal; thus most direct pyrochemical methods are not applicable and processes often involve modern aqueous extractive metallurgy techniques. In this section the more important features of the extractive processes of uranium will be illustrated with emphasis on the chemical principles involved.

Keywords Resources by production method • Uranium recovery • Uranium processing • Uranium chemistry • Uranium leaching • Solution processing • Ion exchange • Acid leaching • Alkaline leaching • Solvent leaching • Solvent extraction • Gel extraction • Precipitation • Uranium ore • Membrane separation

5.1 Introduction

It was in 2003, when for the first time countries reported their distribution of uranium resources by the method of production used irrespective of the type used [1], for instance, open pit or underground mining, in situ leaching, heap leaching, or any other unspecified methods used. These have been represented by Figs. 5.1 and 5.2 below for the reasonably assured resources and the estimated additional resources category I for uranium production under various categories such as <40, <80, and <130/kgU.

B. Mishra (✉) • N.R. Gubel • R. Bhola
Department of Metallurgical and Materials Engineering, Colorado School of Mines, 1500 Illinois Street, Golden, CO 80401, USA
e-mail: bmishra@mines.edu; ngubel@mines.edu; rbhola@mines.edu

J.S. Morrell and M.J. Jackson (eds.), *Uranium Processing and Properties*,
DOI 10.1007/978-1-4614-7591-0_5, © Springer Science+Business Media New York 2013

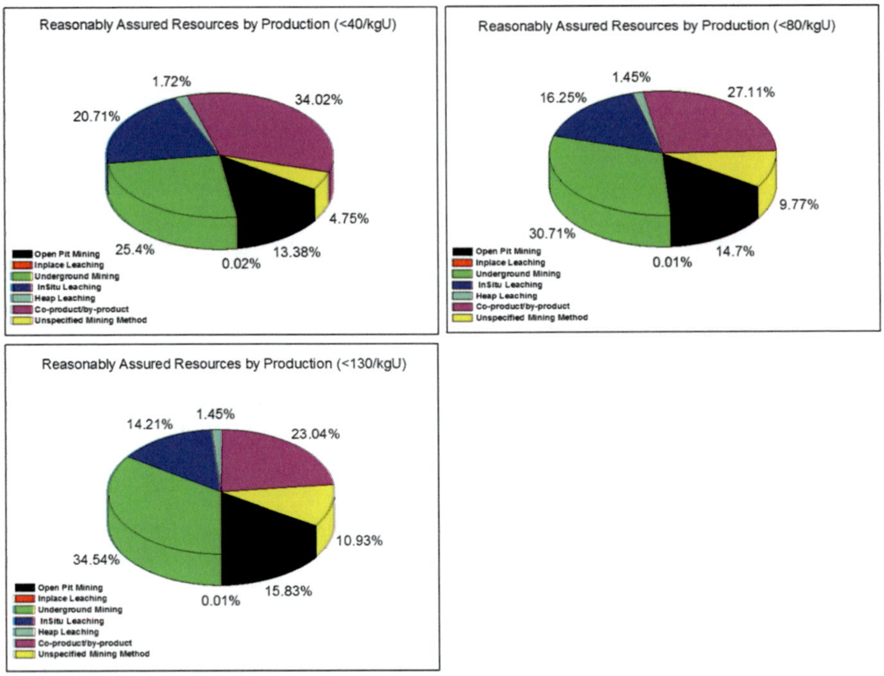

Fig. 5.1 Reasonably assured resources category I for uranium production under various categories such as <40, <80, and <130/kgU

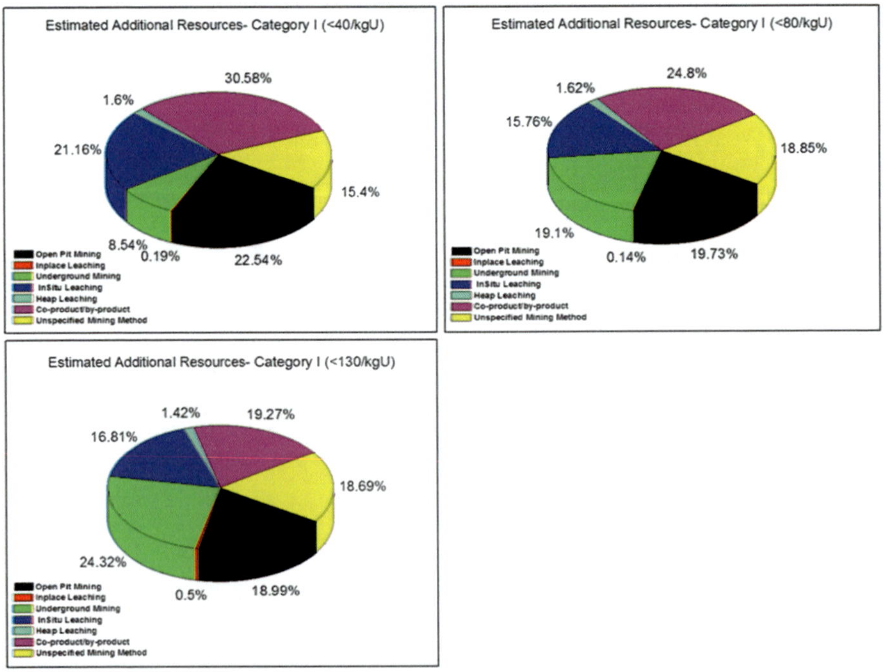

Fig. 5.2 Estimated additional resources category I for uranium production under various categories such as <40, <80, and <130/kgU

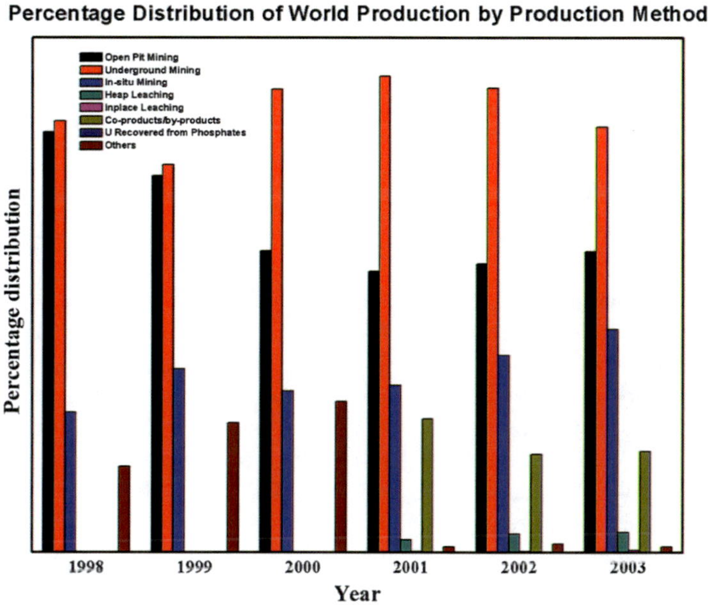

Fig. 5.3 Percentage production of uranium using various production methods

5.1.1 Resources by Production Method

Several countries gave estimates for the production of known conventional uranium resources (KCR = RAR + EAR1) amounting to 1,858,984 tonsU at <$40/kgU which is a 22 % increase compared to 2001 and 2,178,355 tonsU at <$80/kg/U which is a 13 % increase compared to 2001, and these are sufficient to meet the uranium requirement through 2020 [1, 2].

5.1.1.1 Percent Distribution of World Production by Method

The production proportions of uranium using various productions methods since 1998–2003 [1] have been compared in Fig. 5.3 shown above.

5.1.1.2 Projected Production Capabilities

To help in developing the projections of future uranium availabilities, member countries provided their expected projections of production capabilities up to the year 2020. These include projections for existing and committed production centers and for planned and prospective production centers in the <$80/kgU category through 2020 for all countries that are currently producing or plan to produce uranium in the near future [1, 3].

5.2 Leaching: Introduction

The object of leaching operations is to extract the uranium present in the ore into solution, usually aqueous, from which the recovery and purification of the uranium from additional metals can be carried out. Leaching is usually the first of the chemical treatments to which the ore is subjected, and all present chemical processing methods for any type of ore involve some type of digestion of the ore with either acid or alkaline reagents. The extraction of uranium from the majority of the ores is usually more complete by acid leaching than with alternative leaching procedures and is consequently used in most mills.

Due to several factors listed below, it is very hard to concentrate uranium directly from its ore [4, 5]. Firstly, uranium can only be selectively taken into solution from the waste matrix of its minerals. Secondly, uranium usually occurs greatly disseminated in huge rocks, where concentration procedures are usually ineffective and very expensive. Moreover, lower concentrations of the element imply processing large quantities of rocks to quantify a workable amount of the element.

Due to the various difficulties observed in uranium processing, a standard processing protocol has been established [6] which includes various processing steps: (1) leaching to bring the element present in the solid matrix of the ore into the liquid phase, (2) concentration and purification of the liquid to concentrate and purify the uranium extract, and (3) precipitation of the concentrated substrate into a suitable chemical intermediate for final extraction of the purified element.

The dissolution of solids, such as the ore, contains various metals, oxides, hydroxides, sulfides, selenides, tellurides, arsenides, phosphates, silicates, chlorides, and sulfates, i.e., into a liquid media comprising of acids, bases, and water of other solvents form the basis of the leaching process [5, 6]. Depending upon the nature of the mineral, a leaching process can be identified as physical, chemical, electrochemical, or electrolytic and has been summarized below in Fig. 5.4.

The leaching process for uranium with conducting or semiconducting oxides is primarily of the electrochemical type. In chemical nature, uranium minerals may be grouped into simple oxides, multiple oxides, hydrated oxides, simple silicates, and hydrated uranium salts. Thus, when combining the chemical nature of a uranium mineral and its predisposition to leaching, the uranium ores may be grouped under tetravalent ores, hexavalent ores, phosphates and miscellaneous mineral ores, refractory ores, and uranium associated with carbon [5, 7]. As already discussed the dissolution process selection of the uranium mineral depends on the uranium ore classification and has been summarized in Table 5.1.

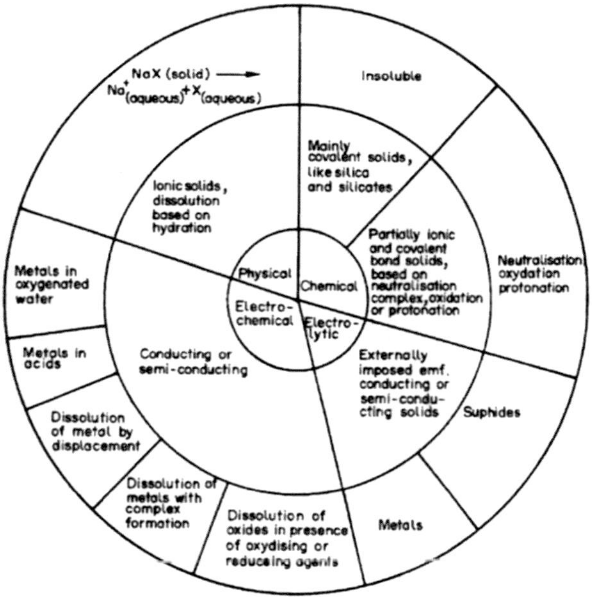

Fig. 5.4 Types of leaching based on type of process [5]

Uranium primarily exists in nature as an oxide and almost never in its free elemental form. It can combine with oxygen in any stoichiometry, but the most common forms are UO_2 (uraninite) and U_3O_8 (pitchblende).

5.2.1 Acid Leaching

Acid leaching involves treating the finely ground uranium ore slurry with sulfuric acid at elevated temperatures, usually in the presence of an oxidant [5, 7, 8]. Sulfuric acid is most effective with most ores containing uraninite and pitchblende and does not require the ore to be broken up into small particles as with carbonate leaching methods. Sulfuric acid attacks other minerals closely associated to uraninite by exposing the uranium-rich ore to the leaching solution. Acid leaching gives a higher yield of the metal due to the refractoriness of host matrix in relation to alkalis. The main requirement for an acid leaching process is that the ore matrix should have low amounts of acids consuming reagents such as lime. Acid leaching has several advantages over other similar techniques such as requiring coarsely ground ore, shorter leaching times, milder reagent concentrations, higher yield, temperatures in moderate range, and subsequent processing of uranium-containing solution using ion exchange and solvent extraction procedures.

Table 5.1 Uranium minerals and their responses toward dissolution in acids [5]

Type	Valency	Natural occurrence	Mineralogical classification (basis: valency)	Solubility
(A) Oxides				
UO_2	4	Uraninite	Primary	Insoluble in dilute H_2SO_4 without oxidant
U_2O_5	5	Not naturally occurring		Partially soluble in dilute H_2SO_4 without oxidant
U_3O_8	4.6	Pitchblende	Primary	Partially soluble in dilute H_2SO_4 without oxidant
UO_3	6	Occurring only with combination with K_2O, V_2O_5	Secondary	Fully soluble in dilute H_2SO_4 without oxidant
(B) Multiple oxides (refractory minerals)				
$(Fe,Ce,U)(Ti,Fe,V,Cr)_3(O,OH)_7$	4[a]	Davidite	[b]	Soluble in strong acid only
$(U,Ca,Fe,Th,Y)_3Ti_5O_{16}$	4[a]	Brannerite	[b]	Soluble in strong acid only
(C) Phosphates				
$Ca_5(PO_4)_3(F,Cl,OH)$	4	Apatite	Primary	Soluble only when matrix attacked
$CaAl_3(PO_4)_2(OH)_5H_2O$	4	Crandallite	Primary	Soluble only when matrix attacked
(D) Carbonates				
$UO_2 \cdot CO_3$	6	Rutherfordine	Secondary	Soluble in dilute acid without oxidant; also water soluble in varying amounts

[a]The uranium in these minerals was probably present in tetravalent state, but now there are also significant amounts of hexavalent uranium due to weathering. Thus, these multiple oxide types contain U(IV) as well as U(VI) together with such elements as Ti, Fe, V, Th, and rare earths

[b]On the valency basis, strict mineral categorization into primary or secondary or mixed cannot be made. However, they are unequivocally identified as refractory uranium minerals

5.2.1.1 Acid Leaching Chemistry Redox

When uranium ore is treated with sulfuric acid, many non-uranium constituents present in the ore will also dissolve and become suspended in solution. It then becomes imperative to separate uranium from most of these impurities to increase the elemental yield. Separation can be accomplished by the selective nature of uranyl ions to form stable compounds with oxyanions such as sulfates, carbonates, and nitrates [9]. If the minerals contain uranium in the 4+ state, an oxidant is added during leaching. Addition of an oxidant may also be desirable if the ore does not contain uranium in 4+ states but contains other reducing agents including metallic iron from extraction processes (mining, crushing, grinding, etc.) of the ore.

Maintaining a proper oxidizing environment during acid leaching is critical to high yields of uranium [2, 5]. The two main oxidizing agents used are manganese dioxide (pyrolusite) and sodium chlorate, although neither of them directly oxidizes the uranium. The reactions involved in the oxidation processes have been summarized below:

$$2Fe^{2+} + MnO_2 + 4H^+ \leftrightarrow 2Fe^{3+} + Mn^{2+} + 2H_2O$$

$$6Fe^{2+} + ClO^{3-} + 6H^+ \leftrightarrow 6\,Fe^{3+} + Cl^- + 3H_2O$$

$$UO_2 + 2Fe^{3+} \leftrightarrow UO_2{}^{2+} + 2Fe^{2+}$$

Subsequent reaction of the uranyl ions with sulfuric acid leads to the formation of uranyl sulfate species shown below:

$$UO_2{}^{2+} + SO_4{}^{2-} \leftrightarrow UO_2SO_4$$

$$UO_2SO_4 + SO_4{}^{2-} \leftrightarrow UO_2(SO_4)_2{}^{2-}$$

$$UO_2SO_4 + SO_4{}^{2-} \leftrightarrow \left[UO_2(SO_4)_3\right]^{4-}$$

The above chemical reactions indicate that by altering the concentration of oxyanions in the uranium-containing solution, uranium behaves amphiphilicly. In the presence of high sulfate concentrations, the thermodynamic stability of the uranium–oxyanion complex is much higher in comparison to the oxides. Thus the presence of these ions increases the driving force for the leaching reactions independent of the presence of protons.

The two important ways for concentration and purification of elemental uranium [10] are ion exchange (strong base anion exchange resins used to treat acid and alkali leach liquors which selectively adsorb anionic uranium complexes and exclude other cations present in the leach liquor) and solvent extraction (used to treat acid leach liquor with both the anionic and cationic extractants, the final outcome of the uranyl oxyanion will depend upon the pH range for the precipitant, >1.2 precipitates uranyl arsenate, whereas >1.9 precipitates uranyl phosphate).

5.2.1.2 Chemical Response of Minerals (pH) and Leaching Parameters

During acid dissolution of uranium ore, satisfactory dissolution of uraninite and pitchblende occurs at a pH range of 1.5 and 2.0 under oxidizing conditions. Preferential dissolution of uranium requires a careful optimization of pH, time, and temperature of the leach liquor due to the various other contaminants acquired during the extraction of the ore [4, 5].

Gummite, shoepite, becquerelite, hydrated salts, and hydrosilicates (coffinite and uranophane) dissolve in the pH range of 1.5–2.0. Torbenite and autenite requires high acid levels, whereas mixed oxide minerals and combinations containing titanium and zirconium are almost unreactive to acids. Moreover, non-uranium minerals such as quartz are nonreactive to conventional acid leach, whereas dolomite, a carbonate mineral, consumes most of the acid during the leaching process. Silicates such as biotite, chlorite, sericite, and various types of clay minerals give variable results near the reaction pH (~1.5–2.0) for dissolution of uranium ore [5].

Several leaching parameters that play a significant role in the recovery of element from its ore are summarized below:

- A suitable acid concentration is required to solubilize the ore to a solution. A satisfactory dissolution of uraninite, pitchblende, and other oxidized uranium minerals occurs at a pH range of 1.5–2.0 with a free acid concentration of 3–7 g/L. Refractory minerals such as davidite and brannerite may require higher free acid concentrations as high as 50 g/L.
- Several oxidizing agents such as MnO_2, $NaClO_3$, O_2 gas, Caro's acid, and H_2SO_5 may be used in industry depending upon the degree of oxidation, substrates used, and processing steps occurring during leaching. The use of oxidants and their effectiveness in leaching operations are routinely monitored using emf measurements. The utilization of oxidant and its cost is an important determining factor in the leaching process.
- Temperature influences the reaction kinetics of the leaching process. High temperatures lead to increased oxidant and acid consumption. The temperature needs to be carefully balanced during the leaching process as there may be an increased amount of heat from the dilution of acids and the exothermic discharge with carbonate minerals. Corrosion kinetics increase with temperature and need to be considered appropriately.
- Time is an independent variable and usually used in conjunction with temperature and other variables to adequately balance the final thermodynamic and kinetic parameters in the leaching process.
- .Particle size is a significant parameter influencing both the degree of liberation and reaction. Finer particles have easier slurry suspension, greater surface area, higher reaction rates, and shorter reaction times compared to more coarse particles.

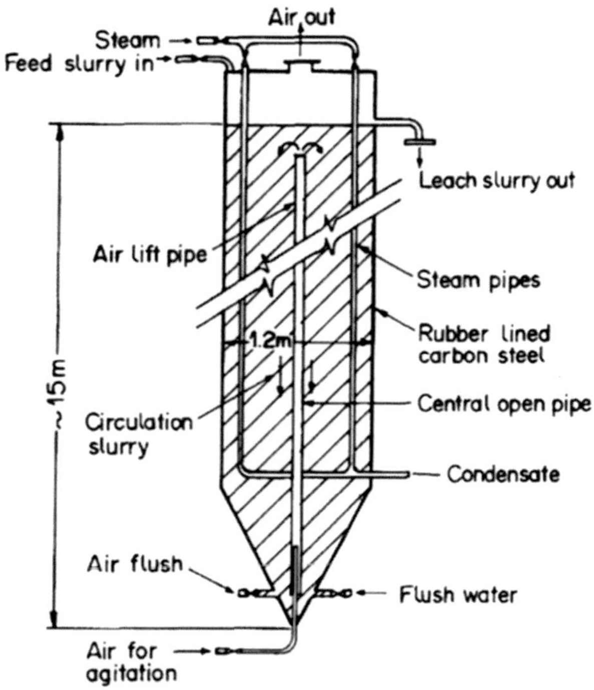

Fig. 5.5 An example of a Pachuca used in pneumatic leach process [5]

5.2.1.3 Leach Process Patterns/Process Diagrams

The contact pattern for the solid–liquid phase in acid leach liquor can vary and has its own advantages and disadvantages that need to be optimized for higher yields of uranium [7]. Uranium resources use two methods for agitation [11], *mechanical* (motor-driven impellers in cylindrical steel or titanium tanks with vertical axis, tanks may have a flat bottom with a jacket for maintaining the optimal reaction temperatures, and optimum solid–liquid ratios in the leach vessels are maintained by a balance between recycling stream and the use of dewatering equipment) and *pneumatic* (compressed air under high pressure is used for agitation); an example of a pneumatically agitated steam vessel is known as *Pachuca* which consists essentially of a wood- or rubber-lined steel cylindrical vessel around 15 m tall with a 60° conical bottom, central vertical tube opening at both ends that provides the compressed air in tank when charged with the pulp causing uniform agitation at low intensity, requiring longer operational times with lower running and maintenance costs. A Pachuca is shown in Fig. 5.5.

Two-stage countercurrent leaching (as shown in Fig. 5.6) has also been developed as a variant to the conventional single-stage leaching with advantages in enhancing uranium extraction, [5, 9] minimizing waste, and reducing reagent consumption. The strong acids and oxidants are added during the second stage of leaching. Following the second stage, the solid–liquid separation is carried out, and the pregnant leach solution containing uranium and unused reagents goes back to the primary stage. Unused solids are discarded from the second stage. During

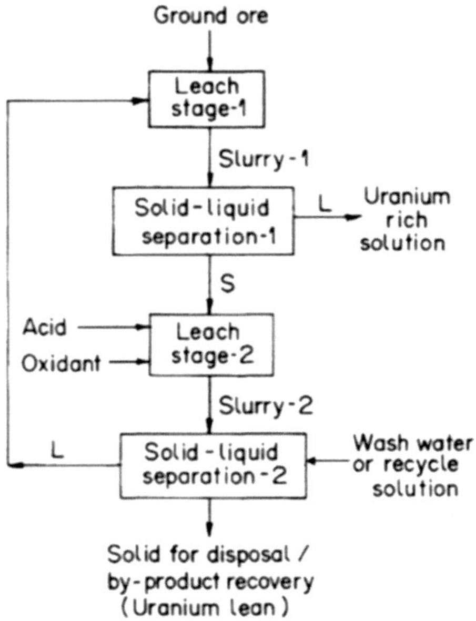

Fig. 5.6 A two countercurrent leach process [5]

the primary stage, fresh solids are brought together without the addition of acids. The following solid–liquid separation step produces the concentrated pregnant solution for further purification via ion exchange or solvent extraction, rerouting the unused solids onto the second leach stage.

Acid curing for levels of high solids is another alternative to two-stage leaching. During this process, dry ore is ground to 2 mm particle sizes, mixed with sulfuric acid, and cured in heaps for some time following which the pugged mass is slurried with water, agitated, and the resulting pregnant solution is separated from the other solids using filtration.

5.2.2 Alkaline Leaching

Alkaline leaching involves treating the ground ore with an alkali, primarily a sodium carbonate solution. The alkaline leaching process utilizes the high stability of uranyl-carbonate compounds to dissolve uranium from the waste matrix and can be utilized for both primary and secondary uranium salts [5]. The alkaline leaching process has several advantages, such as allowing the precipitation of the pure uranium from the leach liquor, reagents consumption is relatively low, and the processing solution is far less corrosive. More over the leach solution can be regenerated for its further use in additional leaching steps.

Alkaline leaching does have certain disadvantages. It requires ore to be ground much finer in comparison to acid leaching and requires elevated temperatures and pressures for effective yields of uranium.

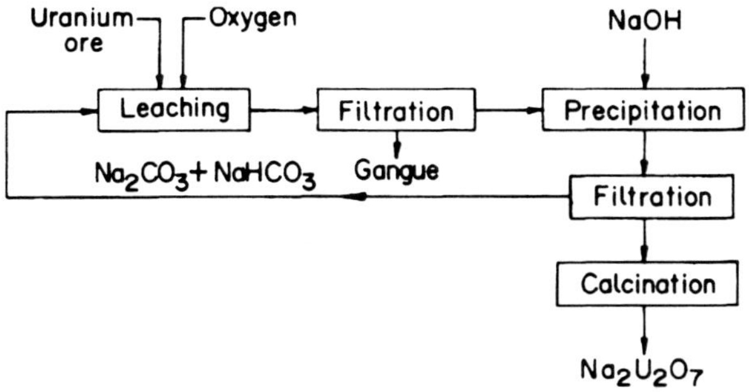

Fig. 5.7 Schematic of alkaline carbonate leach process [5]

5.2.2.1 Alkaline Leaching Chemistry and Pachuca Leaching

Alkaline leaching by carbonate reagent performed in *Pachuca* tanks is very similar to those used in acid leaching described earlier. A schematic of alkaline carbonate leach process is shown in Fig. 5.7. Uraninite ore is ground to 74 μm at a solid density of 55–60 % in a solution containing 30 g/L sodium carbonate and 20 g/L sodium bicarbonate at 71 °C that can provide high uranium yield in about 100 h.

Using deep Pachucas at high pressures (140 kPa) and temperatures (91 °C) can significantly reduce the leaching time and increase uranium recovery [12, 13].

The reaction chemistry occurring during the process can be represented as

$$UO_2 \rightarrow UO_2^{2+} + 2e^-$$

$$UO_2^{2+} + 3CO_3^{3-} \rightarrow \left[UO_2(CO_3)_3\right]^{4-}$$

$$1/2\ O_2 + H_2O + 2e^- \rightarrow 2\ OH^-$$

Summarizing the leach process as

$$UO_2 + 3CO_3^{2-} + H_2O + 1/2\ O_2 \rightarrow \left[UO_2(CO_3)_3\right]^{4-} + 2\ OH^-$$

Due to the generation of hydroxyl ions during carbonate leaching, it is always performed under the presence of bicarbonate ions (sodium bicarbonate) which act as buffers and prevent the increase in pH, preventing the precipitation of diuranate. The entire process can be summarized using the flow chart shown below:

The uranium precipitation reaction can be summarized as follows:

$$2\ Na_4UO_2(CO_3)_3 + 6\ NaOH \rightarrow Na_2U_2O_7 + 6\ Na_2CO_3 + 3\ H_2O$$

The oxidation potential for uranium oxide (UO_2) is 320 mV, and oxygen plays an efficient role in it. The maintenance of oxidizing conditions during carbonate leaching is important to prevent ferrous and ferric ions picked up during various mining procedures to affect uranium production [5].

5.2.2.2 Process Parameters

Uranium salts comprised of arsenates, phosphates, sulfates, vanadates, molybdates, and carbonates are soluble in carbonate solution [14]. Silicates need higher temperatures to solubilize. Vanadates and acidic oxides (P_2O_5) improve uranium solubility if the ore is roasted in a furnace before leaching.

5.2.3 *Pressure Leaching*

As the name implies, leaching performed under high pressure in specialized pressure reactors (autoclaves) is known as pressure leaching [5, 6]. It may be carried out in open or closed reactors at ambient, moderate temperatures or at temperatures much higher and closer to the boiling point of the solution. In either case, oxygen is present as an oxidizing agent and by increasing the partial pressure of oxygen that the leaching rate increases.

For uranium, both acid and alkaline pressure leaching can be performed at higher temperatures to enhance uranium recovery from the solution.

5.2.3.1 Acid Pressure Hydrometallurgy Chemistry

Pressure hydrometallurgy can be successfully employed for uranium ores containing sulfides as it is automatically converted to sulfuric acid in the presence of oxygen at high temperatures and pressure. The expected reactions can be summarized below, which proceed at much faster kinetic rates at elevated temperatures and pressures.

$$2\,FeS_2 + 7\,O_2 + 2\,H_2O \rightarrow 2\,FeSO_4 + 2\,H_2SO_4$$

$$2\,FeS_2 + 7.5\,O_2 + H_2O \rightarrow Fe_2(SO_4)_3 + H_2SO_4$$

$$4\,FeSO_4 + 2\,H_2SO_4 + O_2 \rightarrow 2\,Fe_2(SO_4)_3 + 2\,H_2O$$

$$UO_2 + Fe_2(SO_4)_3 \rightarrow UO_2SO_4 + 2\,FeSO_4$$

Under such high-temperature and high-pressure conditions, elements such as iron, molybdenum, and titanium form precipitates of hydrated ferric oxide or sulfates which further assist in the formation of free acid and reduce the overall consumption of acid.

Both vertical and horizontal autoclaves can be used for pressure leaching. High pressures of 600 kPa at a temperature of 70 °C for 4 h with solids around 40–50 % of the total solution can provide an adequate yield.

Vertical autoclaves are huge steel cylinders equipped with top-mounted turbine-type mechanical agitators. The feed slurry is preheated in steam coil-heated tanks or heat exchangers. The exhaust gases are passed through several recovery vessels to remove excess froth/condensate and recover waste heat.

Horizontal autoclaves, on the other hand, are mechanically agitated steel tanks lined by lead or rubber and partitioned into four interior chambers, each with independent turbine mixers. A continuous flow is required to move the acid slurry from one compartment to another inside the autoclave.

Pressurized tube reactors offer extremely short holding times, high thermal efficiencies, and low capital investments to produce an interesting alternative to the vertical and horizontal autoclaves. The slurry is pumped into an externally heated thick-walled tube and out from the other end of the chamber. The developments of diaphragm-piston pumps that are able to maintain a high pressure of 10,000–20,000 kPa have made such an application for such reactors possible.

5.2.3.2 Alkaline Pressure Hydrometallurgy

Alkaline leaching operations at 700 kPa using ammonium carbonate instead of sodium carbonate have routinely been performed for alkaline pressure hydrometallurgy. The ammonium carbonate decomposes in solution, and the ammonium–uranyl carbonate complex decomposes at atmospheric pressure when heated to 100 °C, causing direct steam stripping of the pregnant leach solution shown by the following reactions:

$$(NH_4)CO_3(aq) \rightarrow 2\ NH_3(gas) + CO_2(gas) + H_2O\ (aq)$$

$$(NH_4)_4UO_2(CO_3)_3(aq) \rightarrow 4\ NH_3(gas) + 3\ CO_3(gas) + UO_3 \cdot 2\ H_2O$$

The gaseous ammonia and carbon dioxide can be recovered and recycled in the leaching process.

5.2.4 Novel Separation Methods

5.2.4.1 Solvent Leaching

In this process a minimal amount of acid is used to wet the ore, and a solvent is used to extract uranium. Tri-n-butyl phosphate (TBP) is used for uranium extraction from ores that have been previously dampened with acid [15]. With the small amount of acid, the ore is pugged and cured at 100 °C for several hours, followed by nitrate addition and further TBP addition under constant agitation. The pregnant extract is then used for uranium recovery, and the solid residue is washed off to recover TBP.

Several other methods involving HCl–acetone system or a similar di-isooctyl phosphoric acid are known to help extract additional uranium from its ore.

5.2.4.2 Bioleaching

Using the potential of single-celled microorganisms to recover element from the ore is known as bioleaching [16, 17]. These single-celled organisms (bacteria) have been organized into various classifications depending upon their growth requirements (autotrophs, heterotrophs), oxygen requirements (aerobic, anaerobic), type of mineral they reduce (sulfide reducing, iron reducing), and shape and size (bacilli, cocci, vibrio, etc.). Many commercial applications of bioleaching in the extraction of uranium have been successful.

Due to the association of uranium with sulfidic and pyritic mineralogy, these bio-reducers play an excellent role in extraction of uranium from ore [9]. The bacteria *Thiobacillus ferrooxidans* is one of the main sources used in the bioleaching and extraction of uranium. It primarily functions through three main mechanisms – direct, indirect, and galvanic – which may operate individually or simultaneously and are summarized below.

The direct mechanism involves the attachment of the bacterium to the mineral and destruction of its lattice by the enzymes produced intrinsically [17, 18]. This enzymatic degradation leaches metal from the ore.

The indirect mechanism involves the dissolution of sulfide ore by the action of an intermediate produced during bacterial metabolism and can be represented as

$$MS + Fe_2(SO_4)_3 \rightarrow MSO_4 + 2FeSO_4 + S^\circ$$

The galvanic mechanism is based on the principle of galvanic coupling in electrochemistry, where two different phases of sulfide (with differences in potential) come in close contact to one another creating a potential difference and via the movement of electrons cause the oxidation–reduction reactions and leaching of pure metal at the anode.

The reactions can be summarized as follows:

$$MS \rightarrow M^{2+} + S^\circ + 2e^- (\text{oxidation} - \text{anode})$$

$$O_2 + 4H^+ + 4e^- \rightarrow 2\,H_2O\ (\text{reduction} - \text{cathode})$$

Thiobacillus ferrooxidans [19, 20] promotes conversion of sulfur to sulfate and allows the anodic dissolution to proceed uninterrupted. For uranium leaching specifically, acidic ferric sulfate acts as an oxidant and is the important bacterial metabolite that brings about the dissolution process. The continuous oxidation of recycled acidic ferrous sulfate leach solution during uranium processing with bacterial involvement is critical to the entire process.

Bioleaching [21–23] is a very slow process, and there are several factors that influence the rate, such as high pyrite content, uranium minerals conducive to bacterial growth and action, permeability of host rock to leaching solutions and gases, and the absence of basic components in the ore and presence of minerals in ore that provide nutrition to sustain living microforms.

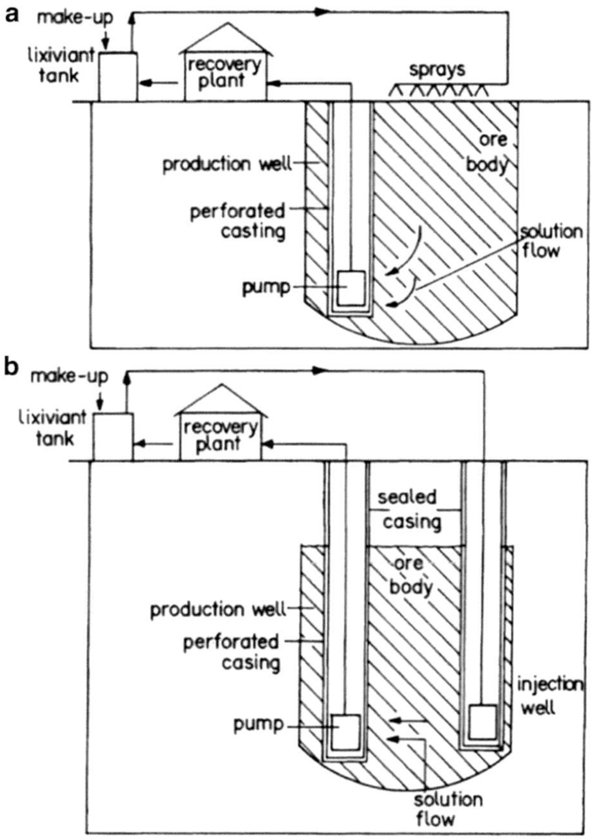

Fig. 5.8 In situ leaching operation for (**a**) an exposed ore body and (**b**) an underground ore body [5]

5.2.4.3 In Situ Leaching

Introduction/Techniques

Solution mining or leaching in place of minerals deposits from within the earth's crust without removal of the host rock or the overlying rocks is commonly employed in industry to extract soluble salt deposits present in low concentration to justify mining costs and expenses, such as potash, sodium chloride, sodium sulfate, and carbonate.

In situ leaching [7, 24] offers several advantages such as minimal surface disturbance, eliminated wastage from conventional mining procedure, and minimal exposure to hazardous mining operations (such as grinding and crushing) [8], and wastes from surface processes are minimal that can be safely handled with cost-efficient procedures. However, lower recovery and risk of groundwater contamination are its major drawbacks.

The in situ leaching primarily involves two techniques (shown in Fig. 5.8). Firstly, spraying over the ore body when exposed or when buried is shown below.

Parameters

Several parameters play an important role for uranium recovery using the in situ technique and are mentioned below [5, 7]. The uranium ore should be in a porous horizontal bed with an impervious nonporous bed underneath it, there should not be any fractured structure in the rock bed, the ore must be leveled below the static water table, the direction and velocity of the natural water flow through the ore must be known, the uranium minerals must be suitable for the in situ leaching process, and the ore body must be of sufficient size to provide justification for the huge expenses involved in the extraction process.

Recovery of uranium using solution mining requires drilling of wells in a geometric pattern spaced every 10–20 m apart. The well patterns may be square, rectangular, honeycomb, or irregular, where corner wells may be used for injecting the leach solution and the middle/central well is used to pump out the pregnant solution for processing.

The wells are cased and cemented with pumps so that they open only at a desired level in the ore. For uranium, the dissolution process using an acid or alkali is primarily the reversal of the natural conditions that led to the formation of ore by promoting the formation of U(+6 state) from the U(+4 state).

ISL Systems

The ISL systems have been used when dealing with different areas with special features such as alkaline leach, acid leach, uranium recovery, and groundwater restoration.

Alkaline leach involves the carbonate bicarbonate leach, used for processing ores containing acid-consuming components. Ammonium bicarbonate is one of the early ISL reagents. Another system uses soda ash and carbon dioxide mixed with groundwater that leads to the formation of a bicarbonate in the pH range 7.5–8.0. The overall reactions at ore body can be summarized as follows:

$$UO_2 + 1/2\,O_2 \rightarrow UO_3$$

$$UO_3 + 2\,HCO^{3-} \rightarrow UO_2(CO_3)_2^{\;2-} + H_2O$$

Similarly, pyrite minerals (FeS_2) can be leached giving ferrous ions sodium sulfate and sodium bicarbonate, where the ferrous ions can serve as an oxidation agent for uranium.

Re-precipitation and coprecipitation are significant in affecting the overall recovery of uranium during the in situ leaching process. Precipitation of contaminant elements can be beneficial in eliminating it from the pregnant solution; however, precipitation of calcium with sulfates or vanadates, molybdates, and arsenates may fix and block the bed preventing uranium movement during recovery [7]. This would therefore require periodic washing of the bed with HCl for sulfate-rich systems and the use of carbon as an active adsorbent for molybdenum and vanadium to increase uranium yield in the solution. Acid leach systems have advantages such as the uranium concentration in the solution is high, oxidant requirement is low, working life and payback period are short, and there is a potential for by-product recovery.

During uranium recovery, dilute uranium solution is collected into a single tank through a channel of field pipes, where it is passed through the activated carbon filter, to an ion exchange system. As soon as the uranium bicarbonate is loaded, chloride is given out in exchange, and subsequently its levels may rise, affecting the exchange process. The liquid effluents from the processing unit are processed along with the groundwater, and a solar evaporation pool is employed for transient storage of the solution.

Groundwater restorations pose significant environmental issues and need to be addressed properly: the prevention and control of leakage of the leach solution into the surroundings (by well-field patterns and use of well monitoring) and restoration of groundwater to acceptable safety standards after completion of the leaching process (pumping of excess water and washing and chemical treatments). Restorations of groundwater involve three main techniques: groundwater sweep (contaminated water is removed from the leach zone and replaced by freshwater from outside), solution processing (solutions recovered from groundwater are processed through electrodialysis or reverse osmosis and reinjected), and chemical precipitation (injecting of suitable chemicals to neutralize and precipitate the contaminants). A combination of single or multiple techniques is used to adequately restore the environment. Once all the mining is complete, the site is decommissioned by returning it into the previous. All wells are plugged, buried pipelines are removed, and surface is landscaped, planted, and monitored for a sufficient period of time before releasing it for public use.

Commercial Considerations

The geological formations where uranium is found usually consist of floodplain, deltaic formations, and shallow marine lagonic deposits. The host rocks are sand and sandstone containing volcanic ash rich in uranium deposits [12]. Gaseous hydrogen sulfide derived from deep-seated hydrocarbons acts as good reducing agents to precipitate uranium from volcanic ash.

The high radioactive ability of uranium and the porous nature of the host bed require an exhaustive survey of several miles of the proposed site which is important, not only for the environment, but also to check the feasibility of the proposed extraction process.

Once the leaching process has been completed, it is necessary to restore the site by pumping out reagents used in leaching and continue pumping until all the reagents used are exhausted from the rock bed. The reagents need to be adequately disposed of, and the well needs to be pumped with fresh, uncontaminated water.

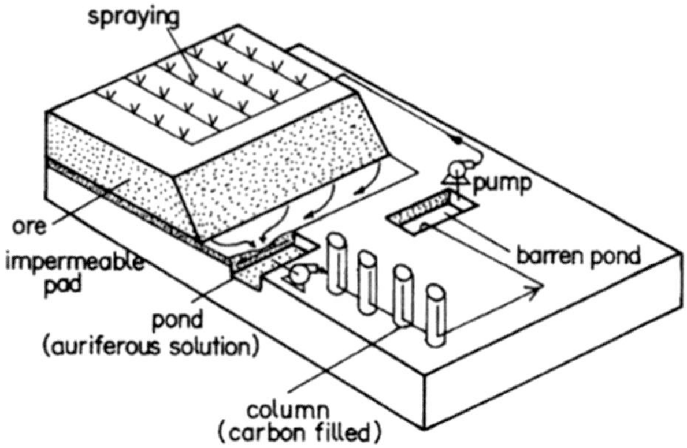

Fig. 5.9 A schematic of heap leaching of a mineral deposit [5]

Penetrative ISL

In penetrative ISL the ore is prepared underground by mine working/explosive shattering prior to injection [7]. Following the blasting operations, the holes are used for feeding, trapping, and draining of working and pregnant solutions. It can be applied to not only explosive shattering but also newer automated, low radiation exposure variants such as hydraulic de-aggregation, using same principles.

5.2.4.4 Heap or Dump Leaching

Heap leaching of uranium with sulfuric acid solution in conjunction with bioleaching has currently revived newer interests [16, 24]. In this process the area is cleared of all vegetation, leveled to an incline, compacted, and covered with an asphalt layer. The crushed ore is then dumped over the asphalt layer, and the leaching solution is sprayed on top of it. The leach solution percolates down and is collected, processed, and refined to extract uranium. The process has been summarized in a schematic sketch as shown in Fig. 5.9.

5.3 Leached Solution Processing/Parameters

The uranium leach solution contains a complex mixture of cations and anions. In alkaline leaching, the solution is relatively pure compared to the acid leaching of uranium which contains a mix of aluminum, iron, titanium, magnesium, vanadium, lanthanides, and large amounts of silica. Regardless of the technique, the uranium concentration is low in the solution [4, 6, 9]. The solution needs to be processed, purified, and concentrated by using an ion exchange technique (resins or solvents). The exchange of ions is very selective and quantitative. Both procedures are comprised of two stages: adsorption phase (where leach solution is brought in

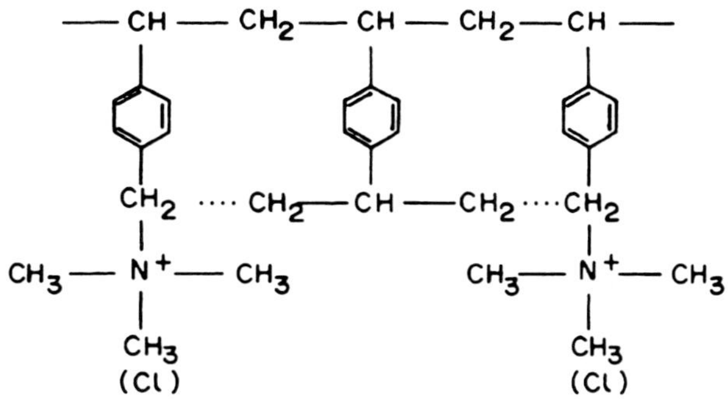

Fig. 5.10 Structure of an ion exchange resin [5]

contact with an organic phase) and elution ion exchange and stripping (where the organic solvent from the previous step is brought into contact with a suitable aqueous media to move the uranium ions from the organic to aqueous phase to selectively extract them). These processes are so selective that they increase the uranium concentration by a factor of 10 in the final, concentrated product which is further processed to get uranium metal. For uranium extraction the ion exchange process is a physicochemical method, whereas solvent exchange is primarily a chemical method and offers several advantages such as reduction of the volume of solution that is chemically treated further to recover uranium and yielding uranium cake with high purity that is very selective in quantitative yield of the element.

5.3.1 Ion Exchange

5.3.1.1 Introduction

The ion exchange process is primarily an ion reversible exchange of the same charge signs between an aqueous solution and an insoluble solid in contact with the solution. Synthetic ion exchange resins have been used since the very beginning of uranium mining to recover the element from both the acidic and alkaline dilute solutions with metal concentration not to exceed a particular range.

Owing to the improvements in resin structures, exchange technology, and resin technology, a continuous exchange sequence is followed in industry to recover adequate quantities of uranium with high purity. Although each step of the process is important, adsorption and elution play a significant role as described below.

5.3.1.2 Adsorption and Elution Principles

The synthetic resin used in uranium industry is a semirigid gel manufactured in the bead form comprising of quaternary amine polymer exhibiting ion exchange functions and shown in Fig. 5.10.

The different forms in which uranium exists in the acid solution are the uranyl ion (UO_2^{2+}), uncharged neutral complexes of sulfate ($UO_2(SO_4)$), and the negatively charged complexes such as $[UO_2(SO_4)_2]^{2-}$ and $[UO_4(SO_4)_3]^{4-}$. Thus uranium behaves as an amphiphilic material that can be adsorbed by both anionic and cationic complexes, anionic complexes being more favorable as only a few ions are absorbed by this method including arsenates, permanganates, chromates, vanadates, molybdates, and ferric ions. These ions can be reduced to some other forms and are not recovered with uranium [25].

Due to the formation of anions, sulfate ion concentrations are paramount in the adsorption of uranium from an acidic sulfuric acid leach system. Under equilibrium the amount of uranium adsorption increases when the sulfate to uranium ratio is less than two. An excess of sulfate ions compete with active sites on the resin resulting in reduced adsorption. The H^+ ion concentration also plays a significant role in the adsorption of uranium on the resin by maintaining the optimum sulfate ion concentration in the solution. The optimum pH for the adsorption is between 2.0 and 2.5. U (+6) is much more readily absorbed compared to U (+4). Carbonates play a similar role in alkaline leach solution as sulfates in acidic leach solution.

Once the ion has been sufficiently adsorbed on to the resin, the resin needs to be regenerated to continue the ion exchange mechanism, and hence, elution comes into the picture. There are situations where elution is done in one step and regeneration of the eluted resin is done in another step. For instance, a weak acid resin is loaded with Cu^{2+} and Ni^{2+} [25].

5.3.1.3 Chemical Reactions

The uranium ion exchange chemistry can be represented through series of chemical reactions shown below [6]. The chemistry begins with the equilibrium reactions of sulfate in acid leach solution represented as

$$HSO^{4-} \leftrightarrow H^+ + SO_4{}^{2-}$$

$$UO_2{}^{2+} + SO_4{}^{2-} \leftrightarrow \left[UO_2(SO_4)_2\right]^{2-}$$

$$\left[UO_2(SO_4)_2\right]^{2-} + SO_4{}^{2-} \leftrightarrow \left[UO_4(SO_4)_3\right]^{4-}$$

Thus several of these ions compete with one another to occupy the active sites on the exchange resin, and thus conditions need to be customized for uranium to be the preferred occupant over the resin:

$$4\,R^+X^- + \left[UO_4(SO_4)_3\right]^{4-} \leftrightarrow (R^+)_4\left[UO_4(SO_4)_3\right]^{4-} + 4\,X^-$$

$$R^+X^- + HSO^{4-} \leftrightarrow R^+HSO^{4-} + X^-$$

$$2\,R^+X^- + SO_4{}^{2-} \leftrightarrow (R^+)_2SO_4{}^{2-} + 2\,X^-$$

where R = fixed ion exchange sites and X = mobile species.

The characteristic elution reactions in uranium processing can also be summarized as follows:

$$(R^+)_4\left[UO_4(SO_4)_3\right]^{4-} + 4\,X^- \leftrightarrow 4\,R^+X^- + UO_2{}^{2+} + 3\,SO_4{}^{2-}$$

$$R^+HSO^{4-} + X^- \leftrightarrow R^+X^- + HSO^{4-}$$

$$(R^+)_2SO_4{}^{2-} + 2\,X^- \leftrightarrow 2\,R^+X^- + SO_4{}^{2-}$$

For alkaline solutions in carbonate leach liquors, the ions involved are sodium, carbonate, bicarbonate, and uranium complexes with carbonates, and main reactions can be summarized as

$$4\,R^+X^- + \left\lfloor UO_4(CO_3)_3\right\rfloor^{4-} \leftrightarrow R_4UO_2(CO_3)_3 + 4\,X^-$$

$$R^+X^- + HCO^{3-} \leftrightarrow RHCO_3 + X^-$$

$$2\,R^+X^- + CO_3{}^{2-} \leftrightarrow R_2CO_3 + 2X^-$$

The capacity of resin for uranium adsorption is significantly reduced when there is an increase in the bicarbonate ion into the solution because bicarbonate ion has a stronger affinity for adsorption versus the carbonate ion. A host of other ions also tend to build up as the carbonate liquor is being recycled and tend to bring down the uranium adsorption for the resin.

The commonly used eluting agent is sodium chloride or ammonium nitrate, and the main elution reaction for the carbonate system can be written as

$$R_4UO_2(CO_3)_3 + 4\,X^- \leftrightarrow 4\,R^+X^- + UO_2{}^{2+} + 3\,CO_3{}^{2-}$$

Elute can be precipitated to uranium by treating it with an alkali or neutralizing with an acid. Poisoning of resin may also occur when some ions are fixed to the resin and eluting is not able to regenerate the resin for further use in uranium recovery, for instance, blockage of resin active sites with vanadates, molybdates, and uranyl sulfate complexes in an acid leach system. An alternative eluting system can be used to remove the incorporated ions from the resin active sites [8].

5.3.1.4 Ion Exchange Materials/Equipment

There are primarily two groups of ion exchange materials: the inorganic category comprising the aluminosilicates often known as zeolites and the organic category consisting of sulfonated coals, synthetic resins of long hydrocarbons with fixed

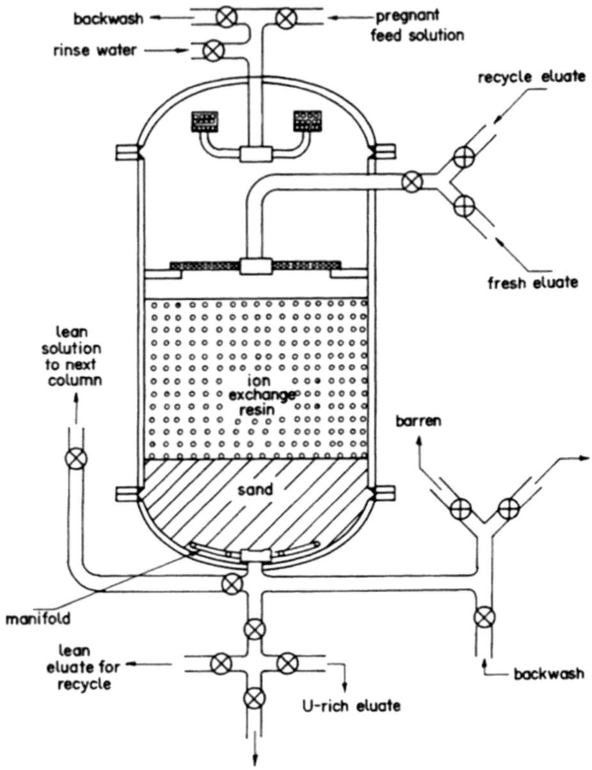

Fig. 5.11 Representation of a typical ion exchange fixed-bed column [5]

ionic groups (weak acid cation exchanger, strong acid cation exchanger, weak base anion exchanger, and strong base anion exchanger) [9].

Since the beginning of uranium processing, various types of ion exchange equipment have been developed and can be represented as fixed-bed columns, resin in pulp basket systems, screen-mix RIP, suspended columns, fluidized bed contractors, and continuous ion exchange systems and are described individually.

Fixed-Bed Columns

In the fixed-bed columns, the feed solution is fed into cylindrical tanks mounted vertically, lined from inside with suitable materials consisting of a packed resin bed resting on a bed of crushed rocks or sand. The solution flows in and out through the columns connected to the cylinder at the level above the resin bed. In the conventional design, the resin is static, and the solution flows and is cycled by suitable valve arrangements. A schematic of this has been shown in Fig. 5.11.

Fixed-bed systems require huge capital investments, large pressure columns, complex valve systems, and initial resin installation inside the column. A newer variant of the traditional design has been successfully used in industry where the

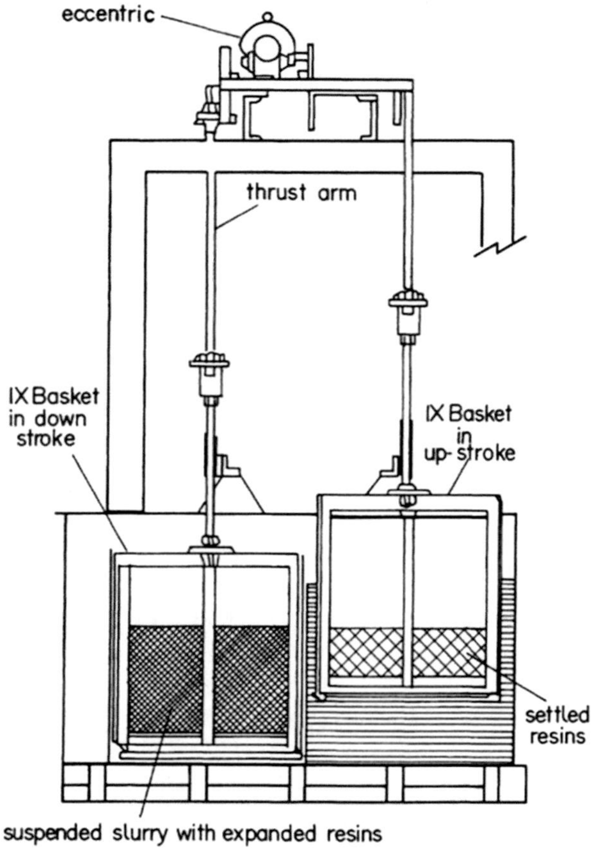

Fig. 5.12 Basket resin in pulp system [5]

resin is fluid and loaded in a separate elution system giving adequate column space and effective regeneration of resin [5].

Resin in Pulp (RIP) Systems

The RIP systems are particularly suitable for treating ores that exhibit poor filtering and settling properties [7]. It can process slurries of medium to high densities and offers several advantages such as low capital and operational costs due to elimination of pre-filtration of slurries before purification and reduction in soluble uranium loses during washing. The different types of RIP systems have been summarized below.

Basket Resin in Pulp

In this technique resin is packed as beads in cube-shaped baskets enclosed by stainless steel screens with a porosity of around 600 μm and suspended in the de-sanded leach slurry. The pictorial representation can be seen in Fig. 5.12.

The baskets are connected through a motor-driven crankshaft and can be raised up and down at a fixed rate 10–12 strokes/min in the rectangular tank containing the slurry. The basket movement not only makes new room in the resin matrix for the solution but also keeps the fine slurry particles in suspension and maintains adequate contact of leach solution with beads [5]. A group of baskets in a tank constitutes a bank, and a number of banks are employed in any processing industry. When the tank is saturated with uranium, either it is drained and rinsed through the rest of the tanks in series or contents are drained out to the feed tank and feed switched to the next tank. The saturated tank is then added to the end of the tank in the elution process which is essentially carried out using a nitrate solution.

Precipitation of uranium using RIP eluate is similar to the columnar system except prior clarification is required in the latter.

Screen-Mix Resin in Pulp

In this process there is a countercurrent flow of leach slurries from the ion exchange resin beds, and the processing involves mixing of ion exchange resin beads into de-sanded leach slurries for adsorption of uranium from the leach solution. Screening of saturated beads in a separate system is followed by recirculation of barren resin into the loading cycle. Incorporation of sand-slime separation with washing circuits has routinely been practiced in industry, where the feed is separated into a fine slurry and coarse slurry. The coarse stream is washed and joins the fine stream to form the RIP feed. During processing the slime slurry and resin move in opposite directions through a series of agitated stages. The discharge from each stage is screened, and countercurrent movement through a number of stages leads to saturation of resin at one end and depleted effluent at the other end. The saturated resins are further eluted in a similar system. Air agitation provides the necessary residence time in each stage and lifts the excessive resin for further screening. The entire process can be visualized through Fig. 5.13 shown below.

Suspended Bed Columns

In suspended bed columns resin loading occurs in an upflow fluidization mode, whereas elution occurs in a downflow mode through the packed bed. The process operates in a circular pattern very similar to the basket-type RIP. The resin column is loaded at the bottom with different layers of sized quartz to control the flow of the leach solution. A compressed air inlet above the quartz bed aids in slime dispersion in the backwash mode. The column has screens on top to prevent loss of resin beads during the upflow mode.

Fluidized Bed Contactors

A fluidized bed contractor consists of a vertical column compartmentalized into stages by horizontally perforated plates and can be visualized in Fig. 5.14.

Fig. 5.13 Screen-mix-type resin in pulp system [5]

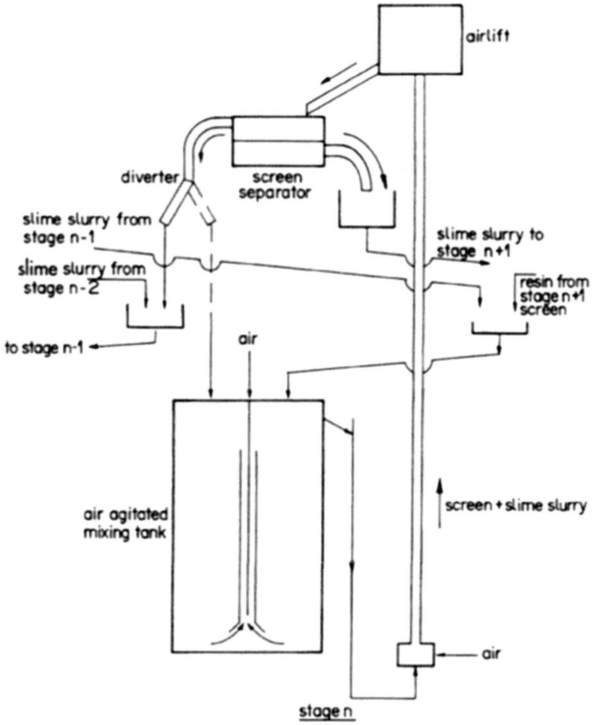

Operation takes place by feeding the slurry in an upflow direction. As slurry moves up, it fluidizes the resin beads and uranium loading occurs onto the beads. On completion of sorption cycle, the forward flow is stopped, and resin then settles onto the plates. The flow is then reversed, and the resin in one stage is transferred to the next stage. The forward-feeding flow is resumed, and a countercurrent flow is created. Freshly eluted resin enters the top of the extraction column, and as it successively moves down, it becomes fully loaded until it leaves the column from the bottom. The loaded resin is eluted separately in a static column.

The processing through such a system depends upon several factors such as the width of the column, resin volume, solution flow rate, void spaces inside the resin column, equilibrium isotherm, uranium concentration in leach solution, temperature, and time of contact with the resin [5].

The Himsley continuous ion exchange system is a variant of the fluidized bed contractor that has been routinely used in industry which involves additional loading of uranium onto a partially loaded resin followed by the elution of uranium from the more highly saturated resin. It displaces impurities from the resin and produces an eluate which can produce a purified yellow disc with reduced eluate consumption. This enables more uranium to be adsorbed to the resin and thus gives a higher yield.

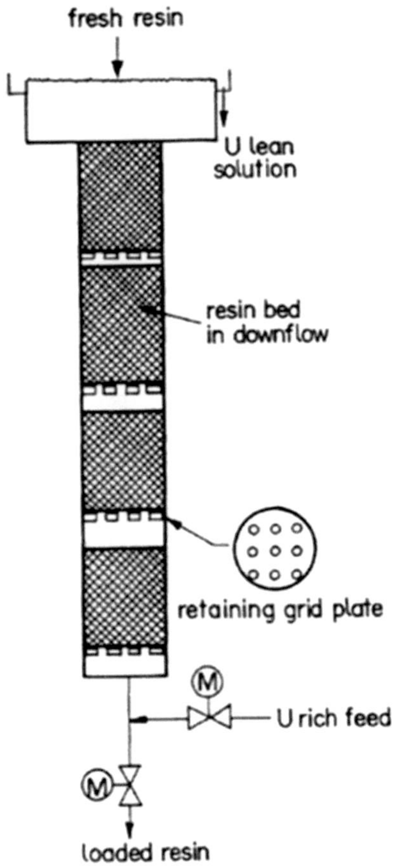

Fig. 5.14 Continuous ion exchange resin system for uranium extraction [5]

5.3.1.5 Uranium Liquor Processing

Ion exchange processing for uranium can be divided into six major steps: (1) adsorption, (2) flushing, (3) backwashing and flushing, (4) elution, (5) backwashing and settling, and (6) standby. Each of these has been described below in detail.

Adsorption: During this process uranium laden liquor is passed through a series of bed of ion exchange resins. Several columns are put in series to saturate the resin from the liquor, and this cyclic process allows continuous sorption of uranium into the used columns. The cyclic processing of the solution can be visualized in Fig. 5.15 below. It further depends on various factors such as uranium concentration in the liquor, flow rate, and the holding capacity of the resin.

Flushing: When the second column in series starts to leak uranium, the feeding loop is stopped, and fresh water is flushed through the second column via the first column

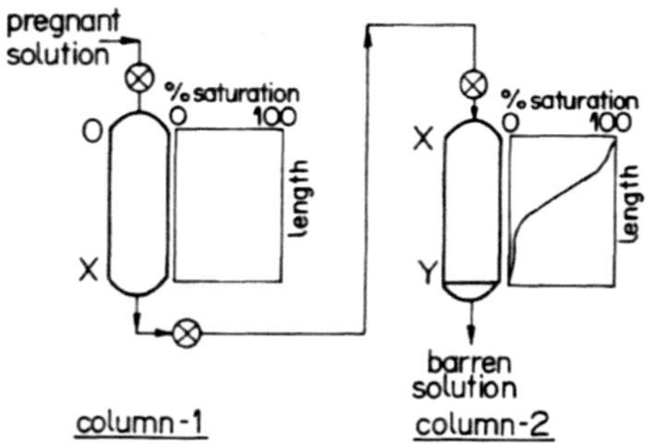

Fig. 5.15 Adsorption profile of uranium during two column processing [5]

at the original flow rate used during adsorption. This not only removes the pregnant liquor filling the void spaces occupying the volume greater than the resin volume but also lowers down the acidity of resin by hydrolysis.

Backwashing and Settling: The ion exchange resin is a strong polyelectrolyte that may flocculate some impurities which bypass the filtration and clarification units that collects as solid on top of the column often causing a pressure drop across the bed, affecting the uranium processing. By forcing water up the column, the bed is hydraulically expanded, and this makes removal of flocculated solids much easier. It should be continued until the overflow effluent is clear followed by allowing the resin bed to settle down before proceeding with the next process of elution.

Elution: Elution of sorbed uranium over resin is carried out using acidified solutions of 1M sodium nitrate or chloride salts. Nitrates are expensive but have higher process efficiency compared to any other salt in elution. The concentration of uranium in initial elute is high and is used to precipitate high-purity uranium economically, and the latter half has dilute uranium and is recycled. The elution is performed in five stages as summarized in the flow chart in Fig. 5.16.

During Stage 1 the column receives eluate from an earlier cycle to displace water containing sulfates, bisulfates, and iron from the void spaces. Stage 2 receives eluate from the previous cycle to displace the major share of desorbed uranium from the resin. Stage 3 receives virgin eluate. Stage 4 receives fresh eluant to achieve maximal uranium removal. Stage 5 receives freshwater to remove eluant ions from the void spaces.

Backwash After Elution: After elution with nitrates, residual ions remaining in the resin void, and those in the top part of the ion exchange column are removed by backwashing. It also aids removal of any solids present in the resin bed.

Standby: It is the last stage in cyclic operations before the change over to the next cycle. Barrens are often passed during this stage into column to ensure conversion of exchangeable active sites into the bisulphate.

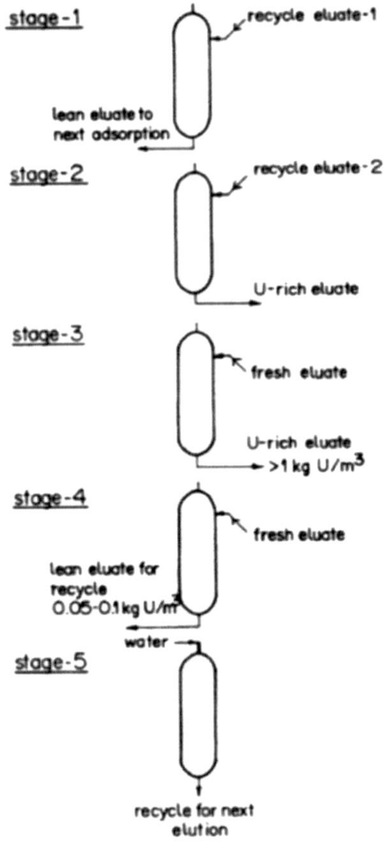

Fig. 5.16 A multistage elution system for a loaded fixed-bed resin column [5]

5.3.2 Solvent Extraction

5.3.2.1 Introduction

Solvent extraction (SX; also known as liquid–liquid extraction or sometimes called liquid–liquid distribution) has been utilized since the beginning of early uranium processing. Solvent extraction, in general, is known to be applied to large-scale copper plants, to some base metals, and to various other cases in metallurgy and chemistry. Continuous advances occurring in industry have all contributed to its development and expansion of the range of applicable areas. Elaborating in the area of copper just mentioned, it may be added that, until the emergence of SX as a commercially viable process, there was no way by which copper occurring in weak sources could be recovered. SX in the scheme of copper extraction involves the leaching of weak sources of copper to ultimately produce a relatively pure and concentrated copper solution, and once concentrated it is easy to win copper by electrochemical means. Therefore, it is often quoted in copper metallurgy that SX

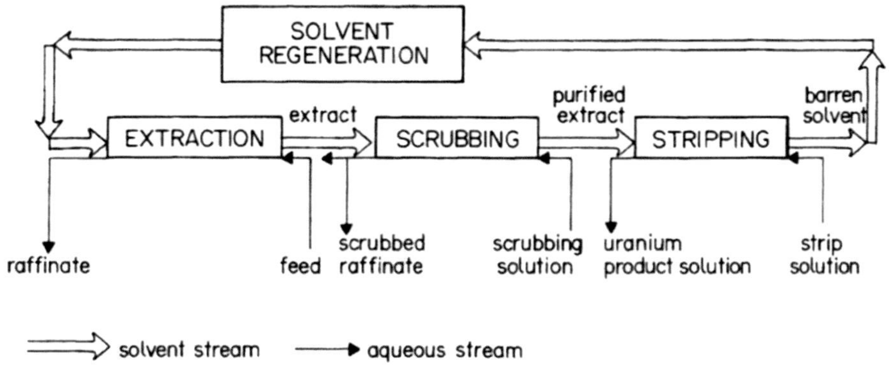

Fig. 5.17 Generalized solvent extraction flow sheet [5]

has bridged the gap between weak copper solution and electrowinning. The benefit has been enormous in that were it not for solvent extraction, much of the copper would have remained unfit for exploitation and remained untapped.

Figure 5.17 illustrates a generalized solvent extraction process. The process entails the contact of two immiscible phases, an organic phase and an aqueous phase. The organic phase is very similar to the resin phase encountered in the ion exchange process, and because of this, the SX process seems to have been analogously addressed as liquid ion exchange. In the first step of extraction, the transport of uranium contained in an aqueous phase to the organic phase occurs when the two phases are brought together into direct contact, for example, by a mechanical mixer. After reaching equilibrium, the two phases are separated in a settler where the difference in phase density leads to efficient separation.

The organic phase may contain traces of co-extracted impurities, which are separated in the "scrubbing" step. During this step, an aqueous scrub solution is mixed with the organic, i.e., the pregnant extract. The impurities are transferred back into the aqueous phase, while the uranium in the organic phase remains intact. The purified extract is then taken to the stripping stage. In the stripping stage, the extract is brought into contact with an aqueous stream of carefully chosen chemical composition so that uranium is transferred into the aqueous phase. The stripping operation results in a uranium product solution and a stripped, clean solvent. The pure product solution is then processed for uranium recovery by typical chemical precipitation processes.

The stripped aqueous stream that results from the extraction stage is usually referred to as the raffinate and is often treated for the recovery of other valuable by-products or discharged as waste after suitable effluent treatment. The scrubbing operation may be bypassed in other SX processes depending on the resulting selectivity of the solvent, impurities present in the feed liquor, or purities required. The solvent resulting from the stripping operation is often subject to regeneration before recycling during the extraction step. Each of the steps of extraction, scrubbing, and stripping usually involves repeated stages which are usually countercurrent [25].

5.3.2.2 Process Chemistry

Extractant Types

Uranium is unique in the fact that it can be extracted from leach solutions either as a cation, as a complex anion, or as a neutral complex. In these forms, uranium occurs in a sulfuric acid leached aquatic medium: UO_2^{2+}, $[UO_2(SO_4)_2]^{2-}$, $[UO_2(SO_4)_3]^{4-}$, or UO_2SO_4. Extraction of either species leads to a shift in the equilibrium so that a larger amount of the extracted species form until an interphase equilibrium is reached. The solvents used in SX divide into three groups depending upon the type of uranium complex extracted.

The principal members of extractants that belong to the first group of the extractants for extracting uranium in its cationic form (UO_2^{2+}) are extractants which can be alkyl phosphoric acids, alkyl phosphonic acids, and alkyl phosphinic acids. The most common is di-2-ethyl hexyl phosphoric acid, generally abbreviated as D2EHPA or DEHPA [26]. Extraction of the uranyl ion involves reactions with two dimers to form a complex via the following reaction:

$$\underset{(aq.)}{\left(UO_2^{2+}\right)} + \underset{(org.)}{2(RH)_2} \leftrightarrow \underset{(org.)}{UO_2(R_2H)_2} + \underset{(aq.)}{2H^+}$$

From the nature of the exchange reaction, it is apparent that as acidity increases, the extraction process decreases. In addition to uranium, other impurities such as iron can be extracted. The acidity of the aqueous phase will affect the separation of uranium relative to iron. The selectivity can also be changed by altering the oxidation state, e.g., the reduced state of iron shows lower extraction rates. Stripping of uranium from D2EHPA is done with strong acids or carbonate solutions. Stripping with HCl converts the uranium into a chloride complex, which tends to have lower extractability. Carbonate stripping involves the reactions

$$UO_2(R_2H)_2 + 3\ Na_2CO_3 \leftrightarrow 2\ NaR_2H + Na_4UO_2(CO_3)_3$$

$$2\ (RH)_2 + Na_2CO_3 \leftrightarrow 2\ NaR_2H + H_2CO_3$$

Sodium carbonate stripping can, at the pH level used, lead to the formation of insoluble hydroxides of other metals and eventually to their precipitation. This precipitation is advantageous, since it eliminates the buildup of impurities in the extractant that could interfere in following extraction steps. The acidic extract can also be stripped with HF solution. Bringing the solvent into contact with HF with which a reducing agent such as $FeSO_4$ has been added leads to precipitation of uranous fluoride. The UF_4 produced may not be of high purity, but the production cost is low.

The second group consists of basic extractants which anionic complex forms of uranium ($[UO_2(SO_4)_2]^{2-}$; $[UO_2(SO_4)_3]^{4-}$) and includes either secondary amines or high molecular weight trialkyl tertiary amines.

The exchange reactions involving amines are similar to those with anion exchange resins. In sulfate media the anions form sulfate and bisulphate salts in the reactions shown below:

$$2\,R_3N + H_2SO_4 \leftrightarrow (R_3NH)_2SO_4$$

$$(R_3NH)_2SO_4 + H_2SO_4 \leftrightarrow 2\,(R_3NH)HSO_4$$

The uranium exchange involves the reactions

$$2\,(R_3NH)_2SO_4 + UO_2(SO_4)_3{}^{4-} \leftrightarrow (R_3NH)_4UO_2(SO_4)_3 + SO_4{}^{2-}$$

$$2\,(R_3NH)HSO_4 + UO_2(SO_4)^{2-}{}_2 \leftrightarrow [R_3NH_2]_2UO_2(SO_4) + 2\,SO_4{}^{2-}$$

The stripping can also be carried out by soda ash solution with the following reaction:

$$(R_3NH)_4UO_2(SO_4)_3 + 5\,Na_2CO_3 \leftrightarrow 4\,R_3N + Na_4UO_2(CO_3)_3 + 2\,H_2CO_3 + 3\,Na_2SO_4$$

Although amine extractants are selected with uranium in mind, some metals such as molybdenum are more easily extracted than uranium, and thus separation is achieved by selective stripping. Uranium is stripped from the solution with chloride, and molybdenum is stripped next with sodium hydroxide. Similar to separation using resins, any $HSO_4{}^-$ ions present may compete for the extractant so the resulting low pH levels tend to lead to decreased uranium extraction.

The principal members that belong to the third group of the extractants are the ethers, trialkyl phosphates, and trialkyl phosphine oxides which extract neutral complexes, which is one of the forms in which uranium exists in aqueous solution. The overall mechanism of extraction can be represented as

$$\underset{\text{(aq.)}}{UO_2(NO_3)_2} + 2\,(TBP) \leftrightarrow \underset{\text{(org.)}}{UO_2(NO_3)_2}\ \underset{\text{(org.)}}{(TBP)_2}$$

A high concentration of nitrate ions will shift the equilibrium to the right and thus enhance uranium extraction. Conversely, stripping of the uranium from the loaded organic can be carried out by contact with water since a low nitrate concentration would shift the equilibrium to the left. A significant advantage of the extraction with tri-n-butyl phosphate is the ease of stripping using demineralized water. Another advantage is the high level of possible selectivity for uranium extraction in comparison with other ions present in the feed solution. The product of the TBP process generally has a high purity. Tri-n-alkyl phosphine oxides are also neutral compounds that are capable of extracting uranium complexes from sulfate leach solutions. However, the extraction performance is lower than that of alkyl orthophosphoric acids. Carbonate solutions can also be used for the stripping of uranium extracted by phosphine oxides [27].

Leach Liquor Characteristics

The most important leach liquor characteristics are uranium valency (or oxidation state), nature of cations, and nature of anions.

The leach solutions may contain uranium in either the hexavalent uranyl form or the tetravalent uraniferous form. Tri-butyl-phosphate (TBP), primary amines, and some secondary amines display higher coefficients for the reduced species, but tertiary and some secondary amine are not sufficiently strong enough for extracting the uraniferous complexes as the uranyl. Alkyl orthophosphoric acids are well known for the ability to extract both forms almost equally from sulfate solutions.

Anions in the aqueous phase determine the type of uranium complex that may be extracted and the type of extractant required.

Uranium leach solutions often contain a number of cationic impurities such as iron, vanadium, titanium, e.g., that result in complex effects of the extraction behavior, with regard to the separation factor and the extraction coefficient. For example, iron and vanadium are easily extracted at lower acidities, whereas uranium extraction from its uncontaminated solutions is less affected by acidity. Leach solutions where uranium is to be separated from iron and vanadium, lower acidity can reduce the extraction coefficient of uranium. An important factor can be the oxidation state of iron. The reduction of ferric iron to ferrous iron limits its extraction potential by alkyl phosphoric acids. The interference of titanium in uranium extraction from sulfate solution by alkyl phosphoric acids can be reduced by combining the leach solution with phosphate or fluoride. Titanium can also be removed from alkyl phosphoric acid extractant by scrubbing with a dilute fluoride solution. Molybdenum extraction requires carbonate regeneration of the solvent to prevent accumulation in the solvent during recycling.

5.3.3 Novel Separation Methods

5.3.3.1 Gel Extraction

Hydrogels

Hydrogels are polymeric materials with great ability to attract and absorb water. Hydrogels have a cellular structure which can swell with water to nearly 1,000 times their mass. In a dry state, these materials have glass-like characteristics, while swollen with water they become elastic. With manipulation of the chemical structure, one can vary the extent of swelling, diffusion coefficient of water, cross-linking density, and size, e.g., hydrogel applications are now found in various fields such as bioengineering, pharmaceuticals, biomedicine, pesticides, and metal sorption. Amidoxime derivatives of copolymeric hydrogels prepared from acrylonitrile and divinylbenzene have been used for uranium recovery from seawater. Hydrogels prepared with acrylamide and itaconic acid (AAm/IA) or acrylamide/malefic acid (AAm/MA) show exceptional adsorption capacity for uranyl ions from nitrate media [28].

When uranyl ions interact with the carboxyl groups of itaconic acid in the hydrogel, swelling in uranyl solutions is 580 %, compared with 1,660 % in water. These interactions of uranyl ions tend to be electrostatic in nature, such as ionic or ion–dipole mechanisms. Uptake of uranyl ions in the range of 42–76 mg/g adsorption has been reported for AAm/IA hydrogels, and 14–90 mg/g adsorption for AAm/MA hydrogels, from solutions containing 20–430 mg/l. The sorption increases with an increase in either the IA or MA content of the hydrogel, but is invariant with a gamma radiation dose in the range of 2.5–6 kGy. The adsorption depends on the anionic species from which the uranyl solution is prepared. Uranyl acetate shows higher sorption than nitrate. The sorption is physical in nature. Hydrogels soaked in uranyl nitrate solutions can be stripped of metal values by contacting with plain water [28].

Hydrophobic Gels

Hydrophobic gels incorporating the solvent extractant molecules offer a novel route for separation of metal ions from aqueous solutions. Gels can be prepared either by soaking low cross-linkage polymer beads in an extract solution or by the use of a suitable gelatinizing agent.

Gel extraction in columns offers efficiencies exceeding those of conventional liquid–liquid columns. An important characteristic of the copolymers swollen with solvents is the hydrophobicity of the external surface. For effective separation the surface needs to be made hydrophilic. This can be done by treating with hot concentrated sulfuric acid which creates a thin outer shell of sulfonated "wettable" copolymer. Experiments on SDVB gels doped with TBP-perchloroethylene have shown successful absorption uranium from nitrate medium. The effective curve of columnar operations is sharp, which indicates high transfer rates at the gel–solution interface and within the gel [29].

The limitations of gel liquid extraction arise from the natural tendency of the gels to lose the active solvent while moving through a flowing medium and the tendency of the gel to degenerate by solutions containing high levels of acidic agents, e.g., 6 N HNO_3. Adsorption of large amounts of solute such as nitric acid can create and introduce changes in gel structure that often lead to ejection of the solvent. Gel separation is very advantageous because of the elimination of entrainment losses of solvent and varying applicability to systems employing a large aqueous to organic ratio in normal liquid–liquid extraction processes [30].

5.3.3.2 Membrane Separation

Processes which utilize semipermeable membranes as separation barriers seem to have become popular novel separation processes. Membrane separation processes tend to be more efficient and more economical than conventional separation techniques. A number of exhaustive reviews have covered the development and applications of membranes [31–34].

Membranes should be thought of as semipermeable barriers between two phases. This barrier is able to restrict the movement of molecules across it in a very precise manner. Membrane separation is a rate process, and the separation is accomplished by a driving force, not by equilibrium between the phases as is typical in traditional techniques. In addition to a solid membrane, immiscible liquids can also serve as membranes between two liquid phases. Different solutes have diffusivities and different solubilities in a liquid, which selective permeability leads to separation. When utilizing liquid membranes, the diffusion coefficients are typically an order of magnitude higher than in solids, so a larger flux can be obtained.

The importance and utilization of liquid membranes (LM) is more significant when the solute concentrations are extremely low and large volumes of solutions are to be processed, without generating any secondary wastes. Although LMs have not yet attracted large-scale industrial applications, several pilot plants have been tested. The LMs use the same basic extraction and stripping reactions that are involved in SX processes [31, 32]. The important difference as LMs are concerned is the greatly reduced amount of organic solvent used. It works essentially as a membrane via which the metal is transferred to the stripping agent solution with the extraction and stripping taking place simultaneously on opposite sides of the membrane. The different types of liquid membranes which may be distinguished are presented in Fig. 5.18.

Bulk Liquid Membranes

Bulk liquid membranes (BLMs) are distinguished by the presence of two miscible aqueous liquids (feed and strip) separated by a third immiscible organic liquid (carrier) as shown in Fig. 5.18. The mass transfer from feed to strip occurs via the carrier. BLMs are used to optimize data for both supported liquid membranes (SLMs) and emulsion liquid membranes (ELMs) [35].

Emulsion Liquid Membranes

Emulsion liquid membranes (ELMs) or surfactant liquid membranes are basically double emulsions, i.e., water in oil in water (w/o/w) system. In this system, the oil that separates aqueous feed and strip solutions is the membrane phase. ELMs are

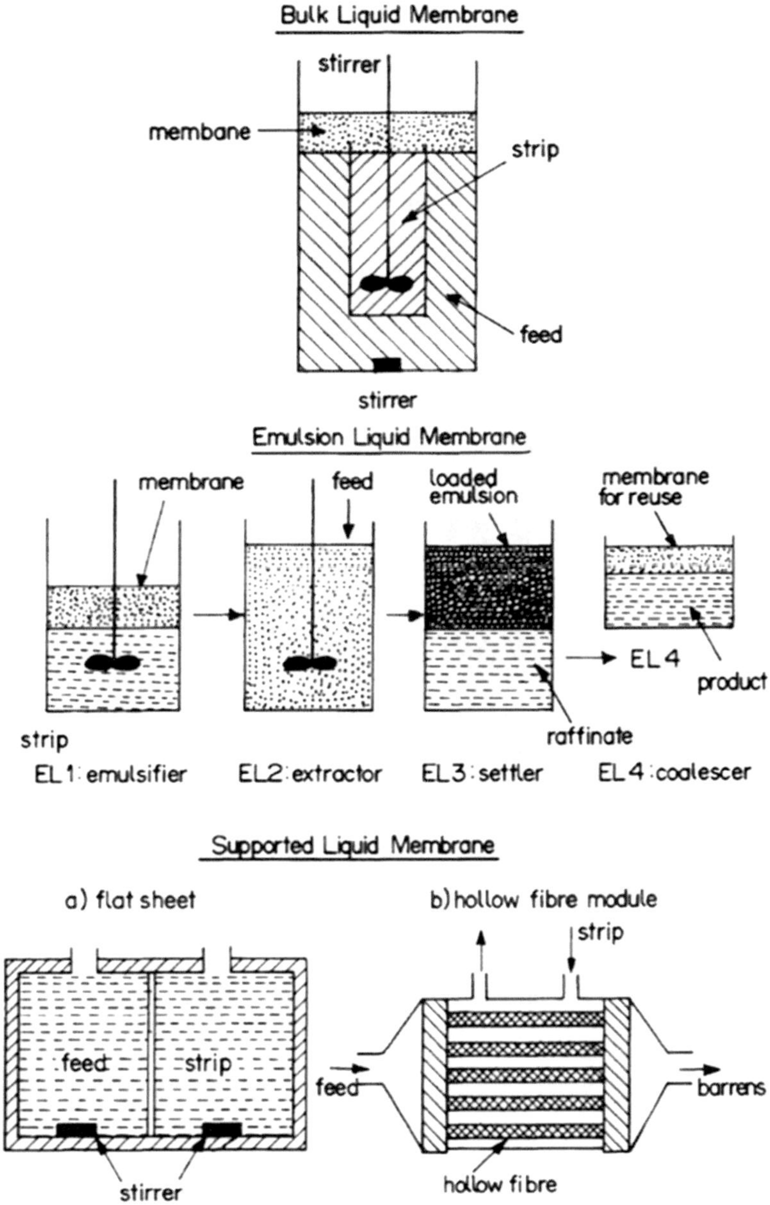

Fig. 5.18 Type of membranes: bulk, emulsion, and supported [5]

prepared by first forming an emulsion between the two immiscible phases (EL1 in Fig. 5.18) making use of a suitable surfactant and then dispersing the emulsion formed in a third continuous phase (EL2 and EL3). Thus, the miscible phases are separated by an immiscible membrane phase. In order to form a stable emulsion, a surfactant is added. Typically, the encapsulated internal droplets in the emulsion are 1–3 μm thick in diameter. When the emulsion is dispersed by agitation in the external continuous phase, many small globules of the emulsion are formed, the size of which depends on the properties and concentration of surfactants, viscosity of internal phase, intensity, and mode of mixing. Usually, the globule size is maintained in the range of 100–2,000 μm in diameter. Each globule contains many internal droplets, and a very large number of emulsion globules can easily be formed to provide a large mass transfer area, adjacent to the external continuous phase [36]

The internal mass transfer area (typically 16 m^2/m^3) is much larger than the external mass transfer area. After extraction, the emulsion and external continuous phases are separated by settling, and the internal phase is recovered by breaking the emulsion via heating, centrifugation, or with an electrostatic coalescer. The main problem associated with ELMs is the emulsion swelling and breakage. The emulsion must be prepared to withstand the shear forces created when mixing during the extraction during removal of the internal phase when separating the emulsion. The ELM process has been comprehensively used to study extraction of uranium from wet-process phosphoric acid [26]. In a specific example, recovery of U(VI) and Th (IV) from a nitrate medium using tri-octyl phosphine oxide (TOPO) as a carrier was reported. It was possible to recover 98 % U(VI) and 82 % of Th(IV) from 0.1 M HNO_3 solution. The liquid emulsion was prepared from a TOPO/Span 80/sodium citrate solution. The recovered products of U(VI) and Th(IV) carried less than 2 % Fe(III) [37, 38].

The major problem of ELM is the emulsion stability. The industrial applications of ELMs call for the further search and synthesis of some new surfactants as emulsifiers for the preparation of membranes.

The objective of using w/o emulsion drops containing magnetic particles in a column operation is to provide the possibility of unique drop motions, such as fluidization or vibrations in the aqueous phase which are favorable for interfacial mass transfer. Moreover, the phase separation of the emulsion from the continuous aqueous stream can be enhanced by magnetic forces [34].

Electrostatic Pseudo Liquid Membrane

Electrostatic pseudo liquid membranes (ESPLIM) consist of a cell filled with an extractant solution separated by a perforated baffle plate into two parts: an extraction cell and a stripping cell. Feed and strip solutions in a cell are as shown in Fig. 5.19. The organic is allowed to flow through the perforated plate, but the aqueous feed and strip are unable. Settlers are located at the bottom of the extraction cell and stripping cell. When an alternating current with a high-voltage electrostatic field is applied to both cells, droplets form out of the feed and strip solutions. The complex formed in the extraction cell diffuses through the baffle plate into the stripping cell, where the product is formed and released. A similar design without the applied electrostatic field has been studied as an emulsion-free liquid membrane (EFLM) [39].

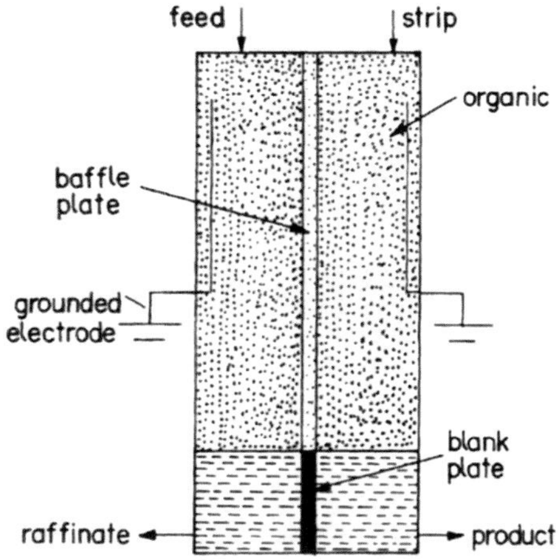

Fig. 5.19 Electrostatic pseudo liquid membrane [5]

Supported Liquid Membranes

Supported liquid membranes (SLMs) utilize a porous polymer support membrane that carries the extractant in its pores. SLMs are formed by doping the pores of a thin, porous polymeric membrane with liquid extractant (carrier) dissolved in a suitable solvent. The stability of the liquid membrane is given by capillary surface forces. The doped membrane is held in a transport cell such that one side is directed toward the aqueous feed solution and the other side is facing the aqueous strip solution (Fig. 5.19). The membrane acts as a common interface between the feed and strip solutions and exhibits selective transport of specific diffusive species through the membrane. This permeation is a simultaneous extraction and stripping operation utilized in a single-stage process under nonequilibrium conditions [40, 41].

Polymeric Inclusion Membranes

The newest, most promising type of membrane system currently in development combines rapid transport with high selectivity. This method is called polymeric inclusion membranes (PIMs). These membranes are made by including a suitable carrier inside the polymer while it is cast as a membrane or via covalently bonding by a chemical method. PIMs are similar to SLMs but are believed to be much more stable.

During the separation of metals, it is essential that multistage extraction is used. However, SLM is inherently a single-stage process. This can be overcome by using multiple or composite SLMs. In this case, the upstaging is done by employing two different methods of metal transport, cotransport and counter-transport. The composition of the first strip solution is designed to promote the back-extraction of the metal ion by a second SLM. Two such membranes are put in series, and the metal of interest can be separated to the desired extent.

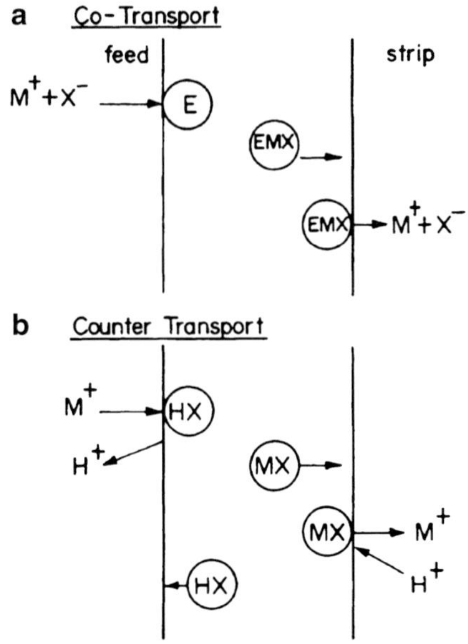

Fig. 5.20 Cotransport and counter-transport processes in membranes [5]

There are other methods of applying SLMs which have future potential. Cellulose acetate RO membranes for decontamination of uranium have been explored. The extractability of Y(III) with PC88 is less than that of Fe(III), but with the use of SLM, it was possible to get selective transport of uranium, as rates of transport and stripping of Fe(III) are low [40, 42].

Transport in Membrane

The transport through the SLM can be performed in two ways: one is by cotransport and the other is by counter-transport.

Cotransport

Metal ions are transported through the SLM, along with the counter ions from the feed solution, as shown in Fig. 5.20. If the aqueous extractant is basic or neutral, the driving force of the system is the difference in diffusion coefficient, D, between feed and strip solutions. This is usually done by maintaining a concentration gradient of the counter ions X^- (e.g., NO_3^-, Cl^-) between the feed and strip solutions. The negative counter ions combine with the metal cations to form a complex with the extractant E in the membrane. Afterward, the complex diffuses to the opposite side of the membrane, and the metal ions and counterions are, as a result, displaced to the strip solution. The chemical reaction for this coupled transport is as follows:

$$M^{n+} + nX^- + E(\text{membrane}) \xrightarrow{\text{Extraction}} EMXD_n(\text{membrane})$$

$$EMX_n(\text{membrane}) \xrightarrow{\text{Strip}} E(\text{membrane}) + M^{n+} + nX^-$$

The liberated extractant molecules then diffuse back toward the SLM–feed interface metal ions, react with counterions, and continue until final equilibrium is attained.

Counter-Transport

As illustrated in Fig. 5.20, the acidic extractant HX forms a complex with metal cations at the SLM–feed interface. Diffusion of the complex to the SLM strip interface takes place while liberating the metal cations to the strip solution and simultaneously picks up H^+ ions from the strip solution. The resulting HX species diffuses back to the SLM–feed interface that picks up more metal ions and the process continues. The extractant molecules continue to transport between feed and SLM strip interfaces.

$$M^{n+} + nHX(\text{membrane}) \xrightarrow{\text{Extraction}} MX_n(\text{membrane}) + nH^+$$

$$M_n^X(\text{membrane}) + nH^+ \xrightarrow{\text{Strip}} nHX(\text{membrane}) + M^{n+}$$

The driving force for counter-transport SLM is simply the pH difference between the feed and strip solutions. This is implicit from the condition of high feed and low strip D values of the system under study, as high feed D would facilitate the extraction via the membrane and low D in the strip would back-extract the metal ions. The difference of D between feed and strip solution is maintained by pH difference. For divalent metal ions and acidic extractant, the interfacial reaction between feed and extractant is

$$M^{2+}(\text{aq.}) + 2\,HX\,(\text{memb.}) \rightarrow MX_2(\text{memb.}) + 2\,H^+(\text{aq.})$$

where aq stands for aqueous and memb for membrane phase and the equilibrium constant K is given by the following expression:

$$K = \frac{[MX_2(\text{memb.})][H^+(\text{aq.})]^2}{[HX(\text{memb.})]^2[M^{2+}(\text{aq.})]}$$

This equilibrium constant is equal on both feed and strip sides of the membrane [41, 43].

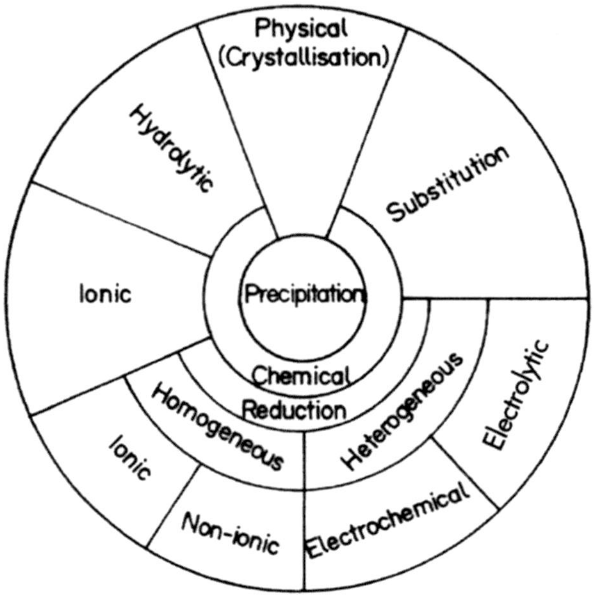

Fig. 5.21 Types of precipitation processes [5]

The SLM technique offers many advantages over traditional solvent processes: (1) elimination of phase separation problems, dispersion-free, (2) negligible organic phase entrainment, (3) single-stage extraction, (4) economic use of expensive extractants due to low inventory, (5) possibility of using high feed to strip ratio to achieve larger concentration of dissolved species, and (6) lower capital and operational costs.

5.3.4 Precipitation

5.3.4.1 Principles

Precipitation is the process in which a dissolved substance is converted into an insoluble form which can then be separated from the bulk of the solution. A chemical or physical change must take place for the transition from a soluble to an insoluble form to take place. The precipitation process is versatile and is universally used as for separation and recovery of metals from solution. The different methods of precipitation are shown in Fig. 5.21. Precipitation processes can be divided into two groups: physical and chemical. Physical methods do not involve the use of reagents, but the concentration and temperature can be adjusted so that crystallization occurs. There are only a few methods considered purely physical methods. In comparison, there are many chemical methods.

Crystallization

Crystallization is strictly a physical process where the concentration and temperature of the solution are changed as to lead to the formation of an ionic solid crystalline phase. The process is perhaps the simplest and oldest method for the recovery of metal from solutions as solid compounds. Typically the temperature will be either raised or lowered until the solubility of the metal salt in solution is exceeded. When the solubility is exceeded, the salt precipitates or crystallizes out of the solution. If several metal salts are concurrently present in the solution, then, as the solubilities of the salts are different, selective crystallization may be utilized. However, complete separation in a single crystallization operation is impossible. For complete separation the precipitates need to be removed from the solution, redissolved, and subsequently recrystallized for increased purity. Separation may also be improved by the selective formation of compounds of different metals. Selective formation of compounds may give rise to large differences in solubilities and result in increased separation of metallic ions during crystallization. The crystallization process can be represented by

$$yM^{x+} + xA^{y-} \rightarrow M_yA_x \cdot n\,H_2O\ (s)$$

Unlike precipitation processes, the crystallization process is usually slow, and the resulting solid is highly soluble in water. The solid most often contains water of crystallization. The basic principles of formation of crystals or precipitates are fundamentally the same: formation of saturated solutions, formation of solid solutions, nucleation and growth, and crystal order. From an industrial production standpoint, crystallization involves a much larger tonnage in comparison to precipitation.

Chemical Methods

Chemical precipitation methods will be discussed in the following sections. These methods involve the addition of reagents to cause precipitation of specific compounds.

Hydrolysis

Chemical precipitation by means of adding water as a reagent is termed hydrolysis. The process most often leads to the precipitation of oxides, hydrated oxides, hydroxides, or hydrated salts. The precipitation of hydroxides can be represented by

$$M^{x+} + nH_2O \leftrightarrow M(OH)_2^{\,-} + nH^+$$

Ionic Precipitation

The previous equation provides clues that precipitation of a hydrolytic product could be enhanced if the H^+ ions produced in the reaction are reacted or neutralized by a base or OH^- ions, shifting the equilibrium to the right side of the equation. This result is shown below:

$$M^{x+} + n(OH) \leftrightarrow M(OH)_x$$

Reactions like those shown above are known as ionic precipitation. They are represented in the following way:

$$yM^{x+} + xA^{y-} \leftrightarrow M_yA_x(s)$$

where A^{Y-} is an anion. Precipitation usually occurs quickly since the compound formed has reduced solubility and the species of the compound are electrostatically bound together.

Without extensive examples, it can be summed up that ionic precipitation is based on the process that if a reagent is added to a solution containing metal ions, a compound is produced whose resultant solubility is very small so that precipitation occurs. In this context, apart from hydroxide, precipitation by hydrogen sulfide is the most commonly used reagent. The ability of hydrogen sulfide to precipitate an extensive list of heavy metals from aqueous solution has long been known by analytical chemists.

Reduction

Precipitation by reduction is the process involving electron transfer from different ions or from a solid, i.e., oxidation–reduction couples. The process can be either homogeneous or heterogeneous. A homogeneous reduction process may be ionic or nonionic. The precipitation of silver from silver nitrate by ferrous ions is a simple example of an ionic reducing agent:

$$Ag^+ + Fe^{2+}(\text{reduced species}) \leftrightarrow Ag + Fe^{3+}(\text{oxidized species})$$

This is an oxidation–reduction process as depicted below:

$$Fe^{2+} \leftrightarrow Fe^{3+} + e^-$$

$$Ag^+ + e^- \leftrightarrow Ag$$

The precipitation process, typically via gaseous hydrogen sulfide, reduction relies on principles of the gaseous reduction of metals from aqueous solutions with hydrogen. With reduction processes, the metals are placed into solution, separated, and precipitated out of solution by a more specific use of the chemical properties of the metal ions in solution.

Reduction of metal species by hydrogen in solution can occur if the hydrogen is at a lower potential than the metal ions at the appropriate metal concentration. The hydrogen potential, and its ability to serve as a reactant, alters considerably with the pH of the solution.

Reduction can also take place where a less valuable metal, X, enters the solution displacing the richer metals as follows:

$$X \leftrightarrow X^{n+} + ne^-$$

$$M^{n+} + ne^- \leftrightarrow M$$

A common example in the above context is from precipitation of copper from copper sulfate solution by metallic iron:

$$Fe \leftrightarrow Fe^{2+} + 2e^-$$

$$Cu^{2+} + 2e^- \leftrightarrow Cu$$

These reactions are considered electrochemical because the transfer of electrons occurs at local areas on the surface of the solid at a distance from where the metal enters the solution.

In electrolytic processes, deposition of a metal from solution is applied by feeding an electric force from a DC source. The process is represented simply by

$$M^{n+} + ne^- \leftrightarrow M$$

Numerous metals such as zinc, copper, nickel, and cadmium are industrially won by the electrolytic process. This exercise has essentially formed a separate branch of reduction chemistry called electrowinning to distinguish it from other extractive metallurgical processes.

Substitution

Substitution is the last of the precipitation processes. Metal ions are extracted by organic solvents by forming a coordination bond. Treatment of this organic phase by hydrogen at high temperatures and pressures, the metal if forced to precipitate into powder form and the organic phase is regenerated. This process is known as precipitation by substitution since no ionic species are involved in the reaction as compared to precipitation by hydrogen ions from an aqueous phase. The reactions can be depicted as

$$H_2(g) \leftrightarrow H_2(org.)$$

$$R_2M\ (org.) + H_2(org.) \rightarrow 2\ RH\ (org.) + M\ (s)$$

where RM = organic solvent and M = divalent metal. This method of precipitation is used for convenience and is also known as reduction by nonionic reducing agents or precipitation of metals from nonaqueous mediums.

5.3.5 *Process Alternatives*

These alternative processes include precipitations based on phosphate, hydroxide, tricarbonate, peroxide, oxalate, and urea.

5.3.5.1 Phosphate Precipitation System

Uraniferous phosphate precipitation from acid leach solutions is commonly practiced in industry. As discussed earlier, selectivity of precipitation products is controlled by pH. There are differences in the pH required for precipitation of ferrous, aluminum, and uraniferous ions. The uraniferous phosphate $U_3(PO_4)_4$ is found to precipitate in the pH range of 1.0–1.8, while precipitation of aluminum occurs at a pH close to 3 and ferrous iron, a pH around 4–5. Precipitation must be carried out under reducing conditions since the ranges for precipitation for ferric, uranyl, and aluminum ions overlap. The phosphate ions supplied for precipitation are most commonly in the form of pyrophosphate, 1.1–2 super stoichiometric. In addition, the pH of the feed solution must be changed prior to the addition of metallic iron to compensate for metal–acid reactions [5, 44].

Uranous phosphate can then be purified by salt fusion using a flux of soda ash, salt, and coal reductant. Fusion produces reduced uranium oxide and allows any impurities to become water soluble.

5.3.5.2 Ammonium Hydroxide System

Purified solutions from acid eluates of ion exchange or solvent extraction strip are generally free of most cationic impurities, except for small amounts of iron, vanadium, and molybdenum after often present. Alkaline strip solutions are similarly pure and can be processed by acidification to expel CO_2 and neutralization or by strong base precipitation. For ammonium carbonate solutions, steam stripping may be completed, to force out ammonia and carbon dioxide, resulting in uranium oxides precipitation.

Precipitation via ammonia in acid and alkali feed solutions is commonly practiced. Solutions can contain nitrates, chlorides, or sulfates with uranium concentrations ranging from 10 to 80 g/l. A two-stage precipitation method is used when the iron level is peculiarly high and control of the sulfate in recycle solutions is desired. Lime (CaO) is used initially to separate iron, titanium, aluminum, and thorium from the basic sulfates and hydroxides. In the presence of phosphate, iron must always be present in sufficient quantities to ensure the precipitation of ferric phosphate; otherwise, uranium can be lost in the form of uranyl phosphate. The iron is then separated and usually recycled by way of an upstream leaching step [45].

From the iron-free solution, uranium precipitation is carried out at pH 7–8. Anhydrous ammonia is vaporized, diluted by air, and injected into the solution. The precipitated product consists of a mixture of diuranates, hydrated oxides, basic uranyl sulfates, and some adsorbed quantities of other ions. Typical reactions are as follows:

$$2\ UO_2SO_4 + 6\ NH_4OH \leftrightarrow (NH_4)_2U_2O_7 + 2\ (NH_4)_2SO_4 + 3\ H_2O$$

$$2\ UO_2SO_4 + 2\ NH_4OH + 4\ H_2O \leftrightarrow (UO_2)_2SO_4(OH)_2 \cdot 4\ H_2O + (NH_4)_2SO_4$$

$$UO_2SO_4 + 2\ NH_4OH + (x-1)H_2O \leftrightarrow UO_3 \cdot x\ H_2O + (NH_4)_2SO_4$$

$$2\ UO_2(SO_4)_3{}^{4-} + 4\ NH_3 + 6\ H_2O \leftrightarrow (UO_2)_2SO_4(HO)_2 \cdot 4\ H_2 + 2(NH_4)_2SO_4$$

Ammonium uranyl sulfate formed by the last reaction requires a third as much ammonia as the first reaction does, but the previous equation is undesirable as it raises the sulfate level of the product. The hydrated oxide has x ~ 2 at temperatures below 60 °C, and above 100 °C, x ~ 4. The compound $2\ UO_3 \cdot 3\ H_2O$ is capable of formation at intermediate temperatures, and hydrated oxides of uranium can exist in several crystalline and allotropic forms, similar to the metallic form of uranium. The precipitated oxides will easily convert to diuranate in the presence of excess ammonia:

$$2\ [UO_3 \cdot xH_2O] + 2\ NH_4OH \leftrightarrow (NH_4)_2U_2O_7 + (2x+1)H_2O$$

Excess sulfate will convert the oxide to basic uranyl sulfate:

$$2\ [UO_3 \cdot x\ H_2O] + (NH_4)_2SO_4 + 4\ H_2O \leftrightarrow (UO_2)_2SO_4(OH_2)_2 \cdot 4\ H_2O + 2\ NH_4OH + 2(x-1)H_2O$$

The basic sulfate will also convert to diuranate:

$$(UO2)_2SO_4(OH)_2 \cdot 4\ H_2O + 4\ NH_4OH \leftrightarrow (NH_4)_2U_2O_7 + (NH_4)_2SO_4 + 7\ H_2O$$

On the other hand, the precipitated ammonium diuranate can either react with the uranyl ion or reform hydrated oxides during solution:

$$(NH_4)_2U_2O_7 + UO_2SO_4 + 3x\ H_2O \leftrightarrow 3\ [UO_3 \cdot x\ H_2O] + (NH_4)_2SO_4$$

$$(NH_4)_2U_2O_7 + (2x+1)H_2O \leftrightarrow 2\ [UO_3 \cdot H_2O] + 2\ NH_4OH$$

The solution chemistry of the ammonia–uranium system is highly complex, and there is a wide series of compounds with various degrees of crystallinity that can form.

The fabrication of coarse crystalline precipitates is favored by slow changes to conditions which encourage precipitation, such as maintaining partial solubility of the precipitate in solution until nucleation and crystal growth initiates. High-temperature dilute feed solutions with relatively low pH values and increasing concentrations of common ions all favor this process. The addition of recirculated crystals, or "seeds," to a solution near the point of precipitation is used often to provide areas for growth to occur in the crystalline product [45].

5.3.5.3 Ammonium Uranyl Tricarbonate (AUT) System

Precipitation of uranium as ammonium uranyl tricarbonate ($(NH_4)_4UO_2(CO_3)_3$) has the advantage of producing a high-purity crystalline powder which is able to rapidly filter and upon drying produces a powder product that can be easily handled with minimal dust hazards. This precipitation process, when used for nuclear fuel fabrication, yields a product that is readily sintered.

Crude uranyl salts (hydroxides, peroxides, urinates, etc.) can be processed by using a two-stage AUT technique. The first stage starts when a feed solution is prepared in a dilute ammonium carbonate solution. Since uranium forms a stable uranyl tricarbonate complex, many impurities form insoluble compounds that can be removed easily by filtration. In the second stage, the filtrate is subjected to uranium precipitation using ammonium carbonate [46].

The soluble complex $[UO_2(CO_3)_3]^{4-}$ reacts with NH_4 to form AUT as follows:

$$\underset{\text{(soln.)}}{4\,NH^+} + \underset{\text{(soln.)}}{\left[UO_2(CO_3)_3\right]^{4-}} \leftrightarrow \underset{\text{(solid)}}{(NH_4)_4UO_2(CO_3)_3}$$

Temperature is very important in the process, and when the solution is heated, AUT yields UO_2CO_3. Dry AUT is unstable in air and will slowly release ammonia at room temperature. To negate this instability, AUT is commonly calcined at temperatures above 350 °C where following reaction occurs:

$$(NH_4)_4\left[UO_2(CO_3)_3\right] \rightarrow UO_3 + 4\,NH_3 + 3\,CO_2 + H_2O$$

When calcination is carried out under inert or mildly reducing atmosphere at temperatures exceeding 550 °C, the reaction product is uranium dioxide.

An interesting AUT precipitation process combines precipitation with the typical stripping phase of solvent extraction. This is often the case when organophosphorus reagents are used for extractive purposes. For example, a loaded organic phase containing tri-n-butyl phosphate is placed in a diluent, stripping solution with a concentrated amount of ammonium carbonate, the following reaction occurs:

$$\underset{(\text{org.})}{UO_2(NO_3)_2 \cdot 2\ TBP} + \underset{(\text{aq.})}{6\ NH_4^+} + 3\ CO_3^{2-}$$

$$\leftrightarrow \underset{(\text{solid})}{(NH_4)_4UO_2(CO_3)_3} + \underset{(\text{aq.})}{2\ NH_4NO_3} + \underset{(\text{org.})}{TBP}$$

This process results in a three-phase system where parameters affecting the particle size distribution as well as the formation rate of AUT crystal seeds include the organic/aqueous phase ratio, time in the stripping mixer, and the intensity of agitation. Strong agitation can lead to the generation of fine grains due to mechanical abrasion.

5.3.5.4 Peroxide System

Uranium peroxide is prepared as a crystalline product with a high degree of selectivity from acidic solutions. The reaction is as follows:

$$UO_{2+}{}^{2+}H_2O_2 + 2\ H_2O \leftrightarrow UO_4 \cdot 2\ H_2O + 2\ H^+$$

The uranium peroxide product is easy to handle in production, precipitation yield tends to be much higher than AUT, and the substitution of ammonia by a nonpolluting chemical is a tremendous advantage.

Cationic impurities like iron can lead to an accelerated decomposition of hydrogen peroxide in some cases. These impurities are removed ahead of the peroxide precipitation by precipitation of ferric ions. Since the bivalent iron ion does not separate, oxidization of all the iron to the trivalent form is essential.

The precipitation reaction with peroxide is followed by the release of hydrogen ions so the pH tends to decrease. Therefore, careful control by the addition of dilute ammonia solution or acid is necessary in plant operations.

The nature of the precipitate formed has been the subject of much research, and a number of species have been suggested. The dihydrate represented in the overall reaction results only after precipitation at about 100 °C. The precipitate obtained at lower temperatures has higher levels of hydration, and the resultant peroxide is a uranyl peruranate ($UO_8(UO_2)_2 \cdot x\ H_2O$) formed as a result of the reactions

$$UO_{2+}{}^{2+} + 3\ H_2O_2 \leftrightarrow UO_8H_4 + 2\ H^+$$

$$UO_8H_4 + 2\ UO_{2+}{}^{2+} + x\ H_2O \leftrightarrow UO_8(UO_2)_2 \cdot x\ H_2O + 4\ H^+$$

The efficiency of peroxide precipitation reactions is heavily influenced by the neutral salts found in solution. Sulfate and other anions capable of complexing with uranium tend to suppress the precipitation. The presence of other ions such as magnesium, sodium, or ammonium sulfate in solution can increase the rate of crystal growth of the peroxide precipitate. Hafnium, thorium, vanadium, and zirconium can form peroxides and coprecipitate with uranium. These impurities in precipitate are removed by washing.

In comparison with other precipitation systems, the peroxide system has higher direct costs because of the costly hydrogen peroxide used, but process improvements and environmental considerations make it a very attractive route for yellow cake recovery.

5.3.5.5 Oxalate System

Precipitation by formation of insoluble oxalate complex $(C_2O_4)^{2-}$ is often used in the cases of nitric acid feed solutions. Impurities are retained in solution

$$UO_2(NO_3)_2 + H_2C_2O_4 \leftrightarrow UO_2(C_2O_4) + 2\ HNO_3$$

when compared with HCl dissolution and reduction of uranium prior to $(C_2O_4)_2$ precipitation, this route is simpler and requires much less oxalic acid. The hydrated oxalate formed is $U(C_2O_4)_2 \cdot x\ H_2O$ where $x = 3–4$. Calcination will then give a dioxide:

$$UO_2(C_2O_4) + 4\ H_2O \leftrightarrow UO_2 + 2\ CO_2 + 4\ H_2O$$

5.3.5.6 Urea Precipitation

Urea or carbamide $(CO(NH_2)_2)$ precipitates uranium homogeneously in solution. Urea is cheaper than caustic soda and is much easier to handle in industrial settings. Precipitation at 100 °C will yield sufficient recovery of particles that will form, settle, and filter more quickly than caustic precipitated materials. Due to complex formation during very slow homogeneous precipitation, the sulfate concentration in the product tends to be higher. A drawback of urea precipitation is that sulfate complexes that form are not decomposed by hot washing.

References

1. IAEA (2003) Uranium 2003: Resources, Production and Demand. International Atomic Energy Agency, Vienna
2. IAEA (1993) Uranium extraction technology. Tech Rep Ser 359. International Atomic Energy Agency, Vienna
3. Otway HJ, Vander-Horst L, Higgins GH (1973) Socioeconomic aspects of a plowshare project. J Nucl Technol 17:58–68
4. Gupta CK, Mukerjee TK (1990) Hydrometallurgy in extraction processes, vol I & II. CRC Press, Boca Raton, FL
5. Gupta CK, Singh H (2003) Uranium resource processing: secondary resources. Springer, New York

6. Katz JJ, Rabinowitch E (1951) The chemistry of uranium, national nuclear energy series. McGraw Hill, New York
7. Akin H (1995) Exploring the future of in situ leach uranium mining. Uranium Institute, London
8. Barlet RW (1992) Solution mining. Gordon and Breach Science Publishers, Langerhorns, PA
9. Clegg JW, Foley DD (1958) Uranium ore processing. Addison Wesley Publishing, Boston, MA
10. Craig WM (1982) Shortcut to yellow cake. Nucl Active 27:3–7
11. Campbell MC (1985) Canadian developments of metallurgical projects in Uranium. In: Proceedings of a technical committee meeting, ST1/PUB/738, ISBN: 92-0-141187-1, Vienna, pp 231–232
12. James HE (1980) Hitting up the centuries. Nucl Active 22:17–21
13. Jensen J (1984) Ice cold uranium. Nucl Active 31:26–32
14. Milde WW (1989) Rabbit lake project: Milling and Metallurgy. CIM Bull 69–75
15. Mann S (1996) Biomimetic materials chemistry. VCH Publishers, New York
16. Barret J, Hughes MN, Karavaiko GS, Spencer PA (1993) Metal extraction by bacterial oxidation of metals. Horwood, London
17. Dwivedy KK, Mathur AK (1995) Bioleaching- our experience. Hydrometallurgy 38:99–109
18. Natrajan KA (1998) Microbiological applications and techniques in biogeochemistry. In Proceedings of the International Workshop Environ Biogeochem, JNU New Delhi, 59–79
19. McCready RGL, Wadden D, Marchbank a (1986) Nutrient requirements for the in place leaching of uranium by Thiobacillus ferrooxidans. Hyrdometallurgy 17.61–71
20. Munoz JA, Gonzalez F, Blazquez ML, Ballester A (1995) A study of the bioleaching of a Spanish uranium ore, Part II: orbital shaker experiments. Hyrdometallurgy 38:59–78
21. Munoz JA, Gonzalez F, Blazquez ML, Ballester A (1995) A study of the bioleaching of a Spanish uranium ore, Part I: a review of bacterial leaching in the treatment of uranium ores. Hyrdometallurgy 38:39–57
22. Livesey GE (1980) An Oscar for bacteria. Nucl Active 20:8–11
23. Rossi G (1990) Biohydrometallurgy. McGraw Hill, New York
24. Box JC, Prosser AP (1986) A general model for the reaction of several minutes and several reagents in heap and dump leaching. Hyrdometallurgy 16:77–92
25. Sole KC, Cole PM, Feather AM, Kotze MH (2011) Solvent extraction and Ion exchange applications in Africa's resurging uranium industry: a review. Prog Hydrometal Appl 29:5–6
26. Bock J, Valint PL (1982) Uranium extraction from wet process phosphoric acid: a liquid membrane approach. Ind Eng Chem Fundam 21:417–422
27. Shakir K, Aziz M, Beheir G (1992) Studies on uranium recovery from a uranium bearing phosphatic sandstone by a combined heap leaching-liquid gel extraction process: part 1: heap leaching; part 2: gel extraction. Hydrometallurgy 31:29–54
28. Small H (1961) Gel liquid extraction: the extraction and separation of some metal salts using TBP. J Inorg Nucl Chem 18:232–244
29. Sekizuka Y (1973) Analytical applications of organic reagents in hydrophobic gel media. Talanta 18:979–985
30. Shakir K, Beheir SG (1980) Gel liquid extraction and separation of U-VI, Th-IV, Ce-III and Co-II. Sep Sci Technol 7(15):1445–1458
31. Babcock WC, Baker RW, Kelly DJ, Lachpelle ED (1980) Coupled transport membranes: the mechanism of uranium transport with a tertiary amine. J Mem Sci 7:71–87
32. Bautista RG (1993) Liquid membrane separation of metals in aqueous solutions. In Emerging separation technologies for metals and fuels. Minerals Metals Materials Society, Warrendale, Pennsylvania, 1993
33. Hofman DL, Craig WM, Smith JJ (1990) Modeling and application of a pilot plant scale supported liquid membrane unit for uranyl nitrate extraction. In: Sekine T (ed) Solvent extraction. Elsevier Science, Netherlands

34. Huang TC, Huang CT (1986) Mechanism of transport of uranyl nitrate across a solid supported liquid membrane using tri butyl phosphate as mobile carrier. J Mem Sci 29:295–308
35. Bock J, Klein RR, Valint PL, Ho WS (1981) Liquid membrane extraction of Uranium from wet process phosphoric acid—field process demonstration. Paper No. 30b, AIChE Annual Meeting, New Orleans, LA, 8–12 Nov 1981
36. Chaudry MA, Islam N, Mohammed P (1989) U-VI transport through a TBP kerosene and liquid membrane supported in polypropylene film. J Radioanal Nucl Chem Articles 1 (109):11–22
37. Hirato T, Koyama K, Awakura Y, Majima H (1990) Recovery of U-VI from wet process sulphuric acid by an emulsion type liquid membrane technique. In: Sekine T (ed) Solvent extraction. Elsevier Science, Netherlands
38. Tourneux JC, Berthet JC, Cantat T, Thuery P, Mezailles N, Ephritikhine M (2011) Exploring the uranyl organometallic chemistry: from single to double uranium–carbon bonds. Am Chem Soc 16(133):6162–6165
39. Huang TC, Huang CT (1988) Kinetics of the extraction of uranium-VI from nitric acid solutions by bis(2-ethyl hexyl)phosphoric acid. Ind Eng Chem Res 27:1675–1680
40. Chiarizia R, Horwitz EP (1990) Study of uranium removal from ground water by supported liquid membrane. Solvent Extract Ion Exch 1(8):65–98
41. Gill JS, Marwah UR, Misra BM (1994) Transport of Sm-III and U-VI across a silicone supported liquid membrane using D2EHPA and TBP. Sep Sci Technol 2(29):193–203
42. Ramkumar J, Shrimal KS, Maiti B, Krishnamoorthy TS (1996) Selective permeation of Cu^{2+} and lie through a Nafion ionomer membrane. J Mem Sci 116:31–37
43. Huang CT, Huang TC (1988) Kinetics of the coupled transport of Uranium-VI across supported liquid membrane containing bis(2-ethylhexyl)phosphoric acid as a mobile carrier. Ind Eng Chem Res 27:1681–1685
44. Li J, Wang TH, Zhang YX (1997) Effects of impurities on the habit of gypsum in wet process phosphoric acid. Ind Eng Chem Res 36:2657–2661
45. Chegrouche S, Kebir A (1992) Study of ammonium uranyl carbonate reextraction crystallisation process by ammonium carbonate. Hyrdometallurgy 28:135–147
46. Lounis A, Gavach C (1997) Treatment of uranium solutions by electrodial. Hydrometallurgy 44:83–96

Chapter 6
The Uranium–Hydrogen Binary System

G. Louis Powell

Abstract Gaseous hydrogen (H_2) dissolves in uranium metal (U) and subsequently precipitates as uranium hydride (UH_3). Near ambient temperature, this process results in destructive, pitting corrosion. At elevated temperatures dissolved hydrogen permeates the uranium metal and precipitates as UH_3 upon cooling to lower temperatures. Near ambient temperatures, trace amounts of UH_3 reduce the tensile ductility of U and trace amounts of oxidizing species in H_2 and on the U surface impede the reaction remarkably. In this chapter the phase relationships and kinetic processes describing the pure binary system are reviewed. Ternary processes involving oxygen are addressed in Chap. 7, and U alloys with other metals are not addressed.

Keywords Uranium hydride • Phase diagram • Hydrogen dynamics • Kinetics • Dehydriding • Mass spectrometric • Mass spectrometry • Hydrogen charging • Quenching • Pitting • Solubility • Impurities • Diffusion

6.1 Introduction

The phase relationships between H_2, solid (or liquid) U, and solid UH_3 are well established. Sieverts [1] reported measurements on this system as early as 1912 and patents for making uranium hydride from uranium metal by reaction with H_2 were obtained in 1931 [2]. During the Manhattan Project, the high-temperature (>300 °C) phase relationships were established for H_2 and deuterium (D_2) and reported [3–7] along with the preparation of UH_3 at lower temperatures.

These measurements were refined over the next two decades, and the stable (α-UH_3) and the metastable (β-UH_3) forms were identified. Both the U–H

G.L. Powell (✉)
Y-12 National Security Complex, 1 Bear Creek Road, MS #8097, Oak Ridge, TN 37831, USA
e-mail: powellgl@y12.doe.gov

J.S. Morrell and M.J. Jackson (eds.), *Uranium Processing and Properties*,
DOI 10.1007/978-1-4614-7591-0_6,

chemistry [8] and U–H metallurgy [9] have been extensively reviewed at the end of that period. At temperatures below 300 °C, the reaction between H_2 and U, typically a form of pitting corrosion, was incredibly unpredictable until Condon demonstrated that, in the total absence of oxygen and related species in both the gaseous and solid phases, this reaction was consistent with the high-temperature phase diagram [10, 11], and that the reaction rate was a strong function of the residual stress in the U [11]. Therefore this chapter addresses those properties that can be attributed to the uranium–hydrogen binary system in the total absence of any tertiary elements resulting from corrosion or alloying.

6.2 The H–U Phase Diagram

H_2 in the gas phase dissolves in U at concentration $[H]_U$ to achieve equilibrium across the U surface in accordance with Sieverts' Law (6.1) or achieve equilibrium with the UH_{3-X} solid phase where X is the vacancy concentration in the hydride phase (6.2). When both the metal and hydride phases are present, the degrees of freedom allowed by the phase rule are decreased to one leading to the Van Hoff's relationship (6.3) where the P and T cannot change independently:

$$k_s(T) = [\mathrm{H}]_U P^{-0.5} \tag{6.1}$$

$$k_v(T) = [\mathrm{X}]P^{0.5} \tag{6.2}$$

$$K_p(T) = P \tag{6.3}$$

Upon reaction with hydrogen, U is the only metal that forms a trihydride without forming any other intermediate hydride phases or stoichiometry. The solubility of H in U is given in Fig. 6.1, (6.4), and Table 6.1 [12]:

$$k_s = \frac{A\left(1 + B\mathrm{e}^{-C/T}\right)/\left(1 - \mathrm{e}^{-1680/T}\right)^3 \mathrm{e}^{(E-25997.5)/T}}{0.02072T^{-1.75}/\left(1 - \mathrm{e}^{-5986/T}\right)^{0.5}} \tag{6.4}$$

The blue data points are solubility measurements made at Y-12 National Security Complex [12, 13] and are the most accurate measured to date. These data were fit to the bound proton model (6.4) [12–15] using the parameters in Table 6.1 to yield the solid curves. This model describes hydrogen isotope effects very well. The hydrogen Einstein temperature of 1,680 K results from a small isotope effect for D_2 and that may be calculated by modifying the parameters in (6.4) [12–15]. This model deals with the nonlinearity of the thermodynamic functions in $1/T$ and has the property that the low-temperature extrapolation is controlled by parameters A, an integer, and E, the ground-state energy relative to H atoms at rest. The B and C terms represent a very high level of thermal disorder in

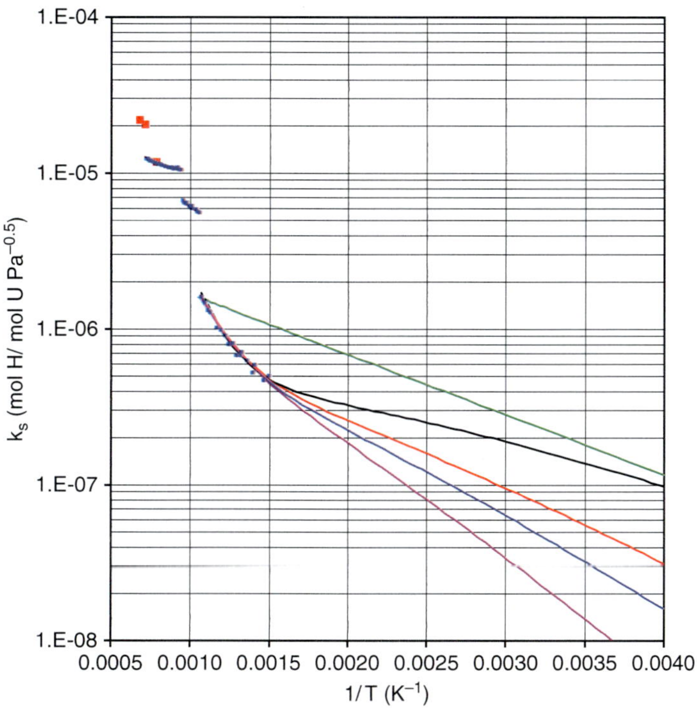

Fig. 6.1 The solubility of hydrogen in uranium (i.e., the Sieverts' Law Constant)

Table 6.1 Coefficients for (6.4) to produce the solid lines

Phase	A	B	C	E
Gamma	3	25	4300	26498
Beta	6	37	4300	25158
Alpha	1	1550	6000	24850
Alpha	2	1500	6200	24390
Alpha	3	300	5000	24120
Alpha	6	200	5000	23680

the alpha phase above 400 °C that causes the alpha-phase solubility to be so large and endothermic. Any of these parameter sets accurately describe the data, and their extrapolations towards ambient temperature form a reasonable basis for the uncertainty at low temperatures. Above 400 °C, the solubility has been described reasonably well since the Manhattan Project as indicated by the red data points representing the solubility in the liquid and gamma phases [7]. The green line is a linear extrapolation by Mallette and Trzeciak [16] and demonstrates the problems

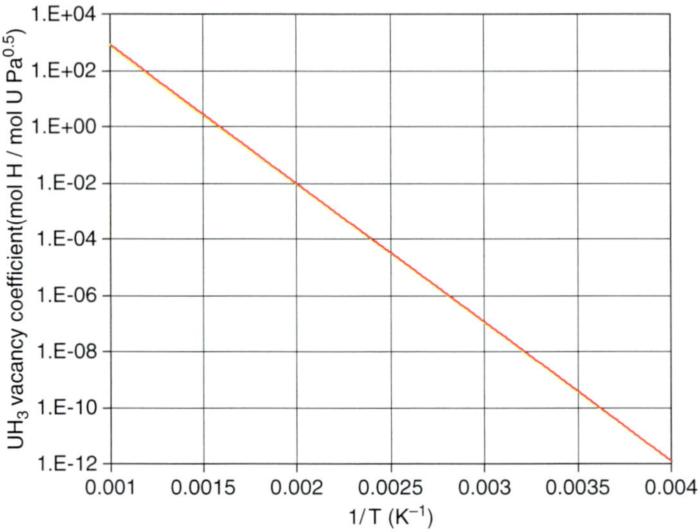

Fig. 6.2 The UH_3 vacancy coefficient as a function of temperature

with extrapolating data that covers a narrow temperature range with some scatter. Davis [17] made very good hydrogen solubility measurements, particularly in the alpha phase. Since this hydrogen solubility has the role of converting measured H_2 pressure into the H concentration used as the boundary condition for applying Fick's Law to H permeation through U, its value cannot be understated.

Figure 6.2 and (6.5) give the equilibrium relationship between the UH_{3-X} and H_2 defining the high composition of the phase diagram as determined by Libowitz and Gibbs [18] that describes the pressure rise required to drive the stoichiometry towards 3.000:

$$k_v = 6.9 \times 10^{30} e^{-11360/T} \tag{6.5}$$

Figure 6.3 and (6.6) give the equilibrium (plateau) H_2 pressure for the H_2–U–UH_3 system when both the metal and the hydride phases are present. These results represent data from Libowitz and Gibbs [18] at higher temperatures and from Condon [19] for temperatures down to 150 °C. Below 150 °C, the equilibrium H_2 pressure becomes too low to accurately

$$k_p = 1.27 \times 10^{30} e^{-14640/T} T^{5.65} \tag{6.6}$$

measure and UH_3 decomposition becomes so slow that it is no longer practical to convert UH_3 to U under vacuum. Equations (6.4) and (6.6) may now be used to solve (6.1) and (6.3) for the low hydrogen composition solvus line and (6.5) and (6.6) to solve (6.2) and (6.3) for the high hydrogen composition solvus line as shown in Fig. 6.4.

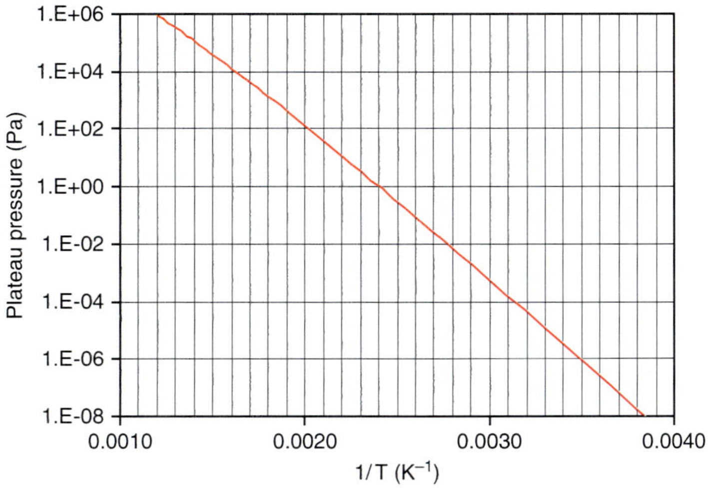

Fig. 6.3 The H_2–U–UH_3 equilibrium "plateau" pressure as a function of temperature

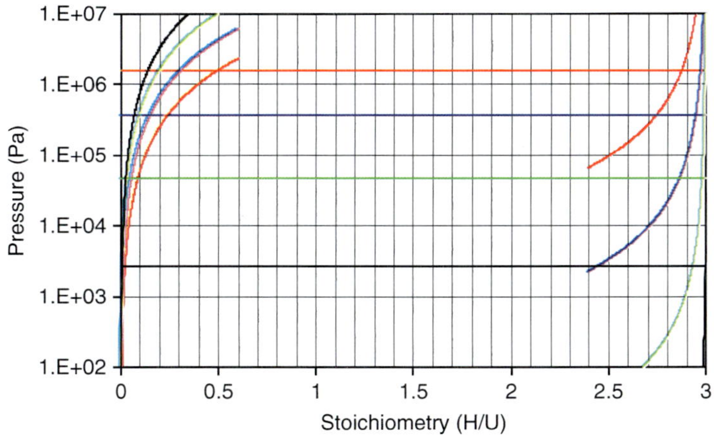

Fig. 6.4 The H_2–U–UH_3 phase diagram (300 °C, *black line*; 400 °C, *green line*; 500 °C, *blue line*; 600 °C, *red line*)

Equations (6.1)–(6.6) were used to construct Fig. 6.4. The single-phase regions of the uranium–hydrogen system are narrow yielding a two-phase equilibrium "plateau" pressure that is broad and flat, with the intersection of the plateau pressure with the solubility curve or the hydride vacancy curve defining the solvus lines for the formation of UH_3 and the disappearance of the U metal.

6.3 Hydrogen Dynamics

Hydrogen dynamics refers to kinetic processes important in the effective achievement of the equilibrium conditions described in the previous section. Figure 6.5 and (6.7) give the diffusion coefficient for H in U determined by thermal desorption experiments derived from mass spectrometric hydrogen analyses [13, 20–22] and generally agree with earlier measurements [16, 17]

$$D = 0.0000019\mathrm{e}^{-5820/T}\left(\mathrm{m}^2\mathrm{s}^{-1}\right) \tag{6.7}$$

at temperatures below 300 °C measurement of D is difficult not only because of small values of D and hydride formation, but also defects such as vacancies, grain boundaries, and impurities trap the migrating H perturbing the diffusion process.

The nonequilibrium dehydriding rate for UH_3, the rate at which UH_3 is converted to U metal under vacuum, is given in Fig. 6.6 and (6.8) where U is the fraction of the material that is the metal phase [10]. This process of converting UH_3 to U becomes too slow to be practical below approximately 150 °C:

$$\mathrm{Rate} = 137.5\left[\mathrm{e}^{-4900/T}\right](1-U)^{1/3} \tag{6.8}$$

Condon [10] initiated a series of thermo-gravimetric experiments under superpure conditions with the goal of correlating the hydriding rate of U powder made by the dehydriding process with the surface area of that powder determined by Ar sorption. The results were that there was no correlation. He had effectively isolated the precipitation process for UH_3 formation at room temperature from diffusion and impurity effects. The U powders were of sufficiently small size that diffusion

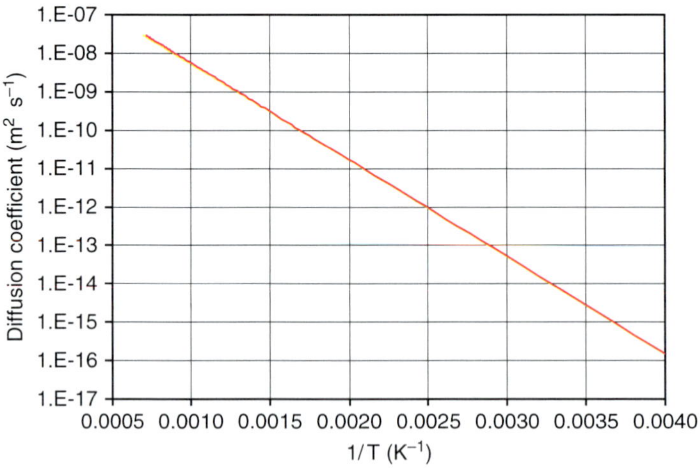

Fig. 6.5 The diffusion coefficient for H in U as a function of temperature

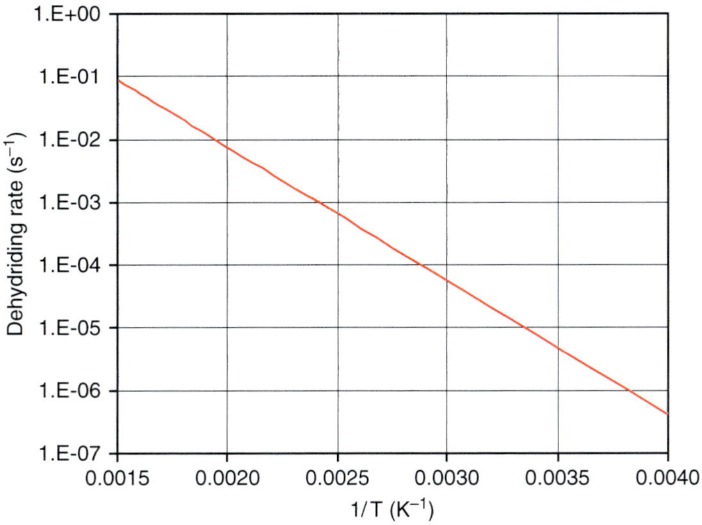

Fig. 6.6 The kinetics for the dehydriding of UH_3 as expressed in (6.8)

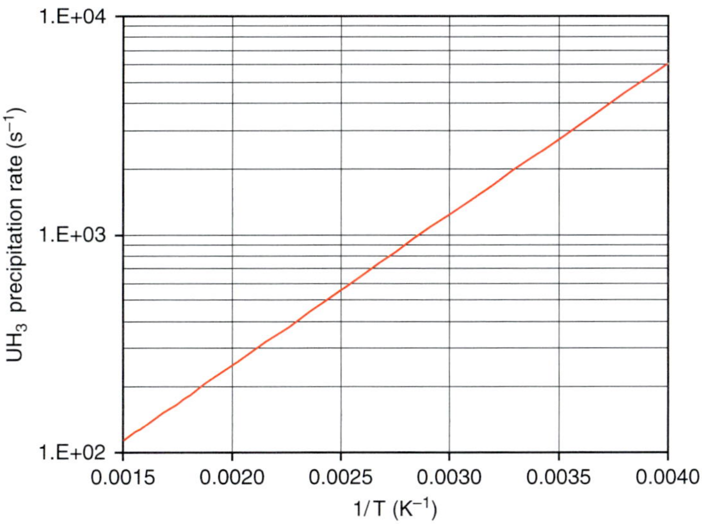

Fig. 6.7 The kinetics for the precipitation of UH_3 in U

through the particles was not a rate-limiting step and surface and gas-phase impurities were not at play. The resulting hydride precipitation rate is shown in Fig. 6.7 and (6.9). As is expected for a condensation process, the precipitation rate increases with decreasing temperature:

$$\text{Rate} = K_H C_H U = K_H k_s P^{0.5} U \quad 0 < U < 1 \text{ with } K_H = 10.4e^{1592/T}\left(s^{-1}\right) \quad (6.9)$$

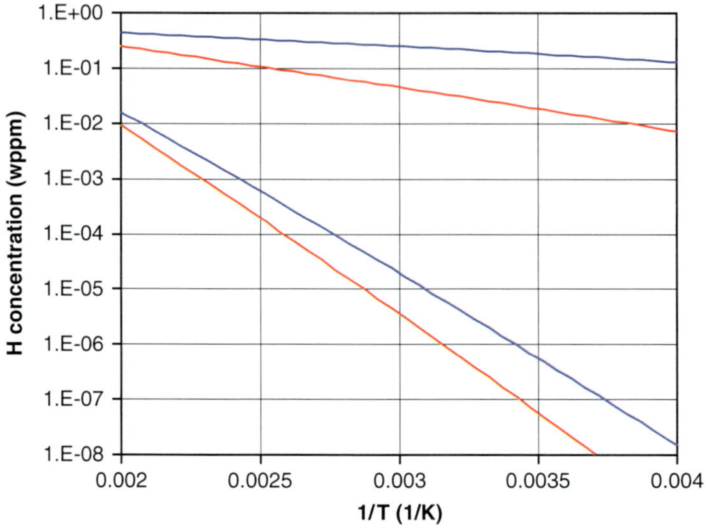

Fig. 6.8 Concentrations of H in U in wppm in equilibrium with the gas phase (*upper curves*) and hydride phase (*lower curves*) at 21 °C with the *red–blue* band representing the extrapolation errors from (6.1)

During these measurements, exposure of the U powder to small amounts of O_2 resulted in the introduction of a sigmoidal character to the hydriding rate that was explained by the oxidizing chemical potential competing with the H_2 chemical potential that resulted in reaction initiation delays. Figure 6.8 shows calculated concentrations of H in U (wppm) exposed to 1 atm. of H_2 showing the concentration of H in equilibrium with the gas phase using (6.4) for the upper curves, and the concentration in equilibrium with the hydride phase using (6.4) solved for the plateau pressure (6.6) in the lower curves with the red–blue curve pairs showing the uncertainty in the calculated solubility values. Equation (6.9) describes the rate at which H is being removed from solution to form the hydride in compliance with the lower curves. Condon [10] postulated that UH_3 precipitation nucleates as the alpha phase upon which the beta phase grows, favoring alpha-phase production at high rates [23] and low temperatures [24].

6.4 Hydrogen Analysis

A method based on mass spectrometric monitoring of hydrogen evolving from a machined pellet of uranium or uranium alloy sample dropped into a quartz furnace at 900 °C was developed to measure the H content of the bulk of the sample at levels well below 1 μg/g. The design followed from that of Davis [17] who controlled the blank results to 0.2 ± 0.1 wppm (weight parts per million) and reasonable correlation between hydrogen content and tensile ductility.

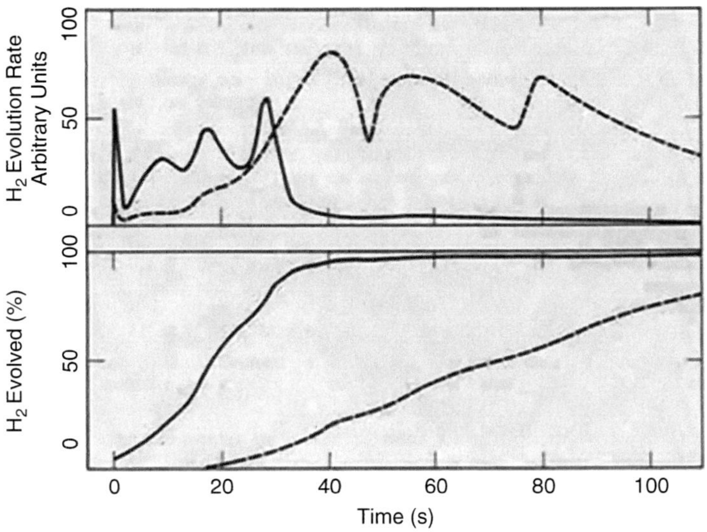

Fig. 6.9 Hydrogen evolution data from two U pellets dropped in a 900 °C furnace. *Broken curve* yielded 16.5 wppm, *solid curve* yielded 0.59 wppm

The mass spectrometric system used ultrahigh-vacuum components, very stable vacuum pumps, and liberal use of liquid nitrogen traps to suppress hydrogen-bearing species in the vacuum system that might form H_2. The system was essentially a temperature-programmed desorption experiment with H_2 evolution monitored [20]. Standards for the analyzers were prepared by annealing U blanks in H_2 at pressures ranging from high vacuum to 1 atm., followed by quenching the specimen in water. The anneal was carried out in a quartz tube followed by breaking the hot end of the tube under water. Specimens could be prepared having hydrogen contents ranging from 0.00 to 15 wppm. Pellets for analysis were machined from this blank to a 6.53-mm diameter by 2.54-mm thickness to achieve a reproducible surface finish, sample thickness, and low surface-to-volume ratio weighing 1.5 g. The samples were ultrasonically cleaned in ethanol before analysis. Eight preloaded samples were individually dropped into the furnace using a rotary mechanical feed-through. Typical results are shown in Fig. 6.9. The instrument operated much like a helium leak detector except that it was calibrated using the expansion of a known volume of H_2 at a measured pressure, the quantity of calibrating gas being typically 1 μg.

The mass spectrometer output was integrated in real time with an analog computer for quantitative analysis. The hydrogen evolved during the first 30 s was interpreted as H_2 produced from surface contamination unrelated to bulk hydrogen content (i.e., dissolved H that precipitated as UH_3 upon quench). The bulk hydrogen reported for the low-hydrogen sample was that evolved after 40 s (0.04 wppm). The blank value for the instrument was therefore determined uniquely for each sample. This particular method resulted in the low-hydrogen standards showing a bias of 0.1 ± 0.1 wppm which was taken as the resolution of the method [22]. This interpretation was confirmed by using deuterium as the annealing gas so that after machining the surface hydrogen was found to be protium. Once the hydrogen solubility had been very accurately measured (6.1) [12, 13], the hydrogen content of the standards could be predicted much more precisely from the annealing pressure than from this analysis. The high-hydrogen specimen hydrogen evolution rate curve shows some surface contamination H_2 for the first 20 s, but it is obscured by the rising evolution of hydrogen due to the increasing dissolution of UH_3 as the specimen temperature rises. The first maximum in this curve arises from the alpha–beta-phase transition during which the incoming beta phase steals hydrogen from the established diffusion path to the surface while the sample isotherms at the alpha–beta transition temperature. Once the alpha phase no longer exists, the evolution increases again until the gamma phase begins to form again stealing hydrogen from the diffusion path to the surface and regulating the specimen temperature. This demonstrates the strong zone refining of H effects of the large increases in solubility with each transition to a higher temperature phase. Once the specimen is fully converted to the gamma phase, the normal diffusion gradients are reestablished and the hydrogen evolution rate decays exponentially with a time constant of $d^2/4\pi^2D$. By analyzing a number of high-hydrogen specimens at temperatures ranging from 1,000 to 400 °C the diffusion coefficient data to arrive at (6.7) was obtained [12, 13].

6.5 Internal Hydrogen Embrittlement

The major consequence of trace amounts of hydrogen in uranium is the loss of ductility. To measure this loss of ductility, blank specimens 12.7 mm in diameter by 100 mm long were annealed in vacuum at 630 °C, after which a measured H_2 pressure blanketed the specimen for sufficient time to achieve the equilibrium solubility (6.1) followed by a vacuum water quench. Small-angle neutron scattering experiments by Spooner et al. [25] demonstrated that this treatment resulted in a dispersion of submicron UH_3 particles. In all 75 specimens were prepared from these blanks ranging in hydrogen content from 0.00 to 1.90 wppm and were tested according to the ASTM E6 specification except for groups of ~11 specimens, each tested at seven levels of cross-head speed from 0.00084 to 6.25 mm/s in an argon atmosphere. Figure 6.10 shows the tensile elongation and reduction in area for the sample set, tested in air at 0.064 mm/s cross-head speed after the yield strength was measured and plotted as a function of the H_2 pressure during annealing. The specimens had yield strengths of 250 ± 34 MPa and ultimate tensile strengths of 817 ± 25 MPa at low-hydrogen contents dropping to 700 MPa at high-hydrogen contents. The ductility is high in the absence of hydrogen, but drops with hydrogen additions to a minimum ductility level (~15 %) at hydrogen contents of 0.35 wppm. Figure 6.11 shows the mass spectrometric hydrogen analysis results for the same material (the hydrogen analysis specimens were wafered from the tensile specimen

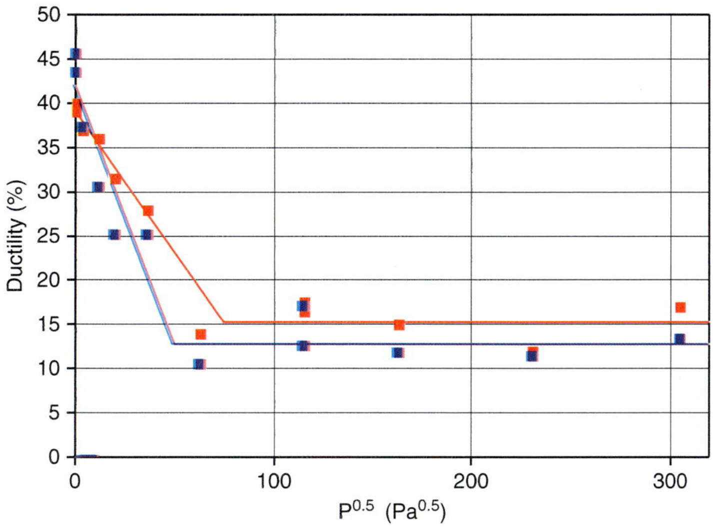

Fig. 6.10 Tensile ductility of hydrogen-charged U specimens as a function of their H charging pressure. Elongation in *red* (*top line*) and reduction in area in *blue* (*bottom line*)

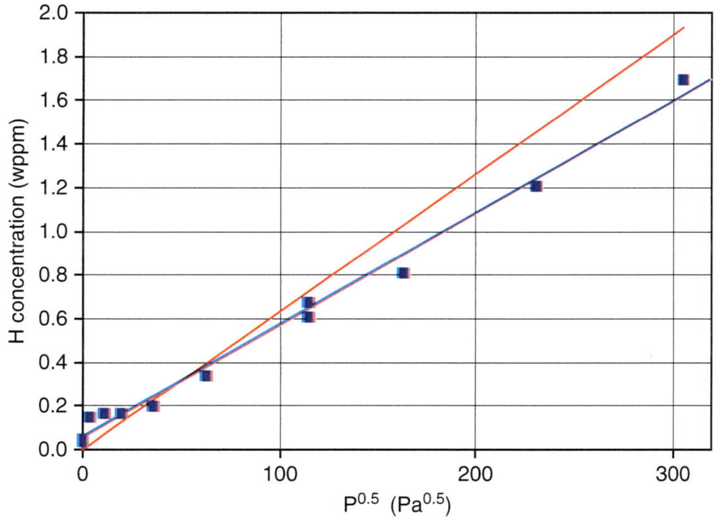

Fig. 6.11 Mass spectrometric hydrogen analysis of hydrogen-charged U specimens as a function of their H charging pressure. Target loading (in *red—top line*) and measured values (in *blue—bottom line*)

shanks) also as a function of the H_2 pressure during annealing. In Fig. 6.11, the linear fit to the data was about 20 % below the target value, and had an intercept of 0.056 wppm and a fit standard error of 0.055 wppm. The ductility increased with decreasing cross-head speed as did the hydrogen concentration for achieving minimum ductility. The ductility sensitivity to hydrogen also increases with increasing yield strength that is achieved in this type of experiment by hydrogen charging and quenching from the gamma phase [26–28].

6.6 Hydriding Kinetics

Hydrogen corrosion of U is a peculiar form of corrosion in that it initiates in what appears to be a chaotic and unpredictable form of pitting corrosion with physical characteristics similar to iron rusting. This unpredictable nature results from the relatively feeble reaction that is the dissociation and dissolution of H_2 at the U surface. The presence of an oxide film or oxidizing impurities in the gas phase or in the oxide film easily shuts down that H_2 dissolution at the U surface [29, 30]. A 250 °C thermal activation of well-annealed bulk U specimens in ultrahigh-vacuum conditions followed by exposure to ultrapure H_2 prepared by filtering the gas through a UH_3 powder bed consistently yields results as shown in Fig. 6.12 [30]. There is consistently a gradual increase in rate to a rate plateau where pitting corrosion is obvious. The rate observed for this plateau is strongly affected by surface preparation; for example, polishing greatly slows this part of the reaction.

Once the pitting process covers the entire surface the rate transitions to a higher value and maintains that value until the specimen is almost consumed whereupon the rate accelerates and then stops. The rate during which time most of the specimen is converted to UH_3 has been successfully modeled by Condon and Kirkpatrick [10, 11, 23, 31, 32] as the establishment of the equilibrium solubility for H at the U surface (6.1), the diffusion of H into the metal according to Fick's Law (6.7),

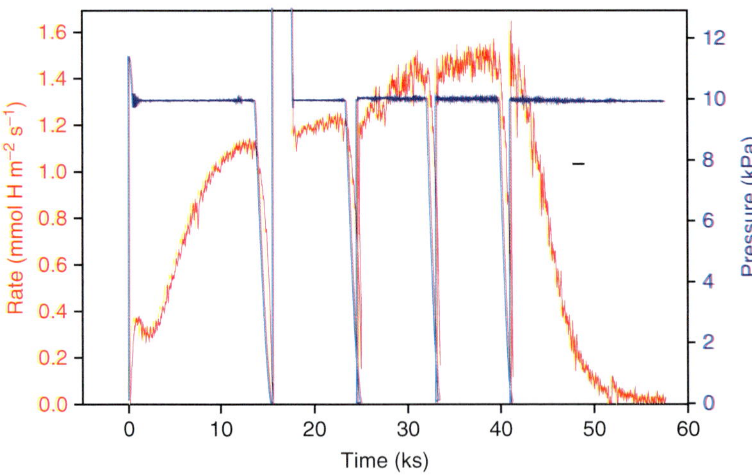

Fig. 6.12 Hydriding rate of well-annealed and thermally activated U foil (0.43 mm thickness) in ultrapure H_2. *Dashed line* is the Condon–Kirkpatrick model prediction for well-annealed U

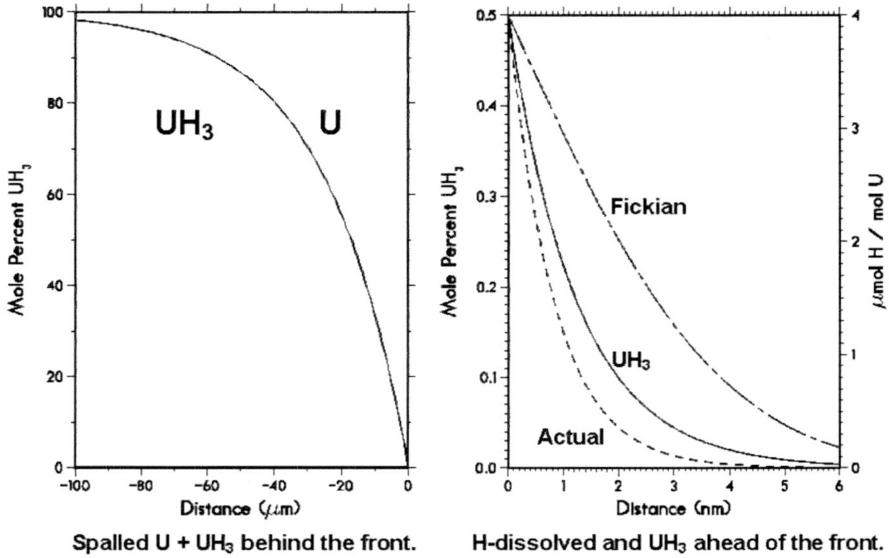

Fig. 6.13 The distribution of H and UH_3 on either side of the reaction front predicted by the Condon–Kirkpatrick model

and the precipitation of UH_3 within the bulk U (6.9). The volume increase associated with UH_3 formation fractures the metal leading to a steady-state process that reduces to a linear reaction rate consistent with experimental observations. The heart of the model is condensed into a simple expression given in (6.10) where the linear rate is simply the geometrical mean of the hydrogen permeability (Eq. 6.1 times Eq. 6.7) and the hydride precipitation rate (6.9) multiplied by a parameter relating the fracture stress for U to the volume expansion due to UH_3 formation (i.e., reaction extent). In the pressure–temperature domain where the pressure dependence is half-order, the models of Wicke and Otto [24] and of Condon and Kirkpatrick [10, 11, 23, 31, 32] are very simply related:

$$\mathrm{Vs} = \left(\left(C_0 k_\mathrm{H} C_0 D / 3{U_0}^{0.5}{U_\mathrm{C}}^{0.5}\right)^{0.5} / -\ln U_\mathrm{C}\right) \tag{6.10}$$

By choosing $U_\mathrm{C} = 0.995$, where $U_0 = 1$, $\mathrm{Rate_{CK}} = (0.51\mathrm{e}^{-3033/T}\,P^{0.5})$ and

$$\mathrm{Rate_{WO}} = \left(0.49\mathrm{e}^{-3020/T}P^{0.5}\right) \tag{6.11}$$

Numerical calculations of the distribution of reactants and products across the spall front for the linear hydriding of uranium yield the profiles given in Fig. 6.13 for 20 °C and 10 kPa. Fickian diffusion is calculated at first spall time and is shown in Fig. 6.13.

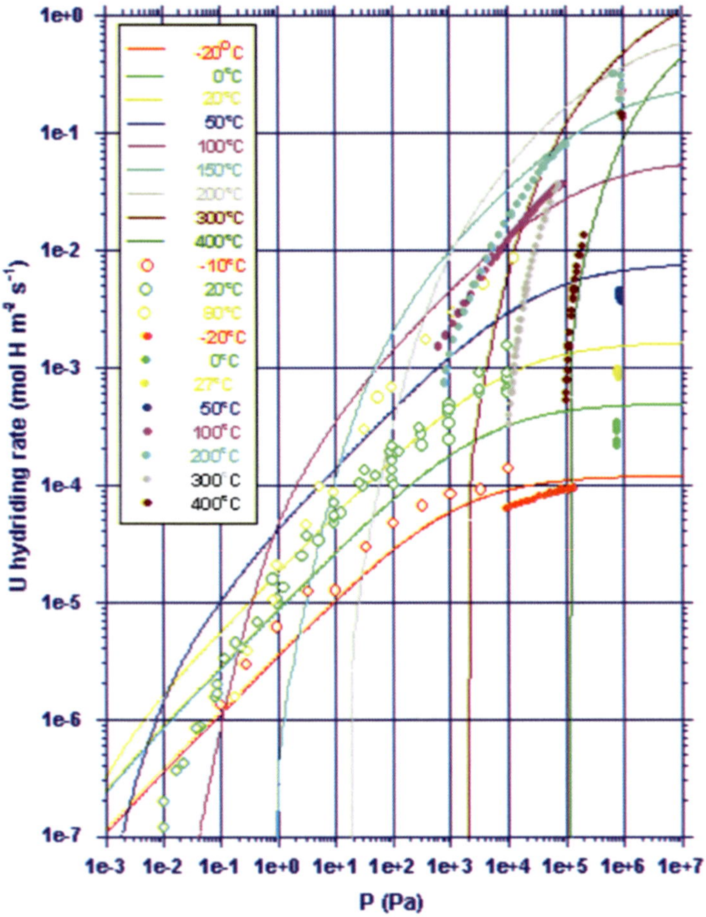

Fig. 6.14 A comparison of the experimental hydriding rates for well-annealed U (*points*) with the predictions of the Condon–Kirkpatrick model (*lines*)

Some refinements of the model to account for dehydriding rates at the low-pressure extremes, concentration saturation effects at the high-pressure extremes, and a slight adjustment to the UH_3 yield required for fracture result in the comparison of model and data shown in Fig. 6.14 [32]. The rates in Fig. 6.12 only apply to well-annealed U specimens. If the specimens contain work as indicated by an increased yield strength, the rate will be remarkably slower, but will recover upon annealing at 630 °C [11].

6.7 Conclusions

Hydrogen reacting with U in a purely binary system does so within the constraints of a relatively simple phase diagram that functions consistently from ambient temperatures to above the melting point. Given a sound understanding of the underlying principles governing phase relationships, chemical dynamics, and the presence and role of impurities in the H–U reaction system, this system of reactions can be controlled and understood. On the other hand, every equation put forth in this chapter is seriously deficient for some conditions and there is a tremendous opening for improving the quality of what is known about the H–U system. There are new areas for research such as small-angle neutron scattering that can measure the concentration of hydride precipitated in U metal and determine the size and shapes of the UH_3 precipitates [25]. Trace hydrogen analysis [21–23] in bulk U is a straightforward analytical technique that can be greatly improved by sample preparation and automation, and reinforced by small-angle neutron scattering. UH_3 appears to have a highly reversible surface phase related to H_2 sorption that accounts for approximately 0.5 % of the stoichiometry; coverage has a logarithmic dependence on H_2 pressure, and probably depends on surface area of the UH_3 [23, 30]. These and many more await those who will meet the incredible challenge of experimental control.

References

1. Sieverts A, Bergner E (1912) Versuche uber die Loslichkeit von Argon und Helium in festen and flussigen Metallen. Ber Deut Chem Ges 45:2576–2583
2. Driggs FH (1931) United States Patent 1,835,024
3. Spedding H, Newton AS, Warf JC, Johnson O, Nottdorf RW, Johns IB, Deane AH (1949) Uranium Hydride I. Nucleonics 4:4–9
4. Newton AS, Warf JC, Spedding FH, Johnson O, Johns IB, Nottdorf RW, Ayres JA, Fisher RW, Kant A (1949) Uranium Hydride II. Nucleonics 4:17–25
5. Warf JC (1949) Chemical properties of uranium hydride. USAEC Report AECD 2997, Iowa State College
6. Svec HJ, Duke FR (1950) USAEC Report ISC-105, Ames Laboratory
7. Katz JJ, Rabinowitch E (1951) The chemistry of uranium, Chapter 8. McGraw-Hill, New York, pp 183–213
8. Libowitz GG (1968) Metal hydrides, Chapter 11. Academic, New York, pp 490–544
9. Inouye H, Schaffhauser SC (1969) Low temperature ductility and hydrogen embrittlement of uranium—a literature review. ORNL-TM-2563
10. Condon JB, Larson EA (1973) Kinetics of the uranium-hydrogen system. J Chem Phys 59:855–865
11. Condon JB (1975) Calculated vs. experimental hydrogen reaction rates with uranium. J Phys Chem 79:392–396
12. Powell GL (1979) The solubility of hydrogen and deuterium in body-centered-cubic uranium alloys. J Phys Chem 83:605–613
13. Powell GL (1982) Internal hydrogen embrittlement in uranium alloys. In: Jessen NC (ed) Metallurgical technology of uranium and uranium alloys, vol 3. American Society for Metals, Metals Park, OH, pp 877–899
14. Powell GL (1976) Solubility of hydrogen and deuterium in a uranium-molybdenum alloy. J Phys Chem 80:375–381
15. Lässer R, Powell GL (1986) The solubility of H, D, and T in Pd at low concentrations. Phys Rev B 34:578–586
16. Mallette MW, Trzeciak MJ (1958) Hydrogen uranium relationships. Am Soc Metals Trans Q 50:981–993

17. Davis WD (1956) Solubility, determination, diffusion and mechanical effects of hydrogen in uranium. Knowles Atomic Power Laboratory, ASAEC Report KAPL-1548
18. Libowitz GG, Gibbs TRP Jr (1957) High pressure dissociation studies of the uranium hydrogen system. J Phys Chem 61:793
19. Condon JB (1980) Standard Gibbs energy and standard enthalpy of formation of UH_3 from 450 to 750 kelvin. J Chem Thermodyn 12:1069–1078
20. Condon JB, Strehlow RA, Powell GL (1971) An instrument for measuring the hydrogen content in metals. Anal Chem 43:1448–1452
21. Powell GL (1972) Mass spectrographic determination of hydrogen thermally evolved from tungsten-nickel-iron alloys. Anal Chem 44:2357–2361
22. Powell GL, Condon JB (1973) Mass-spectrometric determination of hydrogen thermally evolve from uranium and uranium alloys. Anal Chem 45:2349–2354
23. Powell GL, Harper WL, Kirkpatrick JR (1991) The kinetics of the hydriding of uranium metal. J Less Common Metals 172–174:116–123
24. Wicke E, Otto K (1962) The uranium hydrogen system and the kinetics of hydride formation. Z Phys Chem (Frankfort) 31:222–248
25. Spooner S, Bullock JS, Bridges RL, Powell GL, Ludka GM, Barker J (2003) SANS measurements of hydrides in uranium. In: Moody NR, Thompson AW, Ricker RE, Was GW, Jones RH (eds) Hydrogen effects on material behavior and corrosion deformation interactions. TMS (The Minerals, Metals, and Materials Society), Warrendale, PA
26. Powell GL, Condon JB (1976) Hydrogen in uranium alloys. In: Burke JJ, Colling DA, Corum AE, Greenspan J (eds) Physical metallurgy of uranium alloys, Chapter 11. Brookhill Publishing Co., Chestnut Hill, MA
27. Powell GL, Thompson KA (1990) Hydrogen embrittlement in lean uranium alloys. In: Moody NR, Thompson AW (eds) Hydrogen effects on material behavior. The Minerals, Metals & Materials Society, Warrendale, PA, pp 765–773
28. Powel GL (1996) The relationship between strain rate, hydrogen content, and tensile ductility of uranium. In: Moody NR, Thompson AW (eds) Hydrogen effects on material behavior. The Minerals, Metals & Materials Society, Warrendale, PA, pp 355–361
29. Teter DR, Hanrahan RJ Jr, Wetteland CJ (2003) Uranium hydride nucleation kinetics: effects of oxide thickness and vacuum outgassing. In: Moody NR, Thompson AW, Ricker RE, Was GW, Jones RH (eds) Hydrogen effects on material behavior and corrosion deformation interactions. TMS (The Minerals, Metals, and Materials Society), Warrendale, PA
30. Powell GL (2004) Reaction of oxygen with uranium hydride. In: Chandra D, Bautista RG, Schlapbach L (eds) Advanced materials for energy conversion II. TMS (The Minerals, Metals, and Materials Society), Warrendale, PA
31. Kirkpatrick JR (1981) Diffusion with a chemical reaction and a moving boundary. J Phys Chem 1981(85):3444–3448
32. Powell GL, Ceo RN, Harper WL, Kirkpatrick JR (1993) The kinetics of the hydriding of uranium metal II. Z Phys Chem (NF) 181:275–282

Chapter 7
Uranium Corrosion Near Ambient Temperature

G. Louis Powell

Abstract Uranium metal (U) corrodes in air near ambient temperature by numerous mechanisms, rates, and resulting surface degradations depending on the relative humidity (RH) and the residual hydrogen gas (H_2). Experimental observations of this corrosion are presented along with examples of corrosion by the individual components (O_2, H_2, and H_2O) alone and the pyrophoric reaction of O_2 with UH_3. Air below 50 % RH behaves much as in pure O_2 forming a coherent oxide film that may spall after a few years. Higher humidity greatly accelerates the corrosion and destructive effects on the film. The coherent film is super-stoichiometric UO_{2+X} $X < 0.25$ that is impervious to H_2 and little or no barrier at all to H_2O which uses hydriding at the metal interface to propagate UO_{2+X}.

Keywords Ambient temperature • Corrosion • Reflectance • FTIR • Micropitting • Film thickness • Film growth • Blister

7.1 Introduction

Uranium metal is highly reactive with the gases common to air (O_2, H_2O, and H_2) [1–7]. This is expressed as various forms of corrosion near ambient temperature resulting in a plethora of mechanisms and products, including film growth and pitting corrosion, various oxide stoichiometries, and uranium hydride corrosion products, and punctuated by unpredictability and experimental control challenges. This corrosion has been researched, reported in the literature, and review articles written with little data reported near 20 °C. Measurements by Allen, et al is a rare exception and measures the oxide growth on oxide-free uranium at low pressures of

G.L. Powell (✉)
Y-12 National Security Complex, 1 Bear Creek Road,
MS #8097, Oak Ridge, TN 37831, USA
e-mail: powellgl@y12.doe.gov

J.S. Morrell and M.J. Jackson (eds.), *Uranium Processing and Properties*,
DOI 10.1007/978-1-4614-7591-0_7,

O_2 and H_2O [8]. The high density of uranium metal does not lend itself to thermogravimetric measurements since the addition of O_2 or H_2 represents a small fraction of the corrosion products' (UO_2 or UH_3) weight. H_2 and O_2 can be individually controlled and measured as pure gases, but dealing with them as mixtures is a formidable task, and quantitating H_2O with its tendency to adsorb on all surfaces adds to the complexity. Liebowitz et al. [9] reported the U–O_2 reaction below 200 °C with two linear stages and a marked increase in rate at ~400 nm with ~10-nm resolution. In the same temperature range, Baker and Less [10] reported similar results with the first stage described as parabolic followed by an accelerated linear rate at 450 nm. They also reported a half-order dependence on O_2 pressure below 3–5 kPa, increasing with temperature and becoming independent of pressure at higher pressures. The accelerated rate is generally attributed to the disruption of the oxide by stress, and the oxide is generally described as super-stoichiometric UO_{2+X} $X < 0.25$. Baker, Less, and Orman [11] showed that in the case of well-degassed water, the H_2O only reaction in a closed system corrodes U at a rate that is ~10^4 faster than the O_2 only reaction and demonstrated that very pure gases eliminate many of the chaotic properties attributed to U corrosion. They also demonstrated that O_2 and H_2O together in a closed system resulted in the H_2O remaining constant while the O_2 disappeared linearly and H_2 remained at a very low level until the O_2 disappeared, whereupon H_2,consistent with their equilibrium, was produced while the H_2O decayed away. However, the H_2 was measured by extraction through a palladium membrane that guaranteed the gas phase equilibrium. Kinetic equations have been summarized by Ritchie [5] and Colmenares [6] for several corrosion conditions, but they have difficulty coping with corrosion extent-related mechanisms, and little of the data was measured at room temperature.

The discussion up to here yields some insight into the difficulties and complexities of measuring U corrosion and a literature path into the many details. Uranium, being a very practical metal with many uses, the corrosion characteristics around ambient temperature are very important to practical corrosion control. What follows are the results of some exploratory experiments using specular reflectance Fourier transform infrared (FTIR) spectroscopy to measure the UO_2 film growth on uranium metal coupons in room air near 20 °C. The results will be discussed in term experiments reported in the literature at higher temperatures (35–250 °C) already described and their practical implications to corrosion control. That will be followed by reactions by the individual pure gaseous components (O_2, H_2, and H_2O) and some discussion of the reaction of UH_3 with H_2 and O_2.

7.2 Uranium Corrosion in Room Air

Several FTIR spectrometer accessories were developed by Harrick Scientific, Inc., including (1) the Spectropus™ external-reflectance remote-sensing accessory [12, 13] for multiple specimen monitoring in air; (2) the Refractor Reactor™ for specimens in a controlled environment (Fig. 7.1), an evacuable external-reflectance cell that uses tilted, wedge-shaped, KRS-5 windows to effect the 15° deflection

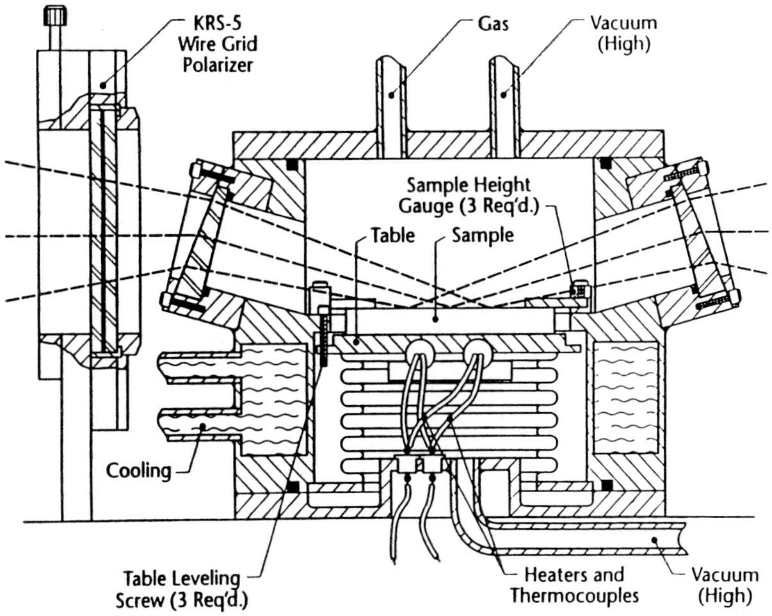

Fig. 7.1 The Harrick Refractor Reactor™ evacuable 75° specular reflection evacuable cell

onto the specimen and to redirect the reflected beam onto the detector; and (3) a hemispherical KRS-5 evacuable diffuse-reflectance cell [14] window that allowed evacuable cells to be used with a Barrel Ellipse™ accessory to measure residual moieties in the oxide film. These accessories, supported by a BIO-RAD FTS-60™ spectrometer, have, as a common characteristic, very high light throughput that, when combined with KRS-5 windows and DTGS detectors, give the necessary signal-to-noise in the 450–650-cm^{-1} range to properly observe the UO_2 band at 575 cm^{-1} [12]. Both external-reflectance accessories have KRS-5 wire-grid polarizers that are lever operated between p and s positions without affecting the spectrometer purge. The external-reflectance measurements were referenced to front-surface gold mirrors for each polarization and the diffuse reflectance was referenced to sand-blasted gold. Later, a Surface Optics SOC 400 portable remote-sensing infrared inspection machine became available with 75° specular reflectance head (Fig. 7.2). These instruments worked much like a top loading balance, a number of samples could be measured at approximately one per minute, and they are adaptable to robotic operation.

The high-purity derby-depleted uranium used for these experiments was cut from the center of a 220-kg billet, ensuring a well-annealed metallurgical condition. The 25-mm by 40-mm by 3-mm coupons were polished to a mirror finish. Later other polished coupons and machined coin-like specimens were analyzed.

These reflectance accessories proved adequate to the task of monitoring changes in the uranium specimens. Figure 7.3 shows the spectral changes resulting from

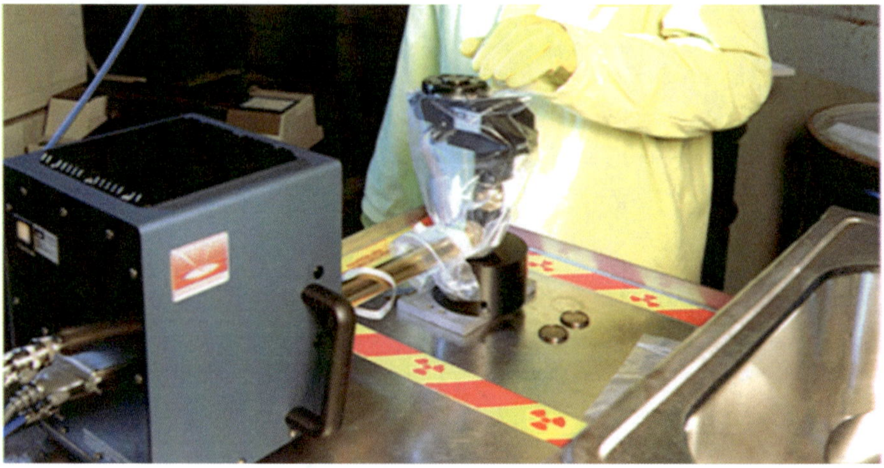

Fig. 7.2 Surface Optics SOC 400 surface inspection machine with 75° specular reflectance head. Note coin-like uranium metal specimens

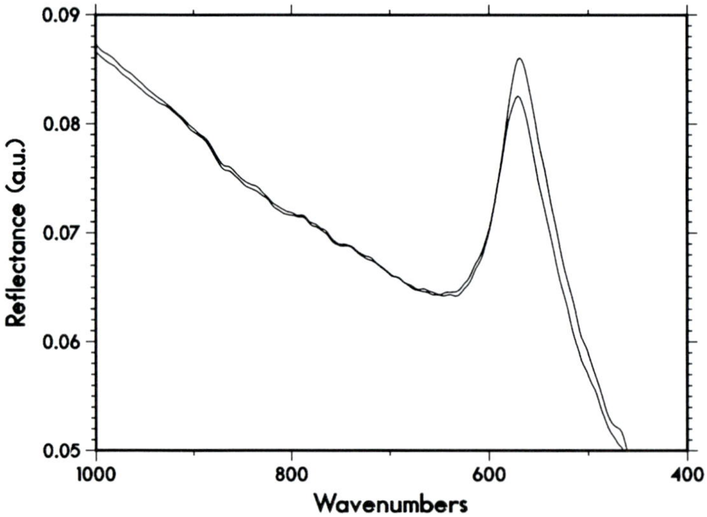

Fig. 7.3 Growth of the 575-cm^{-1} UO_2 band under. *Lower curve*, as loaded; *upper curve*, after 270 ks in vacuum

the oxidation of a polished specimen during a 270-ks period under vacuum (~2 nm), apparently from oxidizing species in the oxide film present after polishing. Specimens exposed to ambient air experienced growth in the 575-cm^{-1} band over time for the low relative humidity (RH) conditions (winter, ~40 % RH room air) experiencing passivation and eventually approaching a linear rate. When conditions shifted to high RH (springtime, ~65 % RH), the corrosion entered an accelerated

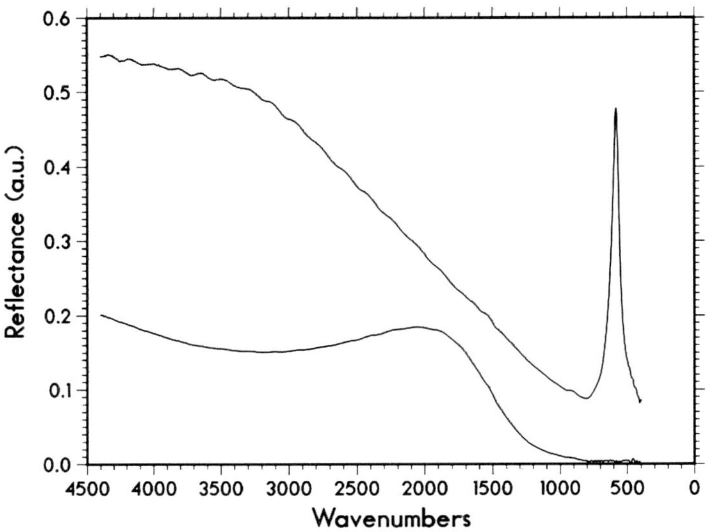

Fig. 7.4 Spectra obtained from ambient air-exposed uranium after experiencing accelerated corrosion. *Upper curve*, p-polarization; *lower curve*, s-polarization

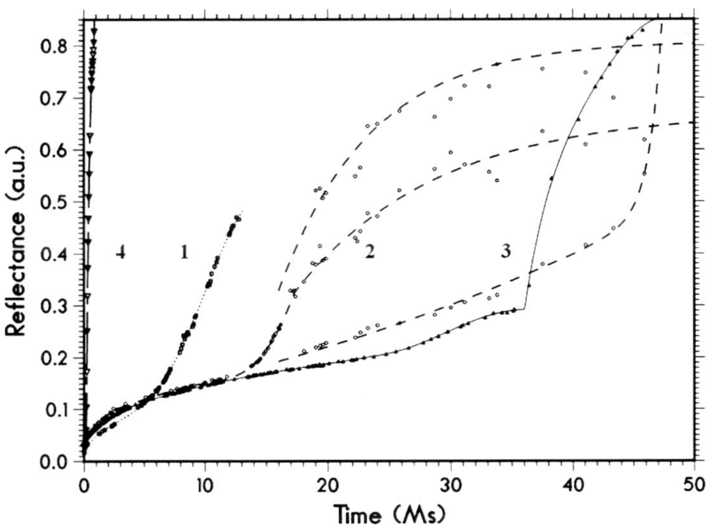

Fig. 7.5 Corrosion of uranium as indicated by the 575-cm^{-1} band of UO_2. *Curve 1*, 1st air; *Curve 2 dashed*, 2nd air; *Curve 3 solid*, O_2; *Curve 4*, stagnant H_2O

mode with a marked deterioration in the quality of the uranium mirror. Figure 7.4 shows the baseline shifted at high energies in the p-polarization curve and a band that appears in the s-polarization curve that gradually propagates from higher to lower energies as corrosion proceeds. Figure 7.5 summarizes the corrosion-related height

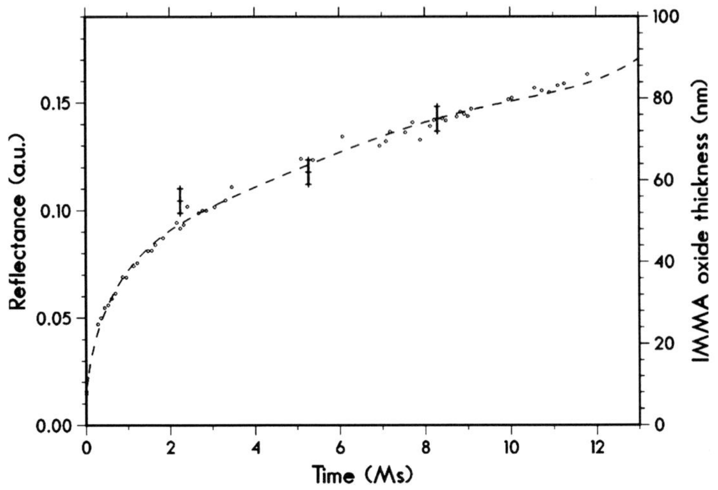

Fig. 7.6 Reflectance data from Experiment 2 compared to three measurements of the UO_2 thickness by IMMA depth profiling

in the 575-cm^{-1} absorption band for four external-reflectance experiments, baseline corrected at 1,000 and 1,375 cm^{-1}, each averaged over 50 cm^{-1}. Experiments 1 and 2 were benchtop experiments starting in winter (low RH) in successive years and continuing into springtime (high RH) experiencing the accelerated corrosion with the seasons. In Experiment 2, three locations on the coupon separated by 10 mm were measured after the onset of accelerated corrosion indicating that local regions on the same coupon corroded at very different rates.

Experiment 3 was in the evacuable cell using research grade O_2 as the corrosive gas (~20 kPa O_2). After 36 Ms, the sample was transferred to high RH room air and immediately experienced the accelerated corrosion. Experiment 4 was isolated in the evacuable cell with 610 Pa H_2O, ice-bath controlled. After 0.2 Ms, UO_2 growth initiated and was off-scale in less than 1 Ms. The pressure in the cell also shot up from H_2 production. During the early stages of Experiment 2, this sample was transferred to an Ion Microprobe Micro-Analyzer (IMMA) where the oxide film thickness was measured based on a sputtering rate for thicker UO_2 films measured using a profilometer. Three such measurements were made at different times during Experiment 2 (Fig. 7.6). Figure 7.7 applies this calibration factor (550 nm/a.u.) to the early part of Experiment 3 along with showing the O_2 pressure to which the U coupon was exposed. Figure 7.5 demonstrates that there is little difference between the corrosion rates in experiments 1, 2, and 3 in low RH.

Before admitting the U coupon to the evacuable cell for Experiment 3, a leak in the evacuable cell bellows was detected and the in-leakage stopped by evacuating the bellows interior. This resulted from a slow decay in the O_2 pressure inside the cell during Experiment 3 that required occasional additions of O_2. The initial O_2 introduction distinctly started the UO_2 film growth, but gross variations of O_2 pressure thereafter had no effect on the film growth.

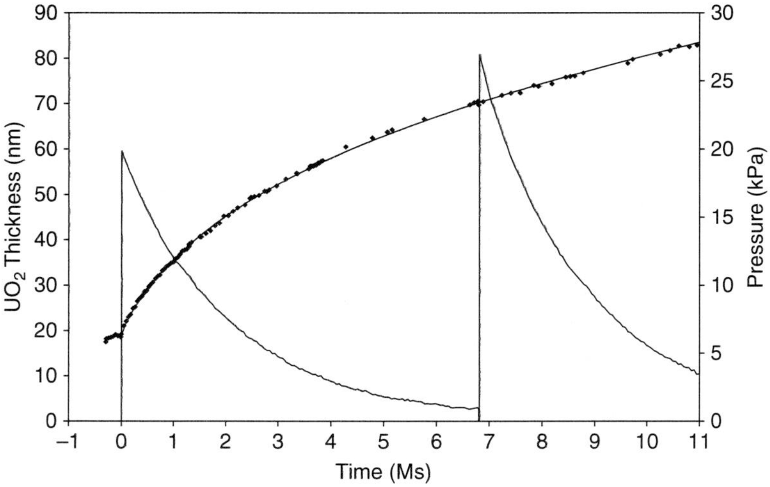

Fig. 7.7 A comparison of UO_2 growth to the O_2 pressure in the evacuable cell

Scanning electron microscope observations of Sample 2 at the end of the experiment (Fig. 7.8) indicated the U mirror degradation process was by nucleation of ~5-μm circular features arrayed linearly along crystallographic directions. Visual and optical microscope observations confirmed the loss of mirrorlike qualities in the specimen commencing with the increase in humidity and reaction rate, but the mirror did not degrade uniformly as can be seen in Figs. 7.5 and 7.8.

These experiments were continued using an SOC 400 surface inspection machine using p-polarized, 75° specular reflection FTIR on similarly polished U coupons. Room temperature was 13 °C. A number of as-machined coupons were also examined to measure the effect of post-machining oil coatings on the oxide film thickness. Figure 7.9 shows some spectra taken during these experiments. There was little difference between spectra from as-machined surfaces and as-polished surfaces, but the calibration factor may differ. One of the polished specimens experienced condensing moisture conditions resulting in visible corrosion. The reflectance spectra contained evidence of uranyl species (UO_3) characterized by the 900-cm^{-1} band, the 575-cm^{-1} band (UO_2, U_3O_8, and UO_3), and a band in between those characteristic of U_3O_8. The polished coupons in room air behaved over time similar to the previous study, and the oil coating apparently arrested the corrosion at a thickness of 20–30 μm, approximately the kinetic break-over point in Fig. 7.10 and (7.1).

Figure 7.10 shows the film growth on three polished specimens that displayed the characteristics observed during the earlier experiments. Over a fraction of an Ms, the rate grew to 20–30 nm and changed to a much slower rate characteristic of a rising exponential function changing to a linear rate (7.1). The solid curves in Fig. 7.10 were fit to this model and also to a parabolic model ((7.2), Fig. 7.11):

$$\text{Film Thickness} = A - B\mathrm{e}^{-Ct} + Dt \tag{7.1}$$

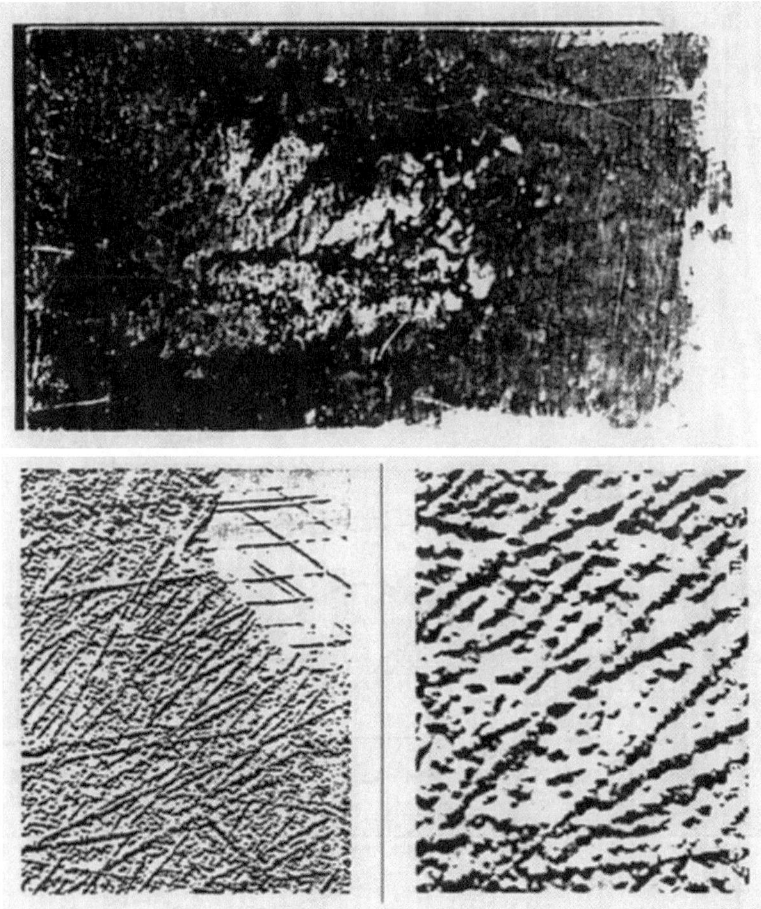

Fig. 7.8 Experiment 2 specimen at the end of the experiment. *Top* micrograph: entire sample 25 mm high. *Bottom left*: ~10× relative to *top*. *Bottom right*: ~10× relative to the *bottom left*

$$\text{Film Thickness} = E + Ft^{0.5} \tag{7.2}$$

The interpretation should be limited until more extensive theoretical evaluation. For example, the parameters from these fits are given in Table 7.1. The initial thickness is $A-B$ for (7.1) and E for (7.2). These models adequately describe the data, but the parabolic fit (7.2) does not involve the initial film thickness in the process. Equation 7.1 fits best at early times with precision better than 1 nm, whereas (7.2) fits better at longer times. Equation 7.1 could be converted to a double exponential rise and better accommodate the slowdown at long times. The increased variability in the film thickness measurement at long times needs also to be studied.

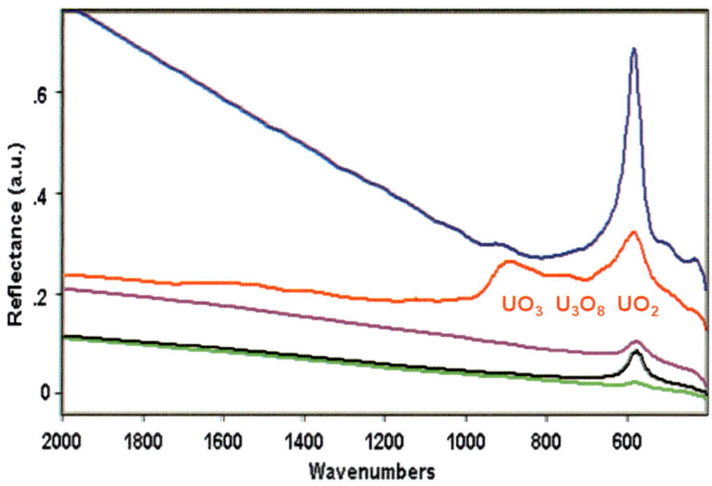

Fig. 7.9 Infrared spectra carry film thickness, oxide stoichiometry, and organic contamination information. From bottom to top: polished U, 600 s in air at 13 °C; polished U, 1.7 Ms in air at 13 °C; as-machined U, oiled post-machining same day, degreased after 15 Ms; pitted corrosion from condensed water with IV to VI oxidation states; and polished, 30 Ms in high RH air

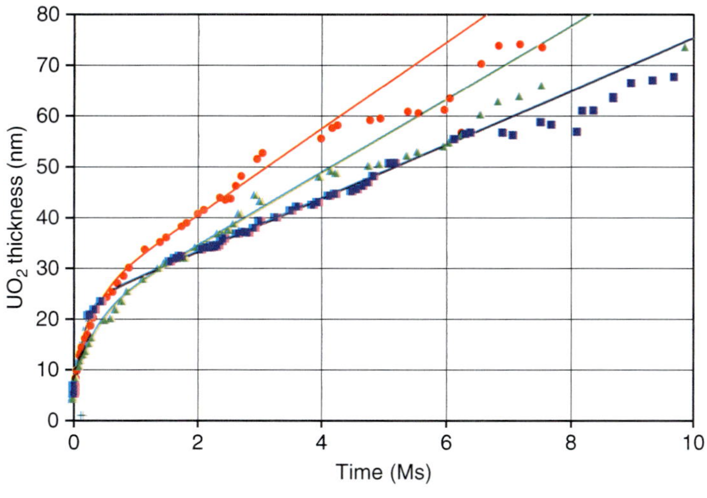

Fig. 7.10 Corrosion of three polished U coupons exposed to room air at 13 °C modeled using a rising exponential to linear fit

Twenty-two as-machined U coupons were oiled for corrosion protection on the day they were machined. After 30 Ms of storage in air, they were degreased and analyzed by this method. The results are shown in Fig. 7.12. The oil appeared to arrest the corrosion at the break point in the corrosion kinetics shown in Fig. 7.10.

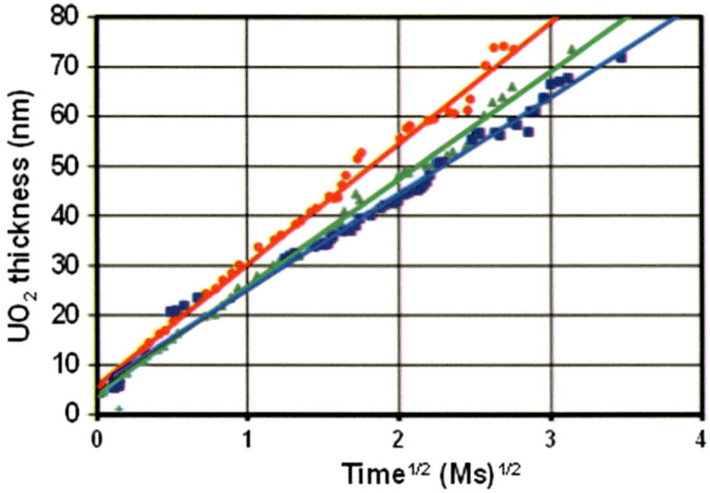

Fig. 7.11 Parabolic plot of corrosion of three polished U coupons exposed to room air at 13 °C modeled using a parabolic fit

Table 7.1 Fit parameters for Figs. 7.10 and 7.11

A (Nm)	*B* (nm)	*C* (1/Ms)	*D* (1/Ms)	(*A*−*B*) (nm)	*E* (Nm)	F^5 (Nm/Ms$^{0.5}$)
22.6	17.26	0.046	0.039	5.4	5.7	19.4
19.9	14.1	0.022	0.027	5.7	4.0	21.6
23.6	17.4	0.054	0.063	6.2	5.9	24.4

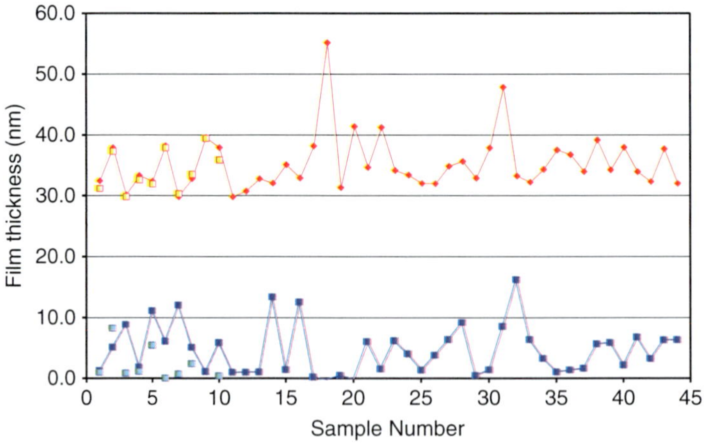

Fig. 7.12 As-machined uranium coupons are readily analyzed after removing a protective oil film after ~30 Ms. Data pairs, starting in sequence, represent opposite sides of each coupon. *Upper curve* UO_2 and *lower curve* hydrocarbon. *Open symbols* are replicate measurements after 0.43 Ms storage in vacuum

Vacuum storage of the degreased U coupons led to little film growth, and the precision of the repeat measurements was good.

Grazing angle p-polarized specular reflection FTIR spectroscopy is a practical tool for monitoring U corrosion. Work presented here is to demonstrate the tool and to draw some insight into this phenomenon at ambient temperature. The method can handle very large arrays of coupons and can map surfaces if properly supported by automation. Such experiments are necessary if the role of U mechanical properties and initial surface preparation in uranium oxidation is to be determined. That polished uranium specimens had oxide thicknesses, at best, no less than 4 nm agrees with the observations of Allen et al. [8] that indicates that the first 4–6 nm of oxide film forms very rapidly before breaking over to a much slower linear reaction, perhaps indicating a minimum oxide film thickness is necessary to initiate passivation.

Some qualitative lessons can be drawn from this work. At low RH and O_2 present, U corrodes slowly as a coherent film, independently of the precise partial pressures of H_2O or O_2, and the corrosion rate is resistant to variations in either. At RH values above 50 %, the corrosion mode changes to a micropitting process that accelerates the reaction. The root cause is probably precondensation of moisture wherein adsorbed H_2O builds to many monolayers, but not yet liquid H_2O. In the case of stagnant H_2O at ~20 % RH equivalent, the accumulation of H_2 resulted in catastrophic production of UO_2, as indicated by the 575-cm^{-1} band shape, after 0.2 Ms. The sharp absorption band at 575 cm^{-1} is an indication of a well-characterized crystal structure, indicating that even the accelerated micropitting or H_2-assisted H_2O corrosion still produces essentially UO_2. That is not to say the stoichiometry is precisely 2.0, because these results indicate strongly that moieties derived from O_2 and H_2O are an intimate part of the corrosion film. In the case where liquid H_2O forms on the surface of U, higher oxidation states of U oxides are formed as part of a very destructive corrosion process. The coherent film formed at low RH stays with the U until stresses cause it to flake off above 500 nm. The micropitting observed at high RH or condensing water products are readily dispersible and thus a radiological contamination source.

7.3 Uranium Corrosion by Hydrogen

From Chap. 6 on The Uranium–Hydrogen Binary System, the pure couple between H_2 and pure U near room temperature and devoid of any oxidizing species or products leads to H atoms dissolving into the U metal, precipitating in the narrow diffusion gradient, rupturing the metal, and achieving a steady-state linear rate conversion of the U metal to UH_3. This is a form of corrosion very concisely described by the Condon–Kirkpatrick Model [15] for the case where no oxide film exists on the U and the UH_3 product as a fine and very reactive powder purifies the reactant H_2 of any trace oxidizing species. Condon and Larsen [16] demonstrated clearly that O_2 exposure of U delayed hydriding by the competition between the

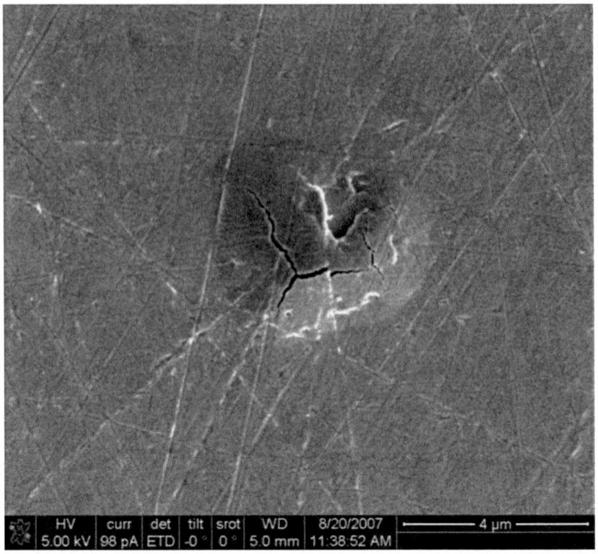

Fig. 7.13 A 4-µm-wide blister on a uranium surface that is just beginning to rupture

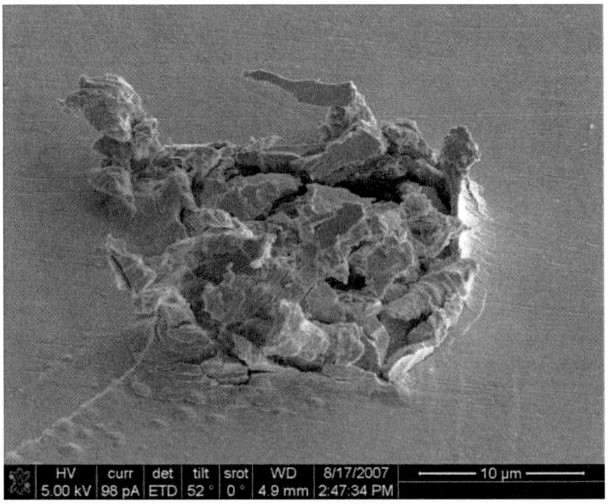

Fig. 7.14 A blister on a grain boundary having ruptured at 15-µm diameter and the surface film is rolling back

chemical potentials of oxidizers and H_2. Condon [17] also demonstrated that nucleation of the hydride phase within the U metal was opposed by the strength of the metal and that residual stress in the metal controlled the hydriding rate. Now consider U with an O-rich UO_2 film, on top of a mechanically worked surface (from machining or polishing) on top of bulk uranium with some residual work, and the H_2 is well supplied with trace amounts of oxidizing species. The H_2 must find its way to the U metal through the oxide film, disassociate, dissolve in the metal, and precipitate as small particles of UH_3. Bingert et al. [18] demonstrated that blisters of

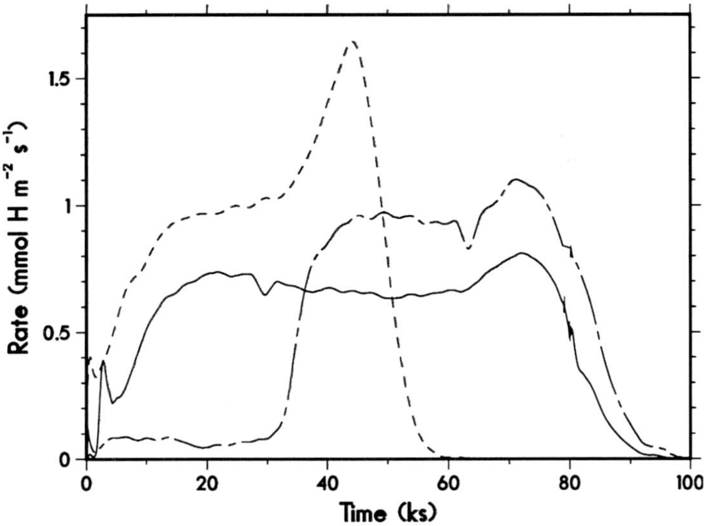

Fig. 7.15 With ultrapure H_2, UHV equipment, a 250 °C vacuum bake, well-annealed U, the three specimens hydrided at 10.0 kPa H_2 near the rate predicted by the Condon–Kirkpatrick Model [14, 16] after a variety of initiation times (pitting processes) for laying bare the bulk U surface. The longest initiation time was for a polished specimen

UH_3 on the order of 10 μm form underneath the U oxide film eventually disrupting the film and rolling it back. These blisters have been shown to be abundant and occur at defects [19] such as grain boundaries as shown in Figs. 7.13 and 7.14. Focused ion beam dissection of these blisters indicates a second phase formed underneath.

In order to form a blister, the H_2 penetrating the defect must overcome tremendous opposition from the oxygen chemical potential associated with the slightly super-stoichiometric UO_{2+X}, $X < 0.25$, as well as OH species, H_2O, and higher oxidation states of UO_X. To promote hydriding, a U metal specimen is typically prepared as well-annealed metal (630 °C, 0.2 Ms) and is then "activated" in an ultrahigh-vacuum system by baking at 250 °C for 0.1 Ms. After such a treatment, exposure to ultrapure H_2 filtered through a UH_3 powder bed results in hydriding rate curves as shown in Fig. 7.15 [20]. The central part of the hydriding rate curves agreed well with the Condon–Kirkpatrick Model prediction [15, 17]. The lengths of the initiation times were much more variable than the hydriding rate after initiation, indicating that differences in the surfaces, most probably mechanical surface work, control the initiation times.

Figure 7.16 shows diffuse-reflectance spectra from a U sample having a high RH corrosion product in an evacuable cell [20]. Light scattered within the fluffy oxide layer carries information of chemical moieties in that film. The initial pumpdown at ambient temperature resulted in removal of water (*A*), subsequent 250 °C bake in vacuum removed more H_2O, perhaps OH, uranyl ~900 cm^{-1} (*B*), for a total loss

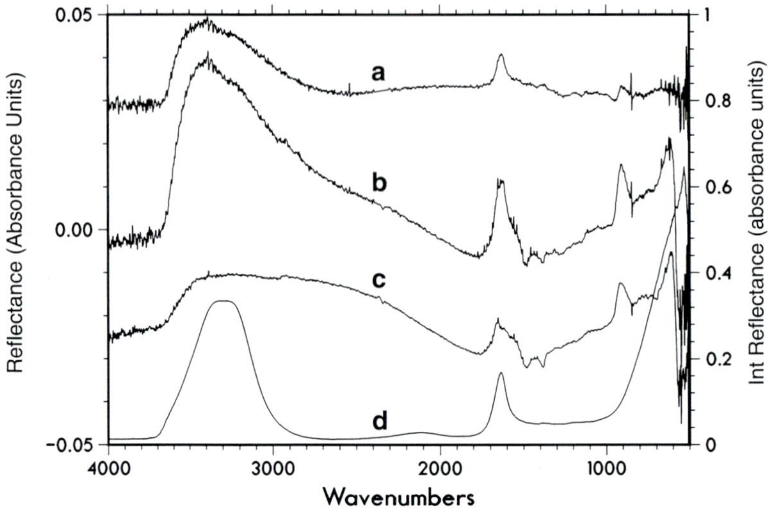

Fig. 7.16 Diffuse-reflectance infrared difference spectra of changes (losses) resulting from (**a**) evacuation, (**b**) 250 °C vacuum bake, (**c**) liquid water spectrum, and (**d**) liquid water spectrum [19]

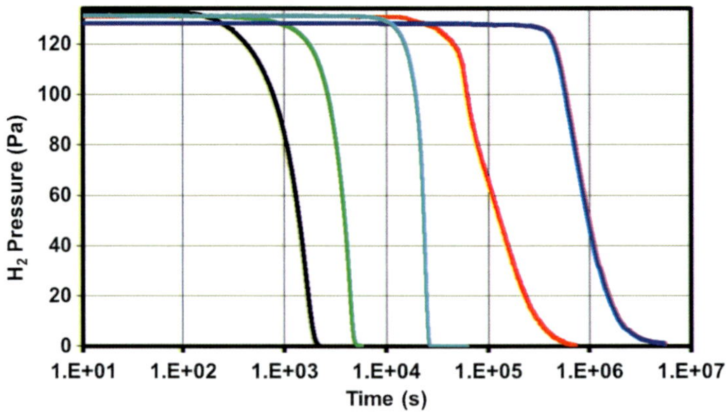

Fig. 7.17 Hydriding initiation times for spherical U powders at room temperature following ultrahigh-vacuum bakes at *right* to *left* 19 °C, 32 Ms; 75 °C, 1.2 Ms; 150 °C, 160 ks; 125 °C, 160 ks; and 300 °C, 65 ks [20]

indicated by (*C*). The vibrational modes of liquid H_2O for reference are (*D*). The UO_2 band at 575 cm^{-1} is inverted after the bake indicating growth of the main oxide band. Thus, the evacuation and bakeout remove many labile oxidizing species from the oxide film. X-ray photoelectron spectra also indicate such chemical changes promoted by baking [21]. Not only does it indicate loss of chemical moieties, particularly oxygen and hydroxyls, but it indicates the U oxide becomes more metallike as the temperature exceeds 250 °C.

Figure 7.17 shows initiation times as a function of ultrahigh-vacuum bake time. The U substrate was spherical U powder (<100 μm) that had an oxide film from aqueous exposure and subsequently kept in an inert environment for 600 Ms followed by 32 Ms at ultrahigh vacuum. Two points to be made from this data. First, the initiation time is decreased by 75 % by the 75 °C bake. Second, the initiation time never goes to zero. This means that the UO_{2+X} film is a most formidable obstacle to H_2, and hydride pitting only occurs at defects that, as formed, are typically protected by an oxidizing chemical potential field. This protective oxidizing chemical potential probably related to the super-stoichiometric oxide does not disappear with extensive inert aging.

7.4 Uranium Corrosion by Water

The corrosion of U by H_2O only is much faster than that of other forms of oxide growth, apparently promoted by the H supplied by the water. Certainly the accumulation of H_2 during the H_2O reaction is a major contributor to fast UO_2 formation. To investigate this mechanism, Condon et al. [22] studied the oxide film growth on using Ion Microprobe MicroAnalysis (IMMA), a secondary ion imaging technique. The technique was used to depth profile the oxide film in much the way it was used for calibrating the 75° specular reflectance FTIR measurements. U specimens were first exposed to $H_2{}^{18}O$ and subsequently to $H_2{}^{16}O$. The results are shown in Fig. 7.18. The last H_2O to be exposed to the sample was found at the oxide metal interface so that H_2O penetrates the oxide film with facility. The slope of the oxide distribution at the interface was interpreted as hydrogen penetrating the metal per the Condon–Kirkpatrick Model, and from that the minimum corrosion rate by water was calculated.

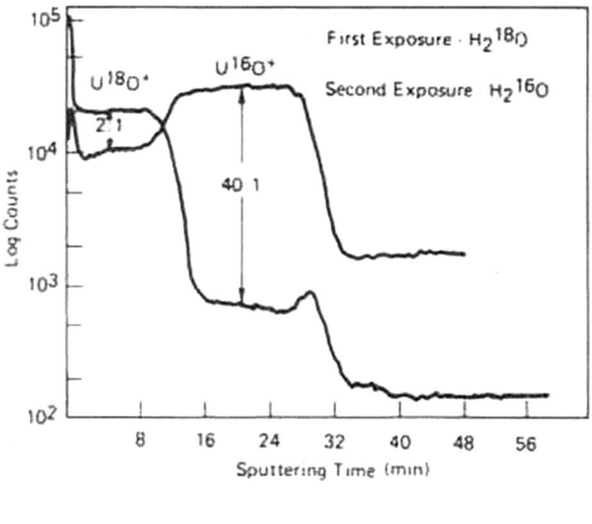

Fig. 7.18 IMMA depth profile of a U coupon exposed first to $H_2{}^{18}O$ and subsequently to $H_2{}^{16}O$ each at 610 Pa for 14.4 ks at 80 °C. The metal interface is at 32 min, and the vacuum interface is at 0 min sputtering time

7.5 Uranium Hydride Corrosion by Oxygen

The intimate connection between U and H_2O corrosion is found as UH_3 in heavily corroded spent nuclear reactor fuel stored for long times in liquid H_2O [23–25]. Upon drying out and some agitation, the UH_3 can ignite when exposed to air. In order to characterize this pyrophoric behavior, a sample of UH_3 was prepared by hydriding a U rod as done in Fig. 7.15 [20, 26]. The specimen was repeatedly reacted in an infrared gas cell with small doses of O_2. H_2 was the only product of the reaction. The surface of the sample was characterized by H_2 sorption [26], and the surface activity was only diminished a small amount with each reaction. With each dose of O_2, the reaction rate was measured. With the first dose after a long evacuation, the H_2 sorption removed all the H_2 product, so the pressure drop could be followed for several decades. Figure 7.19 shows a graphic summary of the sequence of O_2 in and H_2 out. From this it was determined that the appropriate reaction was

$$UH_{3.00} + 1.075\,O_2 \rightarrow UO_{2.15} + 1.5H_2$$

From the O_2 pressure decay, the pressure dependence of the O_2 + UH_3 reaction was determined and is shown in Fig. 7.20.

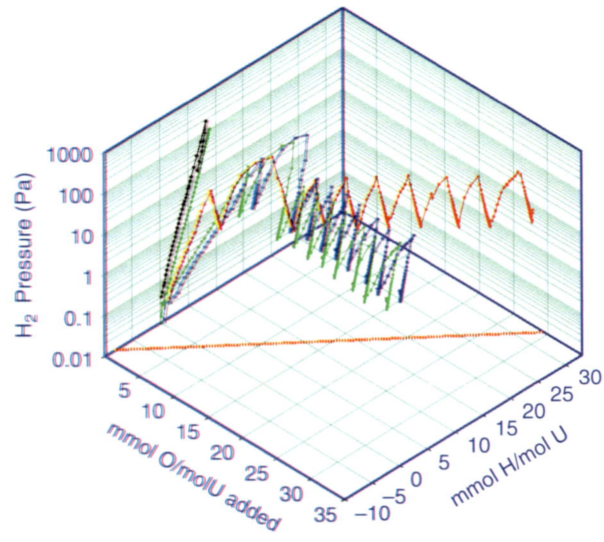

Fig. 7.19 The H_2 surface sorption was characterized by small titers of H_2. Small doses quantitative of O_2 produced H_2 that was quantitatively removed. No H_2O was formed

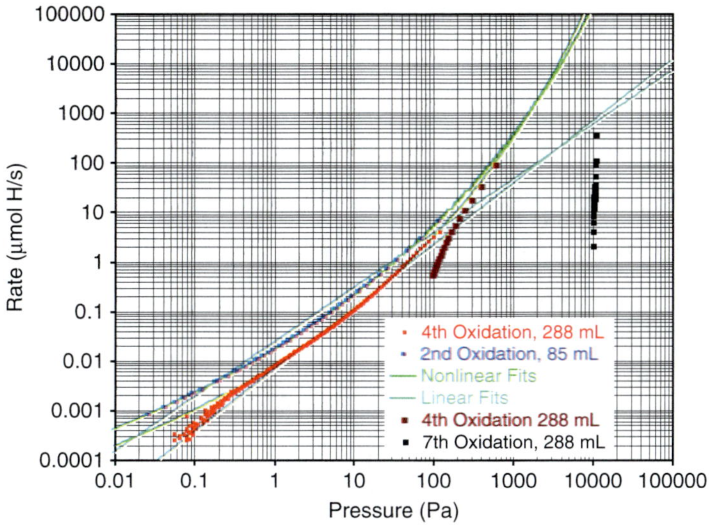

Fig. 7.20 The O_2 pressure dependence of the O_2 + UH_3 reaction

7.6 Conclusions

The corrosion processes for U are many and complex, but with the right combination of measurements, they can be elucidated. The main control for U corrosion is to minimize H_2O and always have some O_2 around. O_2 and its product UO_{2+X} film are the main buffers for controlling oxidation rate and for preventing H_2 attack. Increasing amounts of H_2O easily penetrate the oxide film pushing H ahead of the H_2O-formed uranium oxide. The H_2O-formed oxide is dispersible and, under extreme conditions, may contain UH_3 and be pyrophoric, neither of which is desirable.

References

1. Wilkinson WD (1962) Uranium metallurgy. vol 2. Interscience, New York, p 752
2. Waber JT (1956) A review of the corrosion behavior of uranium. Report LA-2035, Los Alamos National Laboratory
3. Orman S (1976) Oxidation of uranium and uranium alloys. In: Burke JJ, Colling DA, Corum AE, Greenspan J (eds) Physical metallurgy of uranium alloys, Chapter 21. Brookhill Publishing Co., Chestnut Hill, MA
4. Cathcart JV (1976) Gaseous oxidation of uranium alloys. In: Burke JJ, Colling DA, Corum AE, Greenspan J (eds) Physical metallurgy of uranium alloys, Chapter 20. Brookhill Publishing Co., Chestnut Hill, MA
5. Ritchie AC (1981) A review of the rates of reaction of uranium with oxygen and water vapour at temperatures up to 300°C. J Nucl Mater 102:102–182

6. Colmenares CA (1984) Oxidation mechanisms and catalytic properties of the actinides. Prog Solid State Chem 15:257–364
7. Lillard JA, Hanrahan RJ Jr (2007) Corrosion of uranium and uranium alloys. In: Cramer SD, Covino BS, Jr (eds) ASM Handbook, vol 13B, Corrosion: Materials
8. Allen GC, Tucker PM, and Lewis RA (1984) X-ray photoelectron spectroscopy study of initial oxidation of uranium metal in oxygen + water-vapor mixtures, J Chem Soc, Faraday Trans 2:991–1000
9. Leibowitz L, Schnizlein JG, Bingle JD, Vogel RC (1961) Kinetics of oxidation of uranium between 125° and 250°C. J Electrochem Soc 108:1155–1160
10. Baker MMD, Less LN, Orman S (1976) Oxidation of uranium and uranium alloys. In: Burke JJ, Colling DA, Corum AE, Greenspan J (eds) Physical metallurgy of uranium alloys, Chapter 21. Brookhill Publishing Co., Chestnut Hill, MA
11. Baker MMD, Less LN, Orman S (1966) Uranium + water reaction, Part I—the effect of water and other gases. Trans Faraday Soc 62:2525–2530
12. Powell GL, Milosevic M, Lucania J, Harrick NJ (1992) The spectropus system, remote sampling accessories for reflectance, emission, and transmission analysis using Fourier transform infrared spectroscopy. Appl Spectrosc 46:111–125
13. Powell GL, Dobbins AG, Cristy SS, Clift TL, Meyers HM, Lucania J, Milosevic M (1993) The study of the oxidation of uranium by external and diffuse reflectance FTIR spectroscopy using remote-sensing and evacuable cell techniques. Proc SPIE 2089:214–215
14. Smyrl NR, Fuller EL, Powell GL (1983) Monitoring the heterogeneous reaction of LiH and LiOH with H_2O and CO_2 by diffuse reflectance infrared Fourier transform spectroscopy. Appl Spectrosc 37:38–44
15. Kirkpatrick JR (1981) Diffusion with a chemical reaction and a moving boundary. J Phys Chem 85:3444–3448
16. Condon JB, Larson EA (1973) Kinetics of the uranium-hydrogen system. J Chem Phys 59:855–865
17. Condon JB (1975) Calculated vs. experimental hydrogen reaction rates with uranium. J Phys Chem 79:392–396
18. Bingert JD, Hanrahan RJ Jr, Field RD, Dickerson PO (2004) Texture effects on the nucleation of uranium hydride on α-uranium. J Alloys Compd 365:138–148
19. Powell GL, Schulze RK, Siekhaus WJ (2009) Corrosion of uranium by hydrogen at low hydrogen pressures. In: Somerday B, Sofronis B, Jones R (eds) Hydrogen effects on materials. ASM, Materials Park, OH, pp 556–561
20. Powell GL, Ceo RN, Harper WL, Kirkpatrick JR (1993) The kinetics of the hydriding of uranium metal II. Z Phys Chem (NF) 181:275–282
21. Powell GL, Schulze RK, Siekhaus WJ (in press) Hydriding of uranium and thermal decomposition of uranium hydride at low hydrogen pressures. In: Somerday B, Sofronis B, Jones R (eds) Hydrogen effects on materials. ASM, Materials Park, OH
22. Condon JB (1980) Nucleation and growth in the hydriding reaction of uranium. J Less Common Met 73:105–112
23. Totemeier TC (1995) A review of the corrosion and pyrophoricity behavior of uranium and plutonium. Argon National Laboratory Report, ANL/ED/95-2
24. Totemeier TC, Pahl RG, Hayes SL, Frank SM (1998) Characterization of corroded metallic uranium fuel plates. J Nucl Mater 256:87–95
25. Totemeier TC, Pahl RG, Frank SM (1999) Oxidation kinetics of hydride-bearing uranium metal corrosion products. J Nucl Mater 265:308–320
26. Powell GL (2004) Reaction of oxygen with uranium hydride. In: Chandra D, Bautista RG, Schlapbach L (eds) Advanced materials for energy conversion II. TMS (The Minerals, Metals, and Materials Society), Warrendale, PA

Chapter 8
Nondestructive Evaluation of Uranium: Fundamentals and Future

Jonathan Poncelow, David L. Olson, Cameron Howard, Kamalu Koenig, and Craig VanHorn

In theory, theory and practice are the same.
In practice, they are not.—A. Einstein

Abstract Nondestructive evaluation (NDE) capabilities have historically been slow to catch on in the processing of uranium, perhaps due to the difficulties associated with its anisotropy and reactivity or perhaps owing to the limited number of uranium applications. Whatever the case, NDE of uranium components has lagged behind the advances in the field for other material systems, despite some of the actinide's unique properties. A review is presented of the fundamentals supporting NDE techniques as well as the methods themselves which are applicable to uranium systems. Specific features of uranium metallurgy, such as its high degree of anisotropy and low temperature behavior which may be exploited for the purposes of NDE, are described. Finally, case studies in uranium NDE are reviewed briefly in light of the fundamentals established in prior sections.

Keywords Nondestructive • Uranium • Evaluation • NDE • Charge density wave • Solid statephysics • Electrodynamics • Ultrasound • Physical acoustics • Superconductivity • Residualstress • Electropulse processing • Properties

---

J. Poncelow (✉) • D.L. Olson • C. Howard
Department of Metallurgical and Materials Engineering, Colorado School of Mines, 1500 Illinois Street, Golden, CO 80401, USA
e-mail: jponcelo@mines.edu; dolson@mines.edu; cahoward@mines.edu

K. Koenig
Structural Integrity Associates, Inc., 10731 East Easter Avenue Suite 100, Centennial, CO 80112, USA
e-mail: kamalu.koenig@gmail.com

C. VanHorn
URS Corporation 7800 East Union Avenue, Suite 100 Denver, CO 80237, USA
e-mail: craig.vanhorn@urs.com

J.S. Morrell and M.J. Jackson (eds.), *Uranium Processing and Properties*,
DOI 10.1007/978-1-4614-7591-0_8, © Springer Science+Business Media New York 2013

8.1 State of the Art of Uranium Nondestructive Evaluation

Despite its unique properties and importance in the nuclear industry, uranium has not been rigorously targeted for nondestructive evaluation (NDE) research. This failure is perhaps due in part to difficulties in processing the metal or possibly due to a lack of understanding of how the meso- and macroscopic features vary with microstructure—uranium's extreme anisotropy is responsible for both. Whatever the reason, the opportunities for advancement in the field of uranium NDE are numerous, and this text strives to encourage development of nondestructive applications to uranium processing.

8.1.1 Difficulties Associated with Uranium

Most traditional NDE practices are applied to uranium to assess various indications from dimensional changes and inaccuracies, cracking (most likely due to environment), and voids. Uranium, as a metal or alloy, presents practical difficulties for testing procedures such as chemical reactivity which promotes growth of surface scale and hindering practices that require sufficient surface contact, e.g., elastic wave transmission and receiving. The intrinsic radioactivity of uranium makes the practice of radiography difficult as background radiation relative to the signal and personnel safety must be taken into consideration. The fissile nature of certain uranium isotopes and chemically reactive nature of certain alloys make impractical traditional techniques which require the use of water. Additionally, the high degree of anisotropy in physical properties complicates characterization.

8.1.2 The Future of Uranium NDE

Traditional practices are well established in NDE handbooks. For techniques to advance in their applicability to uranium, a reassessment of fundamentals to support new approaches is required.

Uranium and uranium alloy properties and performance are strongly tied to microstructure. Advanced quantitative NDE is the next step in promoting quality in uranium alloy components and fuel. Real-time assessment of processing through microstructural characterization is necessary to increase product yield and quality assurance. Wave analysis techniques (e.g., ultrasonics and eddy currents) allow practices to assess bulk properties. This chapter will present physical fundamentals to thoroughly elucidate the relationships between matter and physical perturbations. Uranium has a significant number of interesting and unique properties which make it of considerable value to study in an NDE context. The chapter will continue with the opportunities afforded by the solid state physical properties of uranium. Finally, recent research efforts in uranium NDE will be highlighted, concluding the discussion.

8.2 Advanced NDE Wave Science and Technology for Nuclear Materials

Various forms of elastic and electronic wave mechanics are essential to the application of future NDE practices at a level that is not often witnessed in traditional practice. The text shall take this opportunity to describe the necessary fundamentals of waves and impart an understanding of the role microstructure plays in their interactions. Each section will begin with derivations of relevant wave equations, continue with a description of observed behaviors, and conclude with phenomenological models of wave mechanics of physical features.

8.2.1 Why Waves?

Before introducing wave concepts it may prove advantageous to the reader to understand why dynamic (i.e., wave-based) measurements are useful in the first place, especially considering the staggering number of static characterization techniques available. One will find, however, that by exciting a particular sample with a *time-dependent* perturbation, an additional "dimension" can be acquired with which that component can be qualified. Such a dimension is best represented by the frequency domain which, being related inversely to time, retains some qualities analogous to reciprocal space in solid state physics.

Perhaps a more revealing argument for the use of dynamic measurements in NDE applications can be made by considering the distinction between *static* and *dynamic* treatments. Consider an analogous field of study: thermodynamics. In stark contrast to its suffix, thermodynamics yields information concerning the static *equilibrium* of a system with emphasis on the energies involved in its components. By equilibrium, of course, time is taken in the infinite limit, where the system is presumably no longer changing. If, however, one considers the *kinetic* aspects of a system, they are rewarded with additional insight into what precisely is occurring at the nano- and microscopic scales.

8.2.1.1 The Static Measurement

What then constitutes a static measurement in the case of other studies? An example is found in elasticity, wherein the elastic modulus—be it shear, Young's or whatever—is obtained by computing the stress response to a static strain. In terms of the underlying phenomena, a measurement made this way reveals the equilibrium of the electrostatic forces between the ions and electrons—a feature which depends on electron density and distribution. What are not made clear by such determination, however, are the "kinetic" aspects of the re-equilibration of electrons to the strained state, viz. how quickly and at what cost the electrons redistribute.

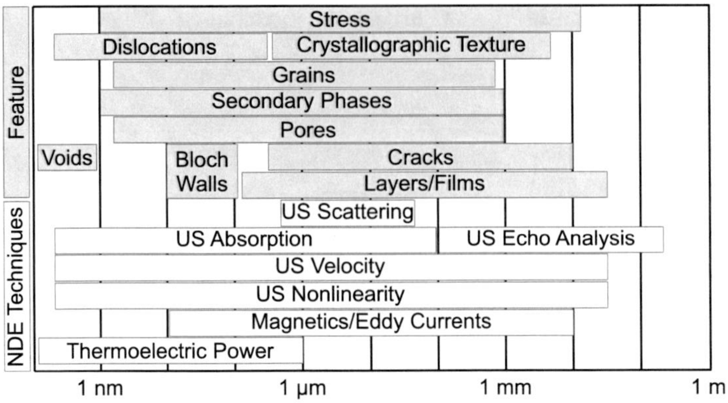

Fig. 8.1 Correlation of dimensionality of lattice defects to the effective range of the various NDE analytical assessment tools (adapted from [18])

8.2.1.2 Time-Dependent Perturbation

If, on the other hand, the strain is *cycled* in time, the opportunity is presented to observe and quantify the time required for electrons (and other nano to micro scale features, as addressed in subsequent sections) to relax to pseudo-equilibrium. Such a philosophy provides the impetus for the use of waves in nondestructive testing: a cyclic perturbation (e.g., elastic, electronic, and magnetic) is applied to a system and the response is measured *dynamically*.

8.2.1.3 The Need for Multiple Measurements

Inherent to the nondestructive characterization of a physical parameter (composition, microstructural morphology, geometry, etc.) is the fact that measurements are dependent on more than just the variable of interest. The dramatic interplay between observable quantities and the microstructural features one wishes to infer demands the use of multiple independent techniques to fully assess a particular specimen. Frequently there is some overlap in available methods to quantify certain mesoscopic details as highlighted in Fig. 8.1.

The necessary advancement in use of physical measurements to assess materials is through the use and correlation of sufficiently distinct physical quantities to extract all the independent material variables. The situation is similar to the algebraic problem of having sufficient equations for a set of given unknowns. The combination of magnetic and elastic property measurements on the same material and with the same thermomechanical history offers a practical opportunity to achieve sufficient correlatable information for assessing the material without the use of material standards. Also, the use of property measurements based on wave analysis (electromagnetic and elastic waves) allows measurements to be made at different frequencies that interact with different specific microstructural details.

Physical measurements performed at different frequencies provide substantial information regarding the exact mechanisms by which the incident wave is affected by the material. Similarly, time domain analysis lends spatially resolved information about a particular specimen.

8.2.2 *The Wave Equation*

For the purposes of developing the necessary mathematical framework for wave theory, continued use shall be made of the particular case of linear elasticity. This case has been chosen for its "generality" in providing longitudinally polarized waves (electromagnetic waves are strictly transverse) and for its intuitive application of Newton's second law, which constitutes the starting point for the wave equation.

Consider an infinitesimal unit cube of an infinite single crystal with mass density ρ fourth-order elastic tensor[1] C, Newton's law gives

$$\frac{\partial \sigma_{ij}}{\partial x_j} = \rho \frac{\partial^2 u_i}{\partial t^2}, \tag{8.1}$$

for a stress σ_{ij} and corresponding displacement u_i in the $i, j \in \{1, 2, 3\}$ directions. Hooke's law relates the stress to the strain, ε_{kl}, by

$$\sigma_{ij} = C_{ijkl}\varepsilon_{kl}, \tag{8.2}$$

where the strain (to first order) is given by

$$\varepsilon_{kl} = \frac{1}{2}\left(\frac{\partial u_k}{\partial x_l} + \frac{\partial u_l}{\partial x_k}\right). \tag{8.3}$$

Combining (8.1) to (8.3) yields

$$\rho \frac{\partial^2 u_i}{\partial t^2} = \frac{1}{2} C_{ijkl}\left(\frac{\partial^2 u_k}{\partial x_j \partial x_l} + \frac{\partial^2 u_l}{\partial x_j \partial x_k}\right), \tag{8.4}$$

over all directions, establishing the differential equation which relates the temporal and spatial variation of displacement. Equation (8.4) is the *wave equation* for linear elasticity, and it has a direct analogue with the wave equation of electrodynamics as derived from Maxwell's equations (Sect. 8.2.6.1). Naturally, a wave-like solution to (8.4) $\left(\text{with amplitude } u_0 = \sqrt{\sum u_{0i}^2}\right)$ is presumed. Thus,

[1] See [34] for a thorough review of the tensorial nature of the elastic constants.

$$u_i(\mathbf{r},t) = u_{0i}\mathrm{e}^{\mathrm{i}(\mathbf{k}\cdot\mathbf{r}-\omega t)}; \quad \mathrm{i} = \sqrt{-1}, \tag{8.5}$$

where $\mathbf{k}$ and ω are the *wave vector* and *angular frequency* (given in radians per unit time), corresponding to the independent variables in the Fourier transforms of the *spatial* and *temporal* domains, respectively (more will be said about the Fourier domains in Sect. 8.2.4.)

Substituting (8.5) into (8.4) and implementing the Kronecker delta function,[2] δ, to ensure separation of the three directions produces three independent equations of the form

$$(\rho\omega^2\delta_{il} - C_{ijkl}k_jk_k)u_l = 0, \tag{8.6}$$

called the Christoffel equation, for which there exists a nontrivial solution only when the determinant of the coefficients is zero, or

$$|C_{ijkl}k_jk_k - \rho\omega^2\delta_{il}| = 0. \tag{8.7}$$

Equation (8.7) is termed the *dispersion equation* and has three roots for $\omega(\mathbf{k})^2$, corresponding to three "branches." For *each* branch, there exist three solutions to the Christoffel Equation (8.6) which relate the wave vector and the angular frequency. These solutions are the *dispersion relations* (i.e., the frequency-wave vector relationship) for the three (orthogonal) *polarizations*; introducing the *phase velocity* $c_p = \omega(\mathbf{k})/k$ gives the velocity of propagation for each polarization which, in general, is dependent on direction of propagation as shown in Fig. 8.2 for single crystal α-uranium, calculated using the elastic constants of Table 8.1.

8.2.2.1 Isotropic Media

For the very special case of a medium which exhibits isotropic symmetry (that is, it is elastically directionally independent), the preceding argument becomes far simpler. Material symmetry reduces the number of independent[3] constants from 21 (for triclinic symmetry) to 2, specifically a bulk modulus C_{11} and shear modulus C_{44} (the Poisson-type modulus $C_{12} = C_{11} - 2C_{44}$ for the isotropic case, but is otherwise independent.) In such a circumstance, the dispersion relation simplifies to

[2] The Kronecker delta function is defined as

$$\delta_{ab} = \begin{cases} 1 & a = b \\ 0 & a \neq b \end{cases} \forall\, a, b \in \mathbf{N}$$

[3] The tensorial symmetry, that is, $C_{ijkl} = C_{jikl} = C_{ijlk} = \ldots$ has already been considered here.

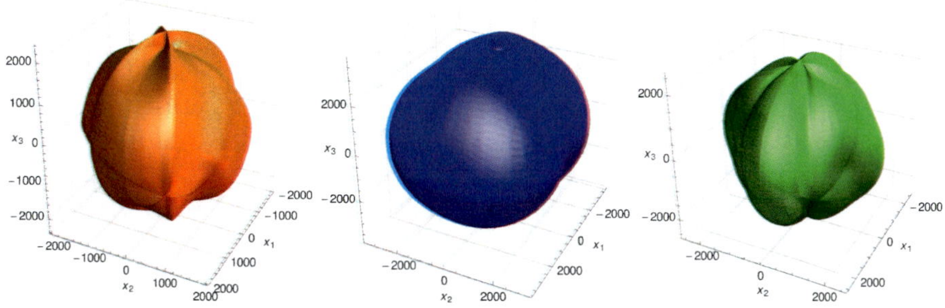

Fig. 8.2 Phase velocity surfaces for single crystal uranium in three dimensions for each branch of the dispersion equation (in units of m s^{-1}). The center figure corresponds to the longitudinal mode

Table 8.1 Single crystal elastic constants obtained by ultrasonic wave velocity at 298 K [23]

Modulus	Value (GPa)
C_{11}	214. 7 ± 0. 3
C_{22}	198. 5 ± 0. 3
C_{33}	267. 1 ± 0. 4
C_{44}	124. 4 ± 0. 1
C_{55}	73. 42 ± 0. 07
C_{66}	74. 33 ± 0. 07
C_{12}	46. 5 ± 0. 3
C_{13}	21. 8 ± 0. 3
C_{23}	107. 6 ± 0. 4

$c_p = \omega/k =$ (constant), and the three solutions to (8.6) collapse to one longitudinal and two degenerate transverse polarizations, with velocities

$$c_l = \sqrt{\frac{C_{11}}{\rho}}$$
$$c_t = \sqrt{\frac{C_{44}}{\rho}}, \tag{8.8}$$

respectively. Thus, the problem is reduced to two types of wave propagation: compressional waves and shear waves. Despite the highly anisotropic crystal structure of uranium, the remaining discussions shall deal primarily with isotropic symmetry for the sake of simplicity, albeit at the cost of generality. Accordingly, the pertinent elastic constants for polycrystalline uranium will be taken as $C_{11} = 222$ and $C_{44} = 84$ GPa [45], yielding longitudinal and shear wave velocities of 3, 410 and 2, 100 m s^{-1}, respectively. (The density of uranium is 19, 040 kg m^{-3}.)

8.2.3 *Frequency-Dependent Properties*

The above equations and relationships hold for situations where the elastic constants are not a function of frequency, but generally this situation is not the case. Indeed there are two broad phenomena in wave theory which manifest the frequency-dependent nature of the elastic constants (or, broadly, any linear functional relating a perturbation to its effect.) In the first of these cases, the constants (moduli) will be generalized to the complex domain, such that the imaginary term becomes an exponential decay term giving rise to attenuation. The second case allows for the moduli to change with frequency, i.e., the degree to which a system changes when perturbed is affected by how quickly the excitation is applied, which is termed dispersion.

8.2.3.1 Attenuation

Suppose that the (linear, for simplicity) proportionality coefficient between the perturbation P and effect E is now not only a function of frequency, but is also complex—contains a real and imaginary term. If a wave is applied, the imaginary terms in the wave perturbation function,

$$P(t) = \mathrm{e}^{\mathrm{i}(kx-\omega t)} = \cos kx - \omega t + \mathrm{i}\sin kx - \omega t,$$

multiplied by the imaginary term in the proportionality coefficient,

$$K(\omega) = K_{\mathrm{r}}(\omega) + \mathrm{i}K_{\mathrm{i}}(\omega),$$

yield a negative (but real) term in the exponential function:

$$E(t) = K_{\mathrm{r}}(\omega)\mathrm{e}^{-K_{\mathrm{i}}(\omega)t}\mathrm{e}^{kx-\omega t}. \tag{8.9}$$

Such a term describes the *attenuation* of the system, characterized by an exponential decay of the signal with time (or, more or less equivalently, with distance). Various symbols are used to denote the attenuation coefficient in literature instead of $K_{\mathrm{i}}(\omega)$ including a, $\mathcal{A}$, and α; the last of which is perhaps the most common and will be used throughout this text. Typical units for the attenuation coefficient are dB m^{-1} or dB μs^{-1}; when natural logarithms are used, as above, the Neper (Np) replaces the decibel, differing by a factor of 8.686.

In addition to the attenuation coefficient, other terms can be used to describe acoustic loss in materials. The quality factor Q and its reciprocal the internal friction Q^{-1} are frequently defined by the full width at half maximum (FWHM $= \Delta f$) for a particular resonant frequency f as $Q = f/\Delta f$ and $Q^{-1} = \Delta f/f$. The quality factor is related to the attenuation coefficient by

$$Q = \frac{\pi f}{\alpha} = \frac{\omega}{2\alpha}, \tag{8.10}$$

when α is given in units of Np s^{-1} and the resonant frequency f (or ω) is in Hz (rad s^{-1}).

8.2.3.2 Dispersion

At the beginning of this discussion, it was suggested that dynamic rather than static measurements could yield more information about a physical system. It was presumed, of course, that the response of the physical system depended on how quickly it was perturbed—a statement which is frequently justified. Suppose that τ is the time taken to change the system from state 1 to 2, then the proportionality constant between the perturbation $P(t)$ and its effects $E(t)$ is $K = K(\tau)$. If P is cyclic and the states 1 and 2 are one half-cycle apart, then τ describes the period and K becomes a function of the frequency of stimulation:[4] $K = K(\omega)$, where ω and τ form a Fourier pair.

Wave dispersion occurs when the speed of propagation (phase velocity) is a function of frequency. Dispersion may be caused to (1) macroscopic geometrical wave propagation effects, (2) material parameters that are rate dependent (due to attenuation as a function of frequency, for example), (3) and other effects beyond the scope of this discussion. The end result is that dispersion is a complication that can provide additional useful information through NDE, or conversely can lead to inaccuracies in determining material properties, such as elastic moduli. The behavior so described introduces the concept of *dispersion* as it is applied to material properties which are *time-dependent*.

8.2.4 Fourier Analysis

An exceptionally powerful device, the Fourier Transform has found countless applications. With the development of Fast Fourier Transform (FFT) algorithms, it has become ubiquitous in signal processing schemes. A review of its properties and usage in NDE is provided, followed by a brief discussion of an "extension" to the Fourier Transform: the Wavelet Transform.

---

[4] If $K = K(\omega, P(t))$, i.e., the proportionality "constant" is *dependent* on the perturbation, the system becomes *nonlinear* (Sect. 8.3).

8.2.4.1 Mathematical Background

Formally, the *Fourier Transform* $F(\omega)$ and its inverse are defined for a function $f(t)$,

$$F(\omega) \doteq \int_{-\infty}^{\infty} f(t)\mathrm{e}^{-2\pi \mathrm{i}\omega t}\,\mathrm{d}t \quad \text{and}$$
$$f(t) \doteq \int_{-\infty}^{\infty} F(\omega)\mathrm{e}^{+2\pi \mathrm{i}\omega t}\,\mathrm{d}\omega, \tag{8.11}$$

where t and ω represent independent variables in the time and frequency domains, respectively, and are easily replaced with the variables x and k to refer to real and reciprocal (wavevector/momentum) as appropriate. The function $f(t)$ can, by virtue of the cosinusoidal nature of the transform, be represented as a Fourier *series* of an infinite number of sine and cosine coefficients:

$$f(t) = a_0 \sum_{n=1}^{\infty} a_n \cos \pi t + b_n \sin \pi t, \tag{8.12}$$

where the coefficients a and b are given by

$$a_n = \int_0^{\infty} f(t) \cos 2\pi n t\,\mathrm{d}t \tag{8.13}$$

$$b_n = \int_0^{\infty} f(t) \sin 2\pi n t\,\mathrm{d}t, \tag{8.14}$$

provided the function $f(t)$ meets the following (Dirichlet) conditions:

- It must have a finite number of discontinuities and extrema over any discrete interval, often taken as $[0, 2\pi]$ or $[-\pi, \pi]$.
- The integral

$$\int_0^{\pi} f(\theta)\,\mathrm{d}\theta$$

must converge (i.e., f must be integrable over the interval.)

Any function which is expressible as a Fourier series can be sampled over a finite interval and transformed according to the *discrete* Fourier Transform (DFT):

$$F(\omega) \doteq \sum_{n=-\infty}^{\infty} f_n \mathrm{e}^{-\mathrm{i}\omega n}, \tag{8.15}$$

which is the form most useful to the field of NDE as signals encountered are of a finite length and are sampled at discrete time intervals Δt.

8.2.4.2 Correlation Theorem

An incredibly useful property of the Fourier Transform is found in the *correlation theorem*, provided here without proof. The cross-correlation operation[5] on two functions $f(t)$ and $g(t)$, denoted by the $\star$ symbol, is defined

$$f \star g \doteq \int_{-\infty}^{\infty} f^*(\tau) g(t-\tau)\,\mathrm{d}\tau = g \star f, \tag{8.16}$$

with * the complex conjugate operator and τ a dummy variable for the independent variable of the functions f, g, and $f \star g$. Cross-correlation can be thought of as the multiplication of two functions, one of which is shifted incrementally in the independent variable. A variety of signal processing applications call for such an operation, as one may wish to correlate two signals, one of which may be a reference or even the signal itself (known as *autocorrelation*).

For the discrete case where m and n shall refer to indices in the ordered lists f and g,

$$f \star g[n] \doteq \sum_{m=-\infty}^{\infty} f^*[m] g[n-m]. \tag{8.17}$$

Equation (8.17) has the undesirable property that its computation requires a summation over all points for *each* point in the correlation, meaning the operation is achieved in $O(N^2)$ time—prohibitive for long sequences! Fortunately, the FFT (which scales as $O(N\log N)$ time) presents a much more expedient algorithm by virtue of the Correlation Theorem:

$$f \star g(t) = F^{-1}[F^*(\omega)\, G(\omega)], \tag{8.18}$$

where the operator F^{-1} is the inverse Fourier Transform. Thus by using the FFT to quickly perform the cross-correlation operation, one gains the capacity for rapid denoising and feature detection for time-domain analysis.

[5] Cross-correlation is very similar to convolution; the latter does not involve the complex conjugate.

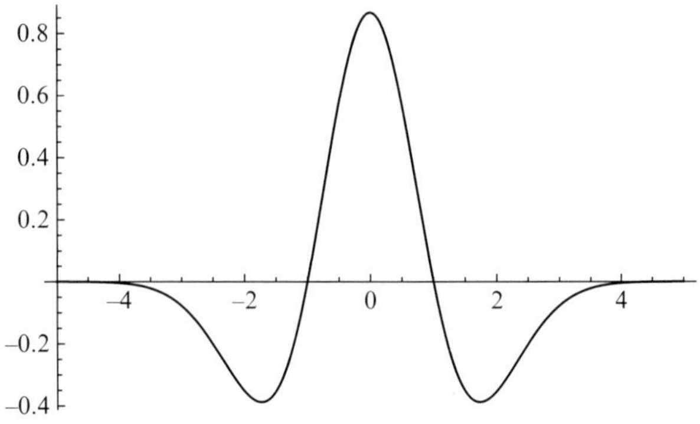

Fig. 8.3 The Ricker wavelet, also known as a "Mexican Hat" wavelet, defined by (8.19)

8.2.4.3 Wavelet Transform

Closely related to the Fourier analysis, the *wavelet transform* offers an extension to spectral analysis to signals whose spectral content changes in time by virtue of its multiresolution capabilities. Just as the Fourier transform uses sinusoidal functions which are "stretched" in the time domain to varying degrees, the wavelet transform stretches and shifts an *arbitrary* function to form its basis functions. The arbitrary function chosen is known as a mother wavelet, and its selection depends on the application. In practical terms, the continuous wavelet transform (CWT) can be considered a convolution of the scaled (stretched in time) mother wavelet with a signal for different widths (thus forming another useful application of the FFT). Each scaled component of the transform represents a distinct spectral component (analogous to frequency in the Fourier transform) and by the convolution gives the variation of the spectral energy in time.

One powerful application of the CWT is the detection and identification of specific features within a signal. If the mother wavelet is chosen such that it accurately replicates the targeted element, such as an echo in the case of pulse-echo experiments, then high values in the CWT correspond to regions in the signal where that feature is present. One common wavelet is the Ricker wavelet, described by the formula

$$\phi(t) = \frac{2}{(3\sigma)^{1/2}\pi^{1/4}}\left(1 - \frac{t^2}{\sigma^2}\right)e^{-\frac{t^2}{2\sigma^2}}, \tag{8.19}$$

for the scale parameter σ; its shape, shown in Fig. 8.3, has given it the colloquial term "Mexican Hat" wavelet.

For an ultrasonic pulse-echo signal, such as is shown in Fig. 8.4, the transit times of an echo may be difficult to extract using ordinary thresholding methods, as the echoes are not clearly distinguishable in the time-voltage domain. The Ricker wavelet CWT of Fig. 8.7 clearly identifies the echoes, and with the aid of a suitable algorithm can yield quantitative results [19]. The CWT is contrasted with the Fourier transform at bottom, which provides the spectral content of the waveform, but does not immediately reveal temporal information.

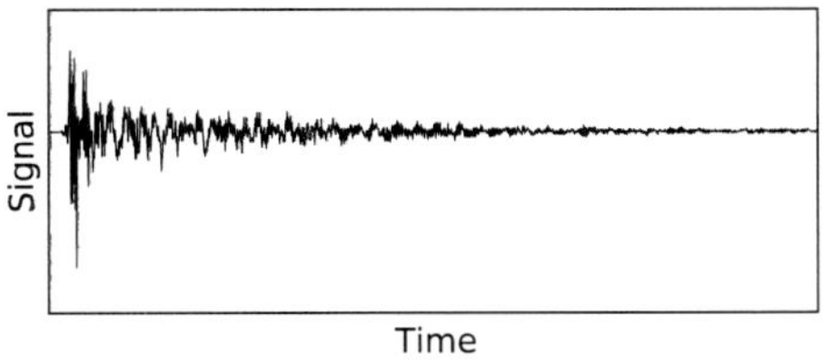

Fig. 8.4 A typical noisy pulse-echo experimental signal. Transits of the 5-cycle burst are unclear

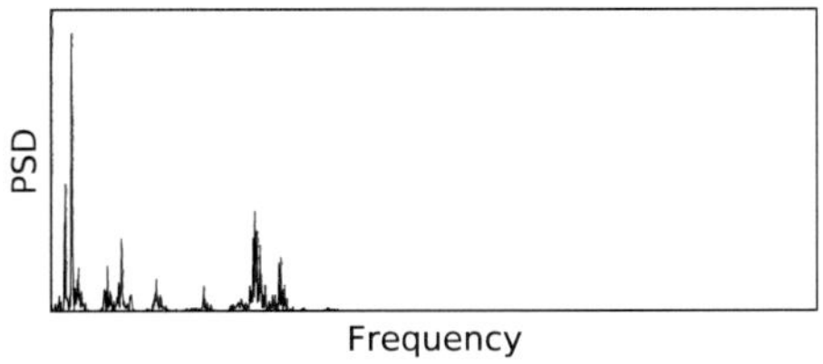

Fig. 8.5 Spectral content (via the Power Spectral Density) for the signal above

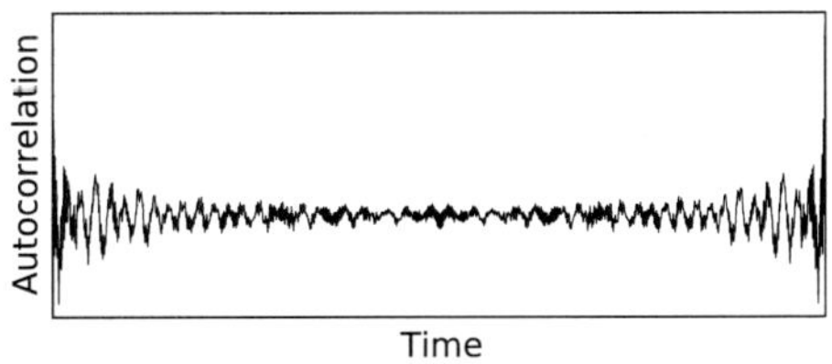

Fig. 8.6 Autocorrelation (*circular*) of the signal in Fig. 8.4

8.2.4.4 Applications in NDE

The Fourier transform is a useful tool in the analysis of various waveforms obtained by wave-based NDE techniques. Often it is the mathematical properties of the Fourier transform which allow for fast computation of various signal processing routines such as denoising and feature recognition, but information about certain waveforms can also be obtained by examining the Fourier spectrum itself. This section highlights a small selection of the many applications of the Fourier transform.

Having defined the Fourier Transform and some of its properties, attention is turned to its practical application in NDE. Returning to the noisy signal and spectrum of Figs. 8.4 and 8.5, the corresponding discrete circular autocorrelation is computed in Fig. 8.6, where the delay times can be observed, albeit with some additional noise (likely mixed-mode ultrasonic waves). The Ricker-based CWT is shown in Fig. 8.7, where the echoes are clearly made visible, due to the abilities of the wavelet transform to isolate and characterize features in a noisy environment.

For samples excited by broadband wave pulses (i.e., packets containing many frequencies), the sample will *resonate* at particular frequencies which are determined by sample geometry and elastic properties. Such a spectrum is provided by the Fourier transform shown in Fig. 8.8.

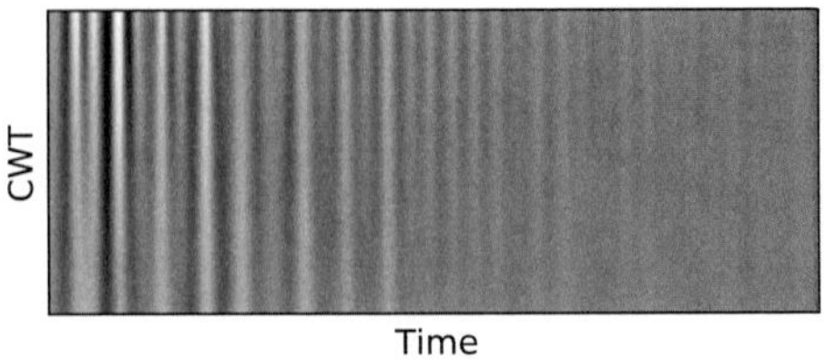

Fig. 8.7 Continuous wavelet transform (using the Ricker wavelet) of the original signal. The ordinate corresponds to different values of σ in (8.19) (i.e., different widths)

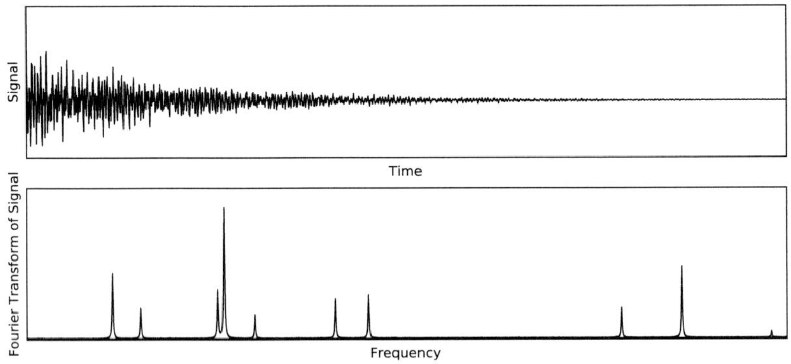

Fig. 8.8 Received signal and corresponding Fourier spectrum for a broadband-excited specimen. Resonant frequencies and quality factor are determined by geometry and elastic tensor, either of which can be inferred by resonant ultrasound spectroscopy

Measurement of the resonant frequencies a sample and quantification of the quality factor, along with prior knowledge of the sample geometry can yield the elastic (and anelastic) properties of the material. Resonant ultrasound spectroscopy (RUS) constitutes one method by which the full elastic and anelastic (i.e., complex) tensors can be inferred [41].

8.2.5 Elastic Waves

In Sect. 8.2.2, the solution (8.5) was derived for the wave equation (8.4) as a plane wave propagating in a definite direction with a velocity c, which varies with elastic properties. As the wave continues through the bulk of the material, it is scattered by discontinuities in physical properties (e.g., inhomogeneities, boundaries) and attenuated by various mechanisms. While these processes appear at first to be detrimental to the application of elastic waves to NDE by increasing the analytical complexity of the problem, they are in fact responsible for providing information regarding the microstructure through their analysis. This section serves to outline some practical aspects of NDE using ultrasound and selected phenomena which may be isolated and characterized to infer the state of materials.

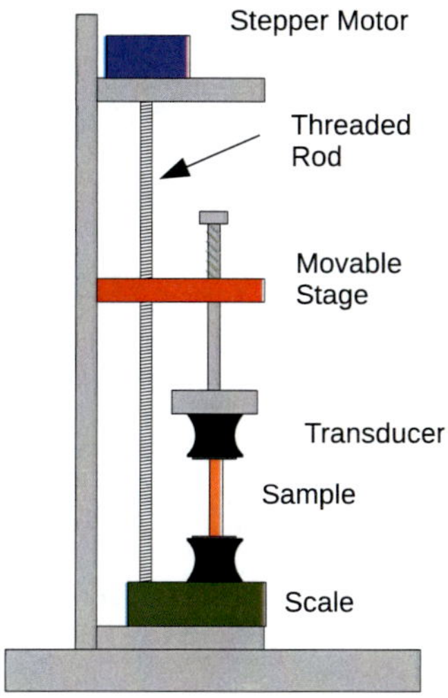

Fig. 8.9 An apparatus used for pulse-echo and resonant ultrasound measurements utilizing piezoelectric transducers. The scale and threaded rod system allow for precisely controlled mountings to reduce variation in coupling effects

8.2.5.1 Generation and Detection

An acoustic (usually ultrasonic) wave can be produced by a variety of methods and mechanisms, but its source falls into one of the two categories: *active* ultrasound, whereby the wave is synthesized and imposed upon the sample under test, and *passive* ultrasound, where the material itself instantiates a wave or series of waves. The former case may consist of a number of techniques and apparatus, each of which will be discussed briefly. For the case of passive ultrasound, only acoustic emission will be mentioned.

High frequency elastic waves can be deliberately introduced into the material and detected by a handful of means, either through direct physical contact between the transducer and sample or indirectly via electromagnetic radiation.

(PZTs) are a contacting method and are a cost-effective means of delivering ultrasonic waves into a material. PZTs are available in a wide variety of form factors, frequency ranges, and require relatively simple equipment for their excitation. A typical setup consists of a function generator, power amplifier, and digitizer (shown in Fig. 8.10), which provides the capacity for waveform selection, or a simple pulse-receiver. A major weakness of a configuration incorporating PZTs is that the transducers must be well coupled to the sample like that shown in Fig. 8.9 which can be inconvenient, inconsistent between loadings, or even destroy the transducer in the case of elevated temperatures or high radiation environments.

On the other hand, high intensity lasers provide a means of introducing ultrasonic waves without touching the sample. Delivery of focused, high-intensity light to the sample surface causes highly localized heating (even ablation) which, for sufficiently short pulses, excites broadband ultrasonic waves in the material. Naturally, such a system is significantly more expensive and difficult to implement, as the laser itself (typically a pulsed Nd:YAG laser) is bulky and costs several tens of

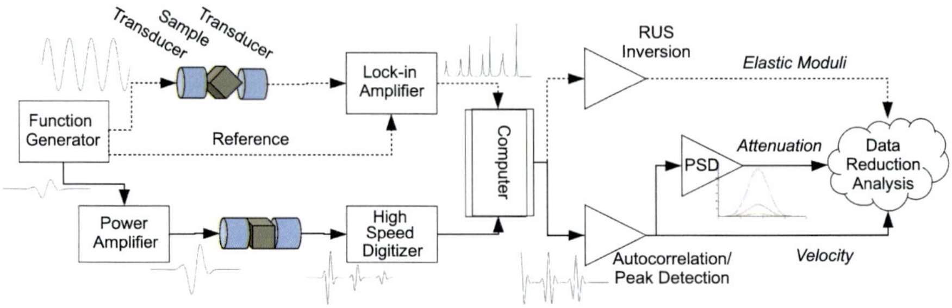

Fig. 8.10 Equipment and functional diagram for RUS (*dotted line*) and pulse-echo (*solid*) experimentation

thousands of dollars. Mode specificity is also difficult with laser-based ultrasonic systems, with mixed shear and longitudinal waves of many frequencies generated.

Strictly speaking, acoustic emission (AE) techniques are passive, as they rely on the detection of acoustic events which are developed within the material itself in response to a stress (compressive/tensile, thermal, corrosive). In practice, however, the applied stress is present in sufficient quantity to generate acoustic emissions only when stress- or proof-testing a material. In this sense, AE constitutes an active form of semi-nondestructive testing. Uranium may well be an ideal candidate for certain (very specialized) applications of AE testing, as the acoustic waves are generated by the movement or activation of slip systems, twins, microcracks, and other microstructural features which abound in the metal.

Two techniques for detecting ultrasonic waves find widespread use in the NDE community. PZTs can be used rather effectively as receivers by simply placing them in contact with the sample. Here again, the physical connection between transducer and material under test is critical, but the low cost of PZTs encourages their use in systems where numerous simultaneous measurements are required. For non-contact applications, laser doppler vibrometers (LDV) provide a means of detecting ultrasonic waves at a distance through interferometry. Once again, the laser-based detection scheme provides a technique for eliminating the transducer-sample coupling variable, and an added benefit: an LDV can be scanned (rastered) across the sample surface to yield spatial information, effectively constituting an array of ultrasonic detectors.

Significant elastic information can be obtained when a traveling wave reflects back onto itself and forms a standing wave within the material, occurring at particular frequencies for a given geometry and stiffness tensor. Termed resonance, this situation allows for the direct inference of the elastic properties of a material through a technique called resonant ultrasound spectroscopy (RUS) when a suitable computational model is available for the prediction of the resonant frequencies [41]. Figure 8.10 illustrates the flow of data and equipment used in a contacting RUS setup.

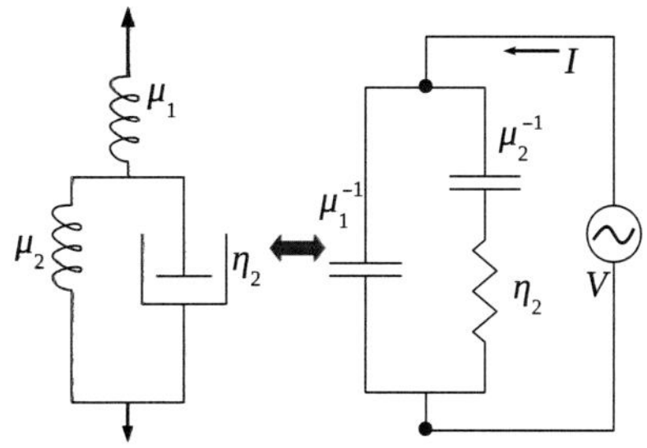

Fig. 8.11 Schematic of Zener's linear elastic solid. The mechanical system (*left*) consists of two springs and a dashpot. The electrical analog (*right*) bears two resistors and a capacitor. Adapted from [39]

8.2.5.2 Interactions with Matter

There are several processes at work which cause a traveling wave to decrease in amplitude with time, some of which are reversible—preserving energy in the form of an elastic wave—and others which are *absorptive*, meaning they convert the energy into heat. Scattering from inhomogeneities (discontinuities in relevant properties) constitutes an example of the former, and its characterization forms a limited (yet fairly well established) means of detecting cracks, grain size, and porosity. Absorptive/dispersive processes, on the other hand, are *material*-dependent phenomena which—if one can accurately isolate them for a given sample—can provide insight into the fundamental physical parameters that cause them.

Broadly, absorptive processes can be divided categorized by their frequency dependence into two distinct categories: relaxation (which includes thermoelasticity, electron- and phonon interactions) and resonant phenomena (primarily dislocation interactions in the field of physical acoustics). A phenomenological description of relaxation behavior will be introduced (owing to its ubiquity in attenuative processes), followed by the introduction of various phenomena and their mathematical relationships with relevant physical parameters. It is proposed that many of the processes discussed may offer substantial opportunities for uranium NDE through careful quantification of the absorption and dispersion as a function of frequency.

8.2.5.3 Relaxation Phenomena

Various ultrasonic loss mechanisms share a common phenomenological root: that of the linear elastic solid as described by Zener [54]. The behavior can be summarized in the following way [39]: imagine a spring with stiffness constant μ_1 in series with another spring (μ_2) and a dashpot of viscosity η_2 in parallel with μ_2, as shown schematically in Fig. 8.11. Upon applying an instantaneous stress, the first spring yields a certain amount of instantaneous displacement. The persistence of this stress will allow time for the dashpot to slowly yield, until the overall (relaxed) spring constant of the system becomes $\mu_1\mu_2/\mu_1 + \mu_2$. Releasing this stress results in the reverse of this process. For a sinusoidally varying stress, one can imagine the electrical analog shown in Fig. 8.11. Thus one can elicit the impedance:

$$\sigma_{12} = \frac{\mu_1\mu_2}{\mu_1 + \mu_2}\left(\frac{1 + \frac{i\omega\eta_2}{\mu_2}}{1 + \frac{i\omega\eta_2}{\mu_1+\mu_2}}\right)\epsilon_{12}, \tag{8.20}$$

which has the appearance of Hooke's Law, albeit for a complex shear modulus (C_{44}.) Note that as $\omega \rightarrow 0$, the modulus reduces to the relaxed modulus, consistent with intuition. Introducing the relaxation times $\tau_S = \eta_2/(\mu_1 + \mu_2)$ and $\tau_T = \eta_2/\mu_2$ into (8.20) yields the expression

$$\sigma_{12}(1 + i\omega\tau_S) = \frac{\mu_1\mu_2}{\mu_1 + \mu_2}(1 + i\omega\tau_T)\epsilon_{12}, \tag{8.21}$$

which can then be rearranged into

$$\frac{\sigma_{12}}{\varepsilon_{12}} = \bar{\mu} = \mu_R\left(\frac{1 + i\omega\tau_T}{1 + i\omega\tau_S}\right), \tag{8.22}$$

where μ_R is the relaxed stiffness constant $\mu_1\mu_2/\mu_1 + \mu_2$. Multiplying by $(1 - i\omega\tau_S)/(1 - i\omega\tau_S)$ gives

$$\bar{\mu} = \mu_R\left(\frac{1+\omega^2\tau_S\tau_T + i\omega(\tau_T - \tau_S)}{1+\omega^2\tau_S^2}\right). \tag{8.23}$$

The internal friction, Q^{-1}, and the attenuation coefficient, α, are then given by the ratio of the imaginary to real parts of the impedance viz. the phase lag from the electrical analogy:

$$\alpha \equiv \frac{\omega}{2Q} = \frac{\mu_1}{\sqrt{\mu_2(\mu_1 + \mu_2)}}\frac{\omega^2\tau}{1+\omega^2\tau^2}; \tau = \left(\frac{\eta_2^2}{\mu_2(\mu_1 + \mu_2)}\right)^{1/2}, \tag{8.24}$$

for which the internal friction Q^{-1} has a maximum at $\tau = 1/\omega$, and the attenuation coefficient is well on its way to a maximum. The more general form of (8.24) is effectively

$$\alpha = \frac{1}{2c}\frac{\Delta\mu}{\mu_0}\frac{\omega^2\tau}{1 + \omega^2\tau^2}, \tag{8.25}$$

where c is the velocity of sound and $\Delta\mu = \mu_1 - \mu_0$ is the difference between the unrelaxed and the relaxed modulus, respectively. For the purposes of semi-empirical parameterization, the constants can be combined to provide an "effective" phenomenological equation,

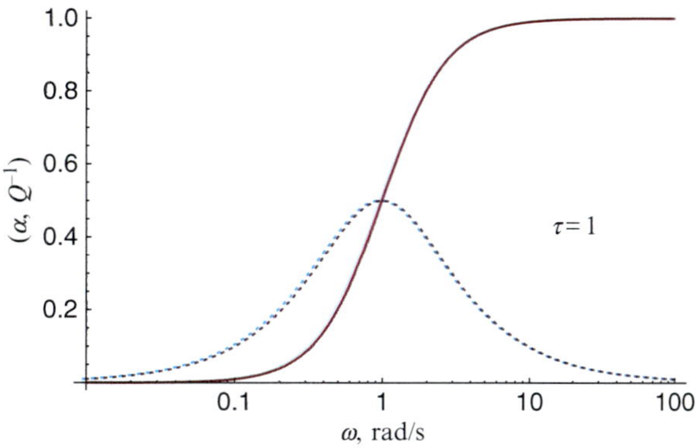

Fig. 8.12 Attenuation, α (*solid line*), and internal friction, Q^{-1} (*dashed line*) as a function of frequency for a loss process that follows a relaxation-type behavior, per (8.26)

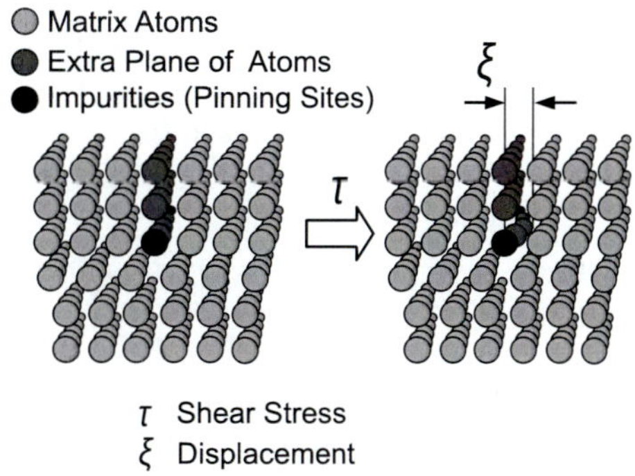

Fig. 8.13 Illustration of an edge dislocation ($\perp$) in a simple cubic lattice, showing an extra plane of atoms. A shear stress τ acts on the slip plane causing the dislocation to bow out while pinned at impurity atoms (defects). The tension of the dislocation (acting out of the paper) is $2Gb^2/\pi(1-\nu)$

$$\alpha = K_i \frac{\omega^2 \tau_i}{1+\omega^2 \tau_i^2}, \tag{8.26}$$

where K_i and τ_i are constants (actually the product of many constants) for some phenomenological process *i*. The general behavior is plotted for the internal friction Q^{-1} and the attenuation coefficient in Fig. 8.12, where the maximum in the internal friction for a relaxation occurs at $\omega\tau = 1$. For some processes, this maximum can be observed at low ultrasonic frequencies at room temperature. Other processes are only observable at very low temperatures (often below 20 K) or at high frequencies (several tens of MHz).

8.2.5.4 Dislocations

Line defects consisting of planes inserted into an otherwise perfect lattice are known as edge *dislocations* and are present in all crystals (Fig. 8.13). These defects are made immobile or "pinned" at locations where they cross each other and at inclusions or impurity atoms which can migrate to the line. At room temperature acoustic energy losses due to dislocations become significant and are observed in the MHz frequency range. Two distinct phenomena have been described [28] which

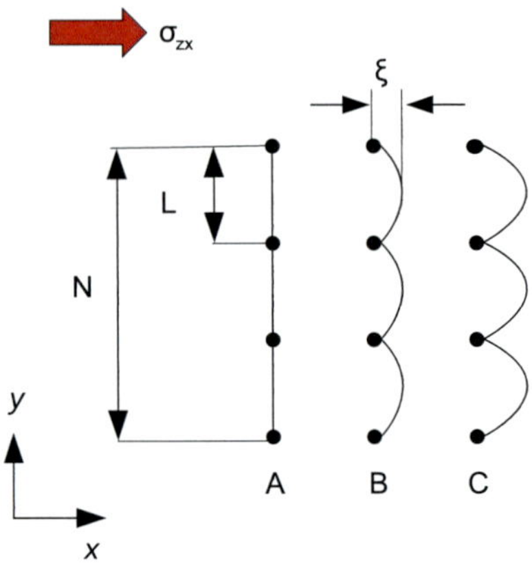

Fig. 8.14 Strain-independent two-dimensional square lattice exhibiting resonance of a vibrating dislocation

contribute a frequency-dependent and strain-dependent (frequency independent) ultrasonic energy loss. For small stresses, the dislocation behaves as a resonating string by bowing out and back from pinning sites, and losses result from phase lag between the defect motion and the cyclic elastic wave. This loss is frequency dependent and is characterized by its resonant behavior. When stresses are sufficient to break the dislocation free from the impurity atoms (but not the *entire* network), there results in an increase in strain without an increase in stress. When the dislocation relaxes back to its equilibrium position (by virtue of the "tension" described below), it completes the stress–strain hysteresis loop of Fig. 8.16. The area enclosed by this loop corresponds to the energy lost per cycle, yielding a frequency-independent loss in the material. Impurity and defect concentrations will affect the length of the dislocation loops, thereby altering their resonant frequencies and changing the breakaway stress: features which could be reliably correlated and identified in the development of a nondestructive tool.

Dislocations exhibit an effective mass and tension, which acts to restore any displacement induced by a stress. This behavior is shown schematically in Fig. 8.14 (A) to (C) for a stress in the direction of dislocation displacement. When an applied stress is cyclic (C) to "(− C)," the system can be thought of as a damped forced oscillator, corresponding to a string with mass $\pi\rho b^2$ and a restoring tension $2Gb^2/\pi(1-\nu)$, with a damping coefficient B and Burgers vector b. Thus as a driven oscillator, a dislocation exhibits resonant behavior, described by the differential equation for the displacement ξ:

$$\pi\rho a^2\frac{\partial^2\xi}{\partial t^2}+B\frac{\partial\xi}{\partial t}-\frac{2Gb^2}{\pi(1-\nu)}\frac{\partial^2\xi}{\partial y^2}=b\sigma, \tag{8.27}$$

which has solution

$$\xi(t)=4b\sigma\sum_{n=0}^{\infty}\frac{1}{2n+1}\sin\frac{(2n+1)\pi y}{l}\frac{e^{\mathrm{i}(\omega t-\delta_n)}}{\left[(\omega_n^2-\omega^2)^2+(\omega d)^2\right]^{\frac{1}{2}}}, \tag{8.28}$$

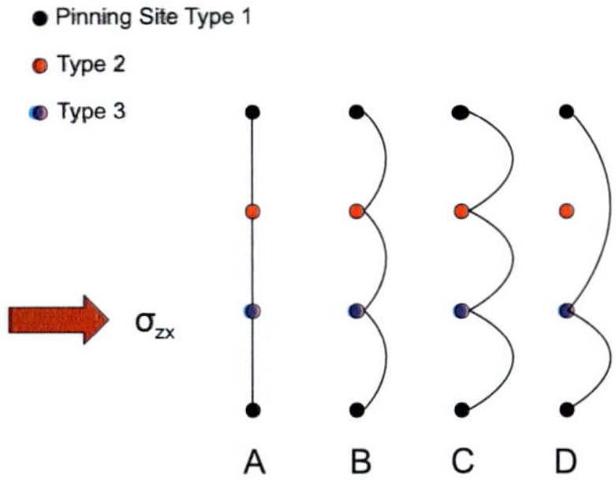

Fig. 8.15 Schematic of the unpinning of a dislocation from impurity sites due to increasing stress. From [28]

where $\delta_n = \tan^{-1}[\omega d/(\omega_n^2 - \omega^2)]$ and $d = B/M = B/\pi\rho b^2$ is the mass-normalized damping coefficient. Deriving the attenuation due to the vibration of the dislocation [6]:

$$\alpha_{\text{disl}} \approx \frac{32}{\pi^5}\frac{NL^5}{Gb^2}\frac{\omega B}{\left(1 - \frac{\omega^2}{\omega_0^2}\right)^2 + \left(\frac{\omega B}{\omega_0^2 M}\right)^2}, \tag{8.29}$$

which has a maximum at $\omega = \omega_0$, the resonant frequency of the dislocation (typically on the order of 10–100 MHz.)

The parameters of interest in (8.29) are the total length of the dislocation loops L and of course the resonant frequency of the dislocation, given by

$$\omega_0^2 = \frac{\pi}{2}\frac{G}{L^2\rho}, \tag{8.30}$$

where L is the length of an individual dislocation loop pinned by intersecting dislocations and impurity atoms. Additional defects will decrease this length, thus increasing the resonant frequency. Parameterization of (8.29) to experimental data can thus yield dislocation information including the dislocation density, viscosity, and loop length; such analysis is provided in Sect. 8.5.1.

At high enough strains, the dislocation can break free from the impurity atoms and the smaller dislocation loops can coalesce (C) to (D) in Fig. 8.15. This occurrence leads to an increase in strain with no additional stress as shown in Fig. 8.16. Upon release of the stress during the second half of the cycle, the dislocation relaxes back to its equilibrium position by virtue of its tension, thus closing a hysteresis loop in the stress–strain relationship—the area enclosed by this loop is the energy lost per cycle.

Absorption due to breakaway of dislocations is not frequency dependent as it does not rely on the resonant nature of dislocation vibration. Clearly the attenuation is strain dependent, however, as sufficient strain is required to break the dislocation free from its impurity pinning points. Furthermore, breakaway from impurity atoms of different defects should occur at different strains owing to differences in their radii and respective precipitate sizes, which will have different pinning binding energies given by

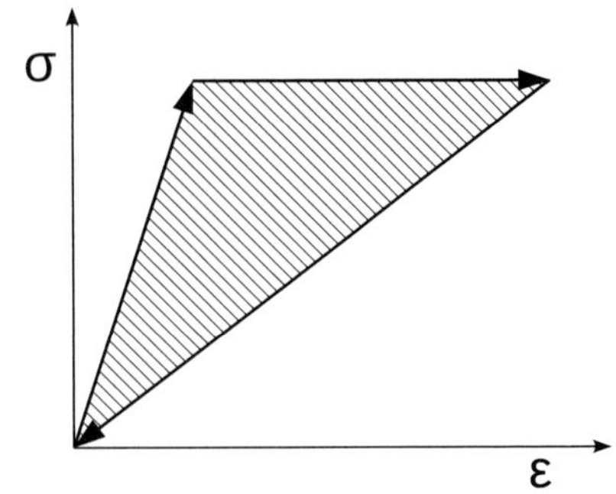

Fig. 8.16 Stress–strain relationship for the hysteretic strain-dependent loss due to dislocation loop coalescence. The *hatched* area corresponds to the energy lost each cycle. From [28]

$$U_{\text{bind}} = \frac{1}{3}\left(\frac{1+\nu}{1-\nu}\right) Gb^3 \frac{r_B - r_A}{r_A}, \tag{8.31}$$

where r_A and r_B are the radii of the matrix and foreign atoms, respectively. It is therefore proposed that by varying the strain amplitude in addition to the frequency of the incident elastic wave, the relative concentrations of the crystal defects may be better determined.

8.2.5.5 Phonons

In the long wavelength limit, where $\omega\tau \gg 1$, phonon interactions play a significant role in acoustic wave losses [5]. In insulators in particular, where phonon conduction plays the dominating role in heat transport, elastic waves attenuate according to the redistribution of phonon modes, a phenomenon which arises from the nonlinear strain dependence of phonon frequencies. For a given elastic wave propagating through a material, the time-varying strain induces a change in the phonon frequency distribution according to the Grüneisen parameter Γ. Specifically,

$$\Gamma(\mathbf{q}, p) = \frac{1}{\omega_0(\mathbf{q}, p)} \frac{\partial\omega(\mathbf{q}, p)}{\partial\epsilon_{jk}}, \tag{8.32}$$

for a given phonon mode $(\mathbf{q}, p)$, where ω_0 is the frequency in the unstrained state. Certain phonon modes in uranium demonstrate a very strong dependence on strain in this regard, as discussed in Sect. 8.4.4.1 on the charge density wave at low temperature.

One wishes to obtain the resulting changes in elastic moduli, as in (8.25). Substituting i for the phonon mode $(\mathbf{q}, p)$ in the sake of clarity, the following relationships for transverse and longitudinal waves,

$$\Delta C_{jj} = \frac{3U_0}{N} \sum_{(i)} [\Gamma_j(i)]^2 \qquad \Delta C_{jj} = \frac{3U_0}{N} \sum_{(i)} [\Gamma_j(i)]^2 - \Gamma^2 C_V T, \tag{8.33}$$

have been obtained [5]. Here the Grüneisen parameter $\Gamma_{j(i)}$ is given by

$$\Gamma_j(i) = -\left[\frac{1}{c_i}\frac{\partial c_i}{\epsilon_j} - \frac{1}{l_0}\frac{\partial l}{\partial \epsilon_j}\right], \tag{8.34}$$

and the thermal relaxation times to be used in (8.25) are determined

$$\tau_{\text{ph,T}} = \frac{3\kappa}{C_P c_0^2} \qquad \tau_{\text{ph,P}} = \frac{6\kappa}{C_P c_0^2}, \tag{8.35}$$

for the transverse and longitudinal modes, respectively. Here N is the number of modes to be summed, U_0 is the expression for the thermal energy from the Debye approximation, l_0 is a characteristic path length, c_0 is the Debye velocity, and κ is the thermal conductivity. Given that phonon transport is the primary means of conduction for the intermetallic compounds, the relative volume fractions of each of these constituents could conceivably be measured using the effects of acoustic wave attenuation by phonon viscosity.

8.2.5.6 Thermoelasticity

Differences in the degree of compression between adjacent regions in a material as a longitudinal wave propagates (or as a rod is in a bending resonance) produce spacial variations in temperature which lead to the flow of thermal energy, ultimately attenuating an elastic wave. There is a strong frequency dependence for this phenomenon, reaching a maximum between two extreme cases: isothermal or adiabatic vibrations. In the low frequency extreme, regions of sufficiently different temperatures are far enough apart that for all practical purposes the sample experiences isothermal conditions. For high frequencies, the regions are not given sufficient time to equilibrate, and the conditions are adiabatic. In the regime between these two extremes, thermoelastic attenuation (as internal friction) reaches a maximum value.

The loss due to thermoelastic effects was determined by Zener [52] via a thermodynamic analysis of heat as a function of stress. Considering the heat generation per cycle δ, the internal friction can be written as $Q^{-1} = 1/2\pi\delta$; for the case wherein temperatures vary on a macroscopic scale, the derivation produces the following relationship for attenuation [5]:

$$\alpha_{\text{th}} = \frac{\Gamma^2 C_V T}{2\rho c^3}\frac{\omega^2 \tau_{\text{th}}}{1+\omega^2\tau_{\text{th}}^2}, \tag{8.36}$$

where Γ is the Grüneisen parameter, C_V the heat capacity at constant volume, T the temperature, ρ the mass density, and the velocity of sound is $c = (C_{11}^S/\rho)^{1/2}$ with $C_{11}{}^S$

the adiabatic elastic constant and $\tau_{\mathrm{th}} = 2\rho C_P l^2/\pi\kappa$ is the characteristic relaxation time. The relaxation time is a function of thermal conductivity κ, the heat capacity at constant pressure C_P, the density, and a characteristic length l. Grain boundaries also exhibit a thermoelastic loss between individual crystallites depending on the degree of anisotropy which is known as the Zener intercrystalline thermoelastic effect and is addressed separately.

It is worth considering some of the ramifications of such an effect. For instance, due to the stipulation that thermoelastic losses must occur in samples wherein there exist regions of compression countered by regions of dilatation, only certain resonant modes should exhibit this phenomenon. Extensional modes and flexural modes, for example, manifest such regions parallel to and orthogonal to the axis of longitudinal wave propagation, respectively. Resonant torsional modes, however, do not exhibit any compression or dilatation, as the deformation is strictly shear. This effect could therefore be separated from other effects by selectively analyzing the attenuation for resonant modes (and their harmonics) which do exhibit thermoelastic losses (as per the FWHM of the resonant peak using RUS) versus those modes which do not. Such an analysis may yield the various relaxation times and coefficients needed to fully characterize experimental data obtained by pulse-echo techniques.

8.2.5.7 Zener Intercrystalline Thermoelastic Effect

In polycrystalline metals, adjacent crystallites can be compressed by an elastic wave by differing amounts, depending on the degree of single crystal elastic anisotropy for the material [44]. This effect is phenomenologically identical to the thermoelastic loss [53] of the foregoing paragraphs, differing only by the length scale involved. Heat transfer in the case of the Zener effect is between microscopic grains, instead of between more macroscopically scaled regions of a sample. The concept is shown schematically in Fig. 8.17 for a two-dimensional rectangular lattice. As before, the lag between the transfer of thermal energy and the elastic wave perturbation leads to the acoustic attenuation provided by (8.37).

$$\alpha_{\mathrm{zener}} = \frac{C_P - C_V}{C_V}\,\frac{\Omega}{2v}\,\frac{\omega^2\tau_{\mathrm{zener}}}{1+\omega^2\tau_{\mathrm{zener}}^2}, \tag{8.37}$$

where C_P and C_V are the heat capacities at constant pressure and volume, $\tau = C_P \langle D\rangle^2/2\pi\kappa$ where $\langle D\rangle$ is the average grain diameter and κ is again the thermal conductivity. The quantity Ω is an elastic anisotropy factor, determined over all of the grain orientations and is given by [52]

$$\Omega = \frac{4}{45}\,\frac{1+\nu}{1-2\nu}\left(\frac{C_{44} - \frac{1}{2}(C_{11} - C_{12})}{C_{11}}\right)^2, \tag{8.38}$$

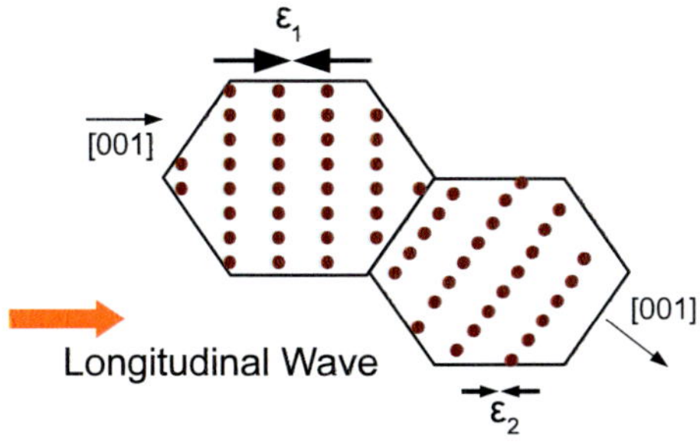

Fig. 8.17 Illustration of the Zener intercrystalline thermoelastic effect. A longitudinal wave propagating in one direction relative to the first grain (in this case parallel to the [100] direction) propagates in a different direction relative to a crystallite of a different orientation (the $[\bar{1}10]$.) The degree of anisotropy determines the extent of the difference in compression between each grain

Adjacent crystallites in uranium exhibit extreme anisotropy ($\Omega \approx 1.5 \times 10^{-4}$), making this effect pronounced in the sub-100 kHz frequency regime, depending on grain size. Additionally, the formation of uranium compounds, which have radically different thermal conductivities and heat capacities from the matrix, could dramatically affect this phenomenon in a way that can be capitalized upon. The development of a nondestructive ultrasonic tool for the measurement of the relative amounts of such compounds should therefore incorporate some analysis of the relevant physical quantities involved.

8.2.5.8 Interstitial Drag

At low frequencies or at temperatures sufficient for interstitial diffusion, dissolved solute atoms can shift position from their equilibrium positions to those otherwise unfavored in response to an elastic wave [39]. This redistribution of atoms lags behind the perturbation; hence, this effect can be thought of as a relaxation process. Because this phenomenon is a thermally activated process, the attenuation is realized by the familiar form

$$\alpha_{\text{int}} = \frac{\omega^2 \tau_{\text{int}}}{1 + (\omega \tau_{\text{int}})^2}, \tag{8.39}$$

where the relaxation time τ_{int} takes on the Arrhenius behavior $\tau^{\circ}_{\text{int}} e^{Q_{\text{act}}/RT}$ (R is the ideal gas constant) and $\tau_{\text{int}}{}^{\circ}$ is on the order of 1×10^{-12} s. Here Q_{act} is the activation energy for an intersticial atomic jump. Q_{act} will be different for each of the dissolved elements and will have a different relaxation time for a given temperature. For the case of carbon in iron, where $Q_{\text{act}} \approx 18$ kcal mol^{-1} and $\tau^{\circ}_{\text{int}} = 4.4 \times 10^{-14}$, $1 / \tau_{\text{int}}$ is unfortunately about 1 Hz at room temperature. If the temperature can be increased to 250 °C, however, $1 / \tau_{\text{int}}$ is on the order of 100 kHz, well in the range of ultrasonic measurements. Attenuation measurements

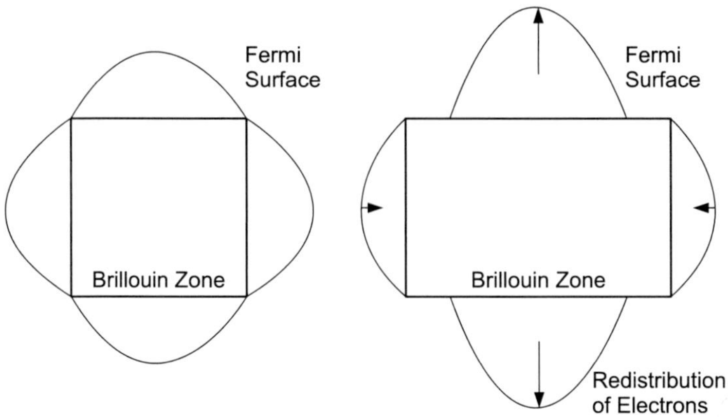

Fig. 8.18 Distortion of the Fermi energy surface of a 2D square lattice due to an elastic wave. The wave in this case is compressing the lattice from right to left, which in reciprocal space is represented by the *stretching* of the Brillouin zone in the same direction

over a range of temperatures and frequencies may therefore provide a means of measuring diffusion-related parameters for the uranium systems in this study.

8.2.5.9 Electrons

As an elastic wave propagates through a crystal of high electron density, typical of metals, electron densities are affected and redistributed away from equilibrium around the ion cores, as described by Morse [43] and Kittel [32]. A schematic representation in reciprocal space is shown for a two-dimensional square lattice in Fig. 8.18. This disturbance creates a localized difference in electric potential whence an electric current is developed, manifested as heat. Considering the model of a free electron gas of "pressure" $P = \frac{2}{5}nE_f$, where n is the electron density and E_f is the Fermi energy, a compressional strain ε_x in the x-direction will result in a pressure difference $\Delta P = \frac{8}{15}nE_f\epsilon_x$. This relationship has the appearance of Hooke's law, giving the difference between the relaxed and unrelaxed modulus $\Delta\mu$ for (8.25). The overall absorption is then

$$\alpha_{\text{el}} = \frac{nE_f}{\rho c^3}\frac{\omega^2\tau_{\text{el}}}{1+\omega^2\tau_{\text{el}}^2}, \tag{8.40}$$

where τ_{el} is related to the electrical conductivity by $\tau_{\text{el}} = \sigma m_e/ne^2$, where m_e is the mass of the electron and e is its charge. It should be noted that measurements made at extremely low temperatures ($\approx$ 20 K) and high frequencies ($\approx$ 1–10 GHz) where the acoustic wavelength is on the order of the electron mean free path, lead to direct phonon–electron interactions such as those discussed in Sect. 8.4.4.1 on the charge density wave in uranium, necessitating a quantum mechanical approach. Since the electronic structure of the lattice is disrupted by solute atoms and other crystal defects, altering the electrical conductivity and electron concentration, the ultrasonic attenuation due to conduction electrons should also be affected in a way that could reliably be correlated.

8.2.6 Electromagnetic Techniques

The nondestructive inspection techniques discussed in this section utilize super low- to medium frequency electromagnetic waves between roughly 30 Hz and 3 MHz to induce eddy currents in conductive test material. Sensors monitor eddy current interactions with defects within the material under test and interpret the resulting electromagnetic effects. Impedance changes of eddy current coil probes are indicative of the time derivative of the magnetic flux density $\mathbf{B}$, while magnetometers such as superconducting quantum interference devices (SQUIDs) and Giant Magnetoresistive sensors (GMRs) monitor the magnetic flux density directly. There is a wide range of sensor geometries and modalities that can be employed that offer varying benefits to the inspector. This section details the eddy current phenomenon, the determining criteria for and distinguishing benefits of various measurement techniques, and practical design considerations of measurement systems.

8.2.6.1 Preliminary Remarks

This section on electromagnetism presents the relevant mathematical relationships for the development of the electromagnetic wave and eddy currents, the latter of which are used extensively in NDE. The discussion begins with Maxwell's equations and concludes with the derivation for the wave equation and eddy current skin depth.

While the derivation of Maxwell's equations (particularly Faraday's law) is an exercise which reveals the underlying elegance of special relativity, it is omitted here for the sake of brevity and clarity.[6] Instead, the familiar equations will be taken as lemmata for the purposes of developing the wave equation for electromagnetics. The differential and integral forms for each of the four equations are listed here with a brief description of their meaning or consequences. Formally, (8.41) and (8.47) are valid only in free space; as such their modification to solid matter is provided summarily.

The first of two laws governing electro/magneto statics is a consequence of Gauss' Theorem (a specific case of Stokes' Theorem):

$$\nabla \cdot \mathbf{E} = \frac{\rho}{\epsilon_0} \tag{8.41a}$$

[6] A thorough discussion is provided in [20].

$$\oint_S \mathbf{E} \cdot \hat{n}\, \mathrm{d}s = \frac{q_{\mathrm{enc}}}{\epsilon_0}, \tag{8.41b}$$

which relates the *electric field* **E** to the total charge q_{enc}, surrounded by a surface S. Specifically, (8.41) states that the electric field is strictly divergent (conservative) and its flux is proportional to the charge enclosed for the case of *electrostatics*.[7]

In the situations of interest to NDE wherein matter is present (as opposed to a vacuum), the electric field must be considered as the sum of the charges present as well as the *polarization* **P** which is the sum of the electric dipoles within the material. This modification to Gauss' law usually considers the polarization as an *effective* charge (with its own charge density) by introducing the *displacement*,

$$\mathbf{D} = \epsilon_0 \mathbf{E} + \mathbf{P}. \tag{8.42}$$

Then the corrected form of Gauss' law is given by

$$\nabla \cdot \mathbf{D} = \rho_{\mathrm{free}} \tag{8.43a}$$

$$\oint_S \mathbf{D} \cdot \hat{n}\, \mathrm{d}s = q_{\mathrm{free,enc}}, \tag{8.43b}$$

where the enclosed charge parameters refer to the free (unbound) electrons.

Magnetic fields are purely rotational (and therefore not conservative), such that the notion of "sources" and "sinks" which exist in electrostatics for charged particles no longer has an analogue. An immediate consequence of the solenoidal nature of the magnetic field is that, at least in the limit of human experience, magnetic monopoles do not exist. Mathematically, magnetic fields must therefore obey the following laws:

$$\nabla \cdot \mathbf{B} = 0 \tag{8.44a}$$

$$\oint_S \mathbf{B} \cdot \hat{n}\, \mathrm{d}a = 0, \tag{8.44b}$$

for *magnetic flux density* **B**, which is Gauss' Law for magnetic fields. Akin to the corrections made to the quantities in (8.43), the magnetic flux density is adjusted by the *magnetization* field **M** in a material:

$$\mathbf{B} = \mathbf{M} + \mu \mathbf{H}, \tag{8.45}$$

---

[7] When magnetic fields are involved (specifically when they are changing), the electric field is *rotational* and is effected even without the presence of a charge.

where **H** is termed the *magnetic field strength*.

In circumstances where the magnetic field is changing *with respect to time*, the once-conservative electric field takes on a rotational character which is directly proportional to the time dependence of the magnetic flux density,

$$\nabla \times \mathbf{E} = -\frac{\partial \mathbf{B}}{\partial t} \tag{8.46a}$$

$$\oint_C \mathbf{E} \cdot \mathrm{d}\mathbf{l} = \frac{\mathrm{d}}{\mathrm{d}t} \int_S \mathbf{B} \cdot \hat{n}\, \mathrm{d}a. \tag{8.46b}$$

Somewhat analogous to Faraday's Law of induced currents, the Ampere–Maxwell law reveals the behavior of a magnetic field in response to a time-dependent electric field (in addition to the static *electric current*, **J**). The curl and contour integral of the magnetic flux density are given by

$$\nabla \times \mathbf{B} = \mu_0 \left(\mathbf{J} + \epsilon_0 \frac{\partial \mathbf{E}}{\partial t} \right) \tag{8.47a}$$

$$\oint_C \mathbf{B} \cdot \mathrm{d}\mathbf{l} = \mu_0 \left(I_{\mathrm{enc}} + \epsilon_0 \frac{\mathrm{d}}{\mathrm{d}t} \int_S \mathbf{E} \cdot \hat{n}\, \mathrm{d}a \right). \tag{8.47b}$$

The two laws given in (8.46) and (8.47) together form the constitutive relation for eddy current generation and detection: a changing magnetic field induces a current in a material which generates an opposing magnetic field, thus conserving energy.

Before continuing to the somewhat special case of eddy currents, Maxwell's equations will be used conjunctively to realize the connection between electricity and magnetism which establishes the generation of many forms of radiation which are at the heart of NDE: the electromagnetic wave equation. Taking the curl (∇) of both sides of the differential form of Faraday's Law (8.46a) and reversing the spatial and temporal derivatives gives

$$\nabla \times (\nabla \times \mathbf{E}) = \nabla \times \left(-\frac{\partial \mathbf{B}}{\partial t} \right) = -\frac{\partial}{\partial t} (\nabla \times \mathbf{B}). \tag{8.48}$$

Application of the curl operator twice upon a vector field **A** is given by the identity

$$\nabla \times (\nabla \times \mathbf{A}) = \nabla (\nabla \cdot \mathbf{A}) - \nabla^2 \mathbf{A}, \tag{8.49}$$

such that (8.48) becomes

$$\nabla (\nabla \cdot \mathbf{E}) - \nabla^2 \mathbf{E} = -\frac{\partial}{\partial t} \mu_0 \left(\mathbf{J} + \epsilon_0 \frac{\partial \mathbf{E}}{\partial t} \right), \tag{8.50}$$

after substitution of the Ampere–Maxwell Law (8.47a). In *free space*, the current $\mathbf{J}$ and charge density ρ are zero, such that those terms drop out and the electric field wave equation is given by

$$\nabla^2 \mathbf{E} = \frac{1}{c^2}\frac{\partial^2 \mathbf{E}}{\partial t^2}, \tag{8.51}$$

where the *velocity* is $c = (\mu_0 \epsilon_0)^{-1/2}$.

The starting point for the foregoing development was Faradays Law (8.46), which relates the electric field to the time dependence of the magnetic field. It is no surprise, then, that the magnetic analogue to (8.51) is derived from the Maxwell–Ampere Law (8.47a) by similar arguments, such that

$$\nabla^2 \mathbf{B} = \frac{1}{c^2}\frac{\partial^2 \mathbf{B}}{\partial t^2}. \tag{8.52}$$

Equations (8.51) and (8.52) are coupled by Maxwell's equations, thus giving rise to *electromagnetic* waves, with $\mathbf{E}$ and $\mathbf{B}$ mutually orthogonal to the direction of propagation, i.e., for an electromagnetic plane wave propagating in the z-direction:

$$\mathbf{E} = \mathbf{E}_0 e^{i(\omega t - kz)} \tag{8.53}$$

$$\mathbf{B} = \mathbf{B}_0 e^{i(\omega t - kz)}, \tag{8.54}$$

where

$$\mathbf{B}_0 = \begin{pmatrix} B_{0x} \\ B_{0y} \\ 0 \end{pmatrix}, \tag{8.55}$$

and

$$\mathbf{E}_0 = \begin{pmatrix} cB_{0y} \\ -cB_{0x} \\ 0 \end{pmatrix}, \tag{8.56}$$

by virtue of the curl relationships in Maxwell's equations. Naturally, (8.53) can be rewritten more generally for an arbitrary direction by replacing z with the *Poynting vector*, $\mathbf{s} = 1/\mu_0 \mathbf{E} \times \mathbf{B}$.

In metals, the current $\mathbf{J}$ is clearly not zero, and (8.51) must be modified to reflect the term in (8.50), such that

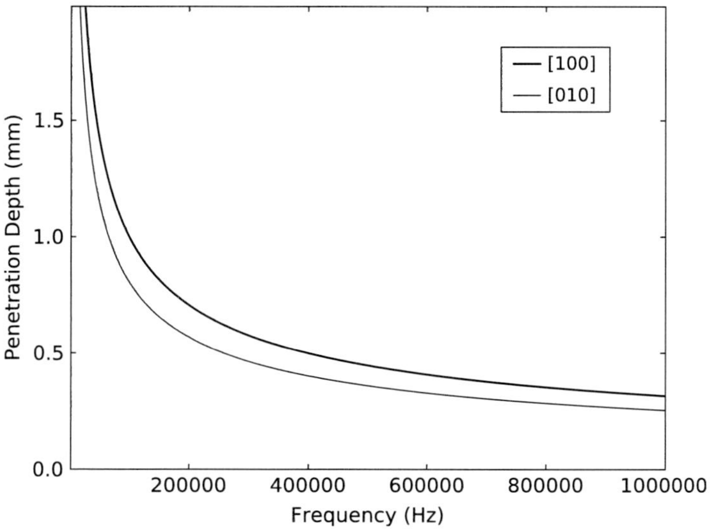

Fig. 8.19 Penetration depths as a function of frequency for the two primary crystallographic directions [100] and [010]. The distinction arises from differences in the respective magnetic and electronic properties in each direction

$$\nabla^2 \mathbf{E} = \mu_m \sigma_m \frac{\partial \mathbf{E}}{\partial t} + \frac{1}{c_m^2} \frac{\partial^2 \mathbf{E}}{\partial t^2}, \tag{8.57}$$

for conductivity σ and velocity c in the medium m. The inclusion of a first order derivative introduces an imaginary term to the electric field, and subsequent substitution yields a wavevector which is *complex*. As before in elastic wave theory, c.f. Sect. 8.2.3.1, the complex wavenumber gives rise to a real exponential decay term for electromagnetic radiation in conductors. This phenomenon is referred to as the skin effect; taking some liberties with substitution one can define the *skin depth*

$$\delta \doteq \left(\frac{2}{\omega \mu_m \sigma_m} \right)^{1/2}, \tag{8.58}$$

to describe the decay of the electric field in the material (again, for propagation in the z-direction)

$$\mathbf{E} = E_{0,x} \mathrm{e}^{-z/\delta} \mathrm{e}^{\mathrm{i}(\omega t - z/\delta)}. \tag{8.59}$$

Values of δ are computed for the two crystallographic directions [100] and [010] based on resistivities of 39. 4 $\times 10^{-8}$ and 25. 5 $\times 10^{-8}$ Ω m [9] and magnetic permeability of 1.257×10^{-5} V s A$^{-1} \cdot$ s^{-1} [47] and shown in Fig. 8.19.

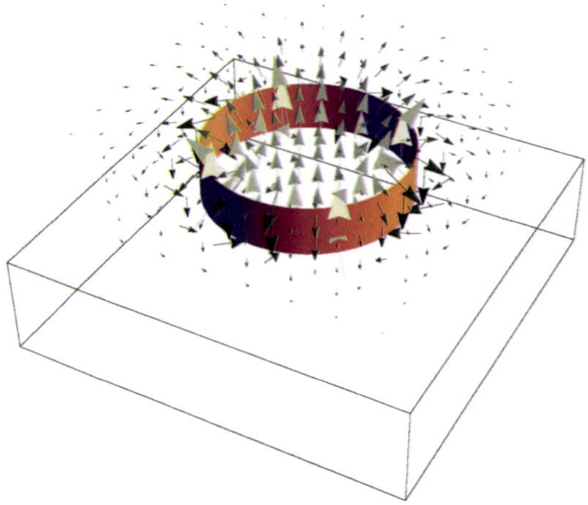

Fig. 8.20 A typical eddy current system. A solenoidal coil produces a magnetic field which penetrates into the sample

Equation (8.59) provides the governing mathematical relationship for the development of eddy currents, the practical aspects of which will now be discussed.

During eddy current inspection, an excitation coil generates a time-dependent magnetic flux that induces eddy currents in the sample beneath the coil. These eddy currents are the mirror image of the excitation currents and produce an opposing secondary flux, as shown in Figs. 8.20–8.22. Eddy current generation via magnetic induction occurs in the reactive near-field of the electromagnetic source.

Eddy current NDE falls under the magnetoquasistatic regime, invoked when the characteristic length of the material under test is much smaller than the source wavelength. The quasistatic condition assumes negligible wavelike electrodynamic effects and drops the displacement current density term in Ampere's Law, resulting in a type of magnetic diffusion in the material. The diffusive wavefield can be cast into the following Helmholtz wave equation in which the square of the wavenumber is purely imaginary [33]:

$$\nabla^2 \mathbf{H} - \mathrm{i}\omega\mu\sigma\mathbf{H} = 0. \tag{8.60}$$

Eddy current density attenuation is caused by dispersive complex material constitutive parameters, electrical permittivity, electrical conductivity, and magnetic permeability (ε, σ, and μ). These constitutive parameters, in addition to the source excitation frequency, define the eddy current depth of penetration, which determines the characteristic measurement length. In the general case for a linear, anisotropic, and dispersive material, the constitutive parameters are complex tensors. Each of the individual values of the permittivity tensor

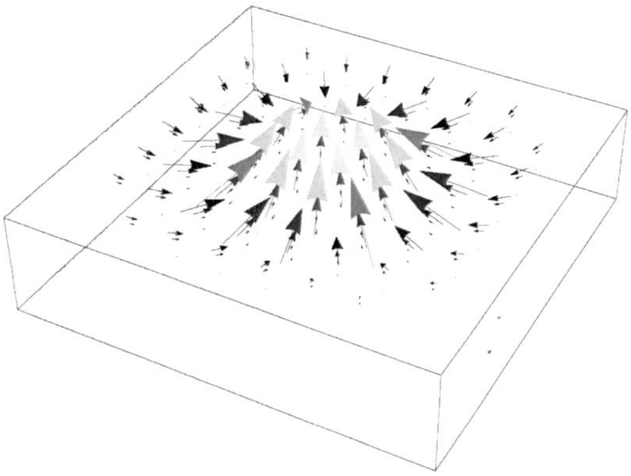

Fig. 8.21 The penetrating field in a conductor decays exponentially with increasing distance and is opposed by the generated current

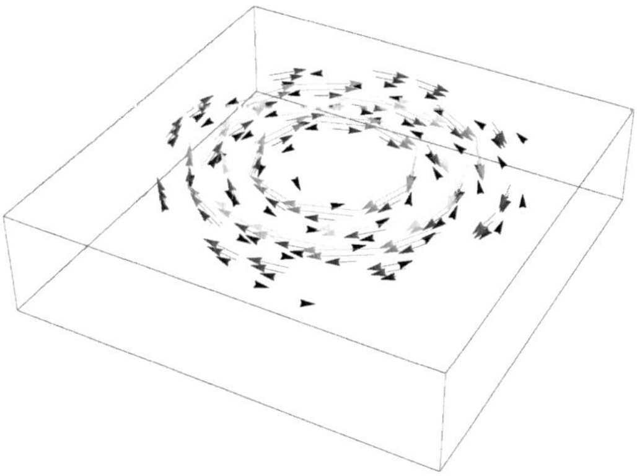

Fig. 8.22 The time-dependent field induces a circular current which quickly decays beyond the coil's region of influence

$$\epsilon = \begin{bmatrix} \epsilon_{xx} & \epsilon_{xy} & \epsilon_{xz} \\ \epsilon_{yx} & \epsilon_{yy} & \epsilon_{yz} \\ \epsilon_{zx} & \epsilon_{zy} & \epsilon_{zz} \end{bmatrix}, \tag{8.61}$$

can be complex. The Kramers–Kronig relations relate the real and imaginary parts of the relative complex permittivity[8]

$$\epsilon_r = \frac{\epsilon}{\epsilon_0} = \epsilon_r{}' + i\epsilon_r'', \tag{8.62}$$

for time-varying fields by

[8] The Kramers–Kronig relations are valid for any causal, *linear* system including elastic solids, in which case they relate the attenuation and dispersion.

Table 8.2 Conductors vs. dielectrics. From [2]

Parameter	Symbol	Exact solution	Good dielectric $\left(\frac{\sigma}{\omega\epsilon}\right)^2 \ll 1$	Good conductor $\left(\frac{\sigma}{\omega\epsilon}\right)^2 \gg 1$
Atten. Coeff.	α	$=\omega\sqrt{\mu\epsilon}\left[\frac{1}{2}\left[\sqrt{1+\frac{\sigma}{\omega\epsilon}}-1\right]\right]^2$	$\approx\frac{\sigma}{2}\sqrt{\frac{\mu}{\epsilon}}$	$\approx\sqrt{\frac{\omega\mu\sigma}{2}}$
Phase Const.	β	$=\omega\sqrt{\mu\epsilon}\left[\frac{1}{2}\left[\sqrt{1+\frac{\sigma}{\omega\epsilon}}+1\right]\right]^2$	$\approx\omega\sqrt{\mu\epsilon}$	$\approx\sqrt{\frac{\omega\mu\epsilon}{2}}$
Impedance	Z_w	$=\sqrt{\frac{\mathrm{i}\omega\mu}{\sigma+\mathrm{i}\omega\epsilon}}$	$\approx\sqrt{\frac{\mu}{\epsilon}}$	$\approx\sqrt{\frac{\omega\mu}{2\sigma}}(1+\mathrm{i})$
Wavelength	λ	$=\frac{2\pi}{\beta}$	$\approx\frac{2\pi}{\omega\sqrt{\mu\epsilon}}$	$\approx 2\pi\sqrt{\frac{2}{\omega\mu\sigma}}$
Velocity	c	$=\frac{\omega}{\beta}$	$\approx\frac{1}{\sqrt{\mu\epsilon}}$	$\approx\sqrt{\frac{2\omega}{\mu\epsilon}}$
Skin Depth	δ	$=\frac{1}{\alpha}$	$\approx\frac{2}{\sigma}\sqrt{\frac{\epsilon}{\mu}}$	$\approx\sqrt{\frac{2}{\omega\mu\sigma}}$

$$\epsilon_r{}'(\omega) = 1 + \frac{2}{\pi}\int_0^\infty \frac{\omega'\epsilon_r''(\omega')}{\omega'^2 - \omega}\,\mathrm{d}\omega' \tag{8.63}$$

$$\epsilon_r''(\omega) = \frac{2\omega}{\pi}\int_0^\infty \frac{1-\epsilon_r{}'(\omega')}{\omega'^2 - \omega}\,\mathrm{d}\omega'. \tag{8.64}$$

The imaginary part of the permittivity defines the equivalent conductivity as

$$\sigma_e = \sigma_s + \omega\epsilon'' \tag{8.65}$$

where the static conductivity for a conductor is defined as

$$\sigma_s = -\mu_e q_e n_e, \tag{8.66}$$

where μ_e is the mobility of electrons (in $\mathrm{m^2\,V^{-1}\,s^{-1}}$), q_e the electron charge, and n_e is the free electron density. (More will be said about free electron theory in Sect. 8.4.1.)

A material can be classified into a good *dielectric* ($\sigma_e\ /\ \omega\varepsilon' \gg 1$) or a good *conductor* ($\sigma_e\ /\ \omega\varepsilon' \ll 1$), and the relevant eddy-current determinable parameters are given in Table 8.2.

Eddy current attenuation is due to electromagnetic interaction within the material resulting in energy losses, such that the attenuation of the amplitude of the magnetic field **H** defined in (8.60) is

$$\mathbf{H} = H_0 \mathrm{e}^{-z/\delta}\mathrm{e}^{\mathrm{i}(\omega t - z/\delta)}, \tag{8.67}$$

for a material ($z > 0$) with a boundary at $z = 0$. The frequency of excitation is the variable in (8.67) that can most easily be adjusted to tune the measurement depth [2].

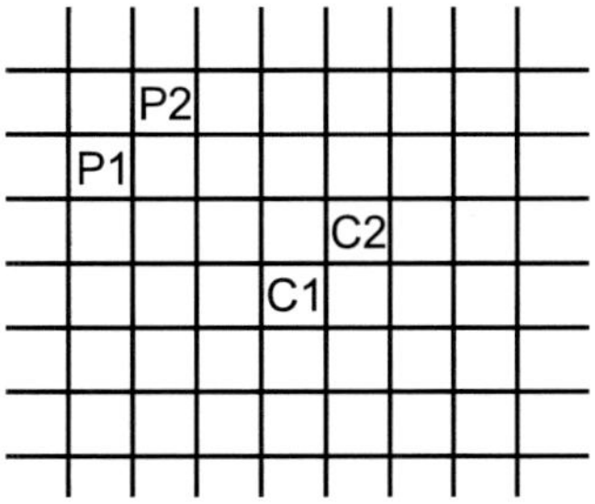

Fig. 8.23 Illustration of the spacial resolution of a measurement array

Material properties such as alloy content, stress intensity and distribution (see Sect. 8.5.2 for one such example), microstructure, carrier content, and radiation damage all affect either the electrical conductivity or the magnetic permeability of a material and can be assessed through electromagnetic means.

The basic requirements of an electromagnetic nuclear material characterization system consist of electronics that control excitation, sensing, signal processing, and data storage. In its simplest form, a single absolute coil generates and measures the eddy current response of the material. Dual coil measurements in the pitch-catch mode have the advantage of increased spatial resolution or increased measurement depth, as they are inversely related. By increasing the pitch to catch distance, deeper eddy current signals are sampled. However, the increased pitch to catch distance results in a decreased spatial resolution for this setup with a larger footprint. In a sensor array, however, the spatial resolution of the pitch-catch system can be retained by selecting successive coil pairs, thereby preserving the spatial resolution of the array. In Fig. 8.23 above, the depth advantage gained from coil pairs (P1/C1) separated by three diagonal grid spaces is the same for the subsequent measurement (P2/C2). However, the collective array spatial resolution is the distance separating the subsequent measurements which is only one diagonal space.

Nuclear Material Electromagnetic NDE systems can be configured to handle various environmental conditions. A robust electromagnetic sensor would have adequate radioactive shielding and waterproof ratings required for the inspection location. Stand-alone measurements would be carried out through a properly packaged self-contained embedded device that would consist of the requisite electronics and software programs. The scope of the automated algorithms employed in processing the acquired data would depend on the information needed. Some of the post-processing of the data may take place outside of the embedded device.

8.3 Departures from Linearity

In the foregoing discussions of Sects. 8.2.2 and 8.2.6.1, it was assumed that cause and effect were related to each other by some *linear* coefficient which itself did not depend on either. Suppose, then, that this relationship no longer holds due to the magnitude of the perturbation, the length scale involved, or because of some other higher-order interaction. Such nonlinear systems are encountered frequently, but given the difficulty of including quadratic (and higher) terms in mathematical relationships are only recently gaining traction in NDE.

8.3.1 The Need for Nonlinear Terms

Processed materials consist of compositional, microstructural, and property gradients which necessitate the modification of various fundamental relationships in materials science. When microstructural inhomogeneities are on the order of nanometers or less, nonlinear thermodynamics and kinetics are required to accurately describe the system, and normal linear relationships between measurable properties break down as a consequence. Cahn and Hillard [10, 11] describe the thermodynamics of nonuniform systems, Hart [30] has addressed inhomogeneous strain issues, and Tu [49] has described the need for nonlinear terms in expressions which relate the evolution of nanoscale thin films.

Bonding interfaces, which can be considered as junctions with steep electronic gradients (characterized by the Debye length), will facilitate the use of electronic and elastic NDE tools to assess material quality. The expressions for describing lattice behavior and the response of the lattice to perturbations will require additional nonlinear terms. The influence of these nonlinear effects on NDE assessment needs to be investigated and the analytical materials science practices need to be developed to allow interpretation of the nanostructural state of many advanced and high performance uranium materials being used.

8.3.2 Features

Assessment of nonlinear behavior in solids is necessarily more involved than the characterization of the "simple" features of ordinary dispersion and attenuation. The complexity arises both in finding general solutions to the nonlinear wave equations and in the practical aspects of measuring nonlinear quantities. In the case of acoustics, two characteristic phenomena have been identified with nonlinear behavior: (1) *acoustoelasticity*, which refers to the stress dependence of the sound velocity and its dispersion relations, and (2) *harmonic generation*, quantified by the nonlinearity parameter β. Both effects have found application in the nondestructive determination of microstructural features in more common materials; uranium offers significant (albeit as yet unapproached) opportunities in this regard.

8.3.2.1 Acoustoelasticity

The strain as given by (8.3) is correct only to first order. In developing the nonlinear wave equation, one must include the mixed derivatives for the strain given by

$$\epsilon_{ij} = \frac{1}{2}\left(\frac{\partial u_i}{\partial x_k} + \frac{\partial u_k}{\partial x_i} + \frac{\partial u_l}{\partial x_i}\frac{\partial u_l}{\partial x_k}\right). \tag{8.68}$$

Numerous systems of constants exist for establishing the relationships between stress and the exact definition of strain, depending on the choice of coordinates. The Huang (or propagation) coefficients, defined by [13]

$$A_{ijkl\ldots} = \rho_0 \left(\frac{\partial^n U}{\partial \epsilon_{ij} \partial \epsilon_{kl\ldots}}\right)_{\partial u/\partial a=0}, \tag{8.69}$$

of order n and material coordinates a, provide a convenient expansion of the internal energy U such that

$$\rho_0 U = A_{ij}\epsilon_{ij} + \frac{1}{2!}A_{ijkl}\epsilon_{ij}\epsilon_{kl} + \frac{1}{3!}A_{ijklmn}\epsilon_{ij}\epsilon_{kl}\epsilon_{mn}\ldots, \tag{8.70}$$

which establishes the *nonlinear wave equation*

$$\rho_0 \frac{\partial^2 \tilde{u}_k}{\partial t^2} = \bar{L}_{ijkl}\frac{\partial^2 \tilde{u}_k}{\partial a_l \partial a_j}, \tag{8.71}$$

where $\tilde{u}_k = u_k - \bar{u}_k$ is the displacement relative to the undeformed state with mass density ρ_0, and the propagation matrix is given by

$$\bar{L}_{ijkl} = A_{ijkl} + A_{ijklmn}\epsilon_{mn} + \ldots. \tag{8.72}$$

Equation (8.71) and the supposed solution $\tilde{u}_k = U_k e^{i(\omega t - \mathbf{k}\cdot x)}$ give rise to the acoustoelasticity equation, analogous to the Christoffel equation (8.6),

$$(\bar{L}_{ijkl}N_j N_l - \rho_0 c^2 \delta_{ik})U_k = 0, \tag{8.73}$$

for propagation vector $\mathbf{N}$ and polarization vector $\mathbf{U}$. Thus (8.73) provides the relationship between the "natural velocity," $c = \omega/|\mathbf{k}|$, and the strains established in the propagation matrix $\bar{L}_{ijkl}$. Measurements of the velocity c as a function of applied stress in a material can therefore yield information about the higher order elastic constants $A_{ijkl\ldots}$, which have been shown to be highly sensitive to microstructure [13].

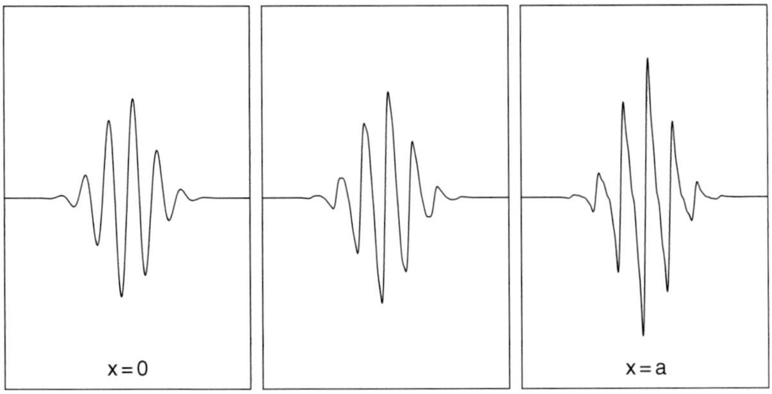

Fig. 8.24 Distortion of a wave burst as it propagates in a nonlinear material in the x direction

8.3.2.2 Harmonic Generation

A very striking and useful nonlinear feature is the generation of harmonics, i.e., waves with frequencies integral factors greater than the fundamental, which are produced following the introduction of a purely sinusoidal wave into the nonlinear medium. As the incident wave propagates through the material, it becomes distorted as shown in Fig. 8.24, and the degree of harmonic generation depends on how far the wave has traveled.

A measure of the nonlinearity of an elastic medium becomes proportional to the ratio of the amplitudes, of the second harmonic generated relative to the fundamental (squared) and normalized to the distance traveled, d. Thus the *nonlinearity parameter* is measured as

$$\beta = 8\frac{P_2}{k^2 P_1^2 d} \tag{8.74}$$

for the wavenumber k. The amplitudes P_1 and P_2 can be obtained from the Fourier spectra as shown in Fig. 8.25. Certain phenomena such as the interaction of acoustic waves with dislocations generate primarily odd harmonics (8.2.5.4), resulting in "anomalous" nonlinearity. Formally, the nonlinearity parameter is defined [14]

$$\beta = \frac{A_{ijklpq} N_j N_l N_q U_i U_k U_p}{A_{ijkl} N_j N_l U_i U_k}. \tag{8.75}$$

Values of β have been shown to be dependent on atomic arrangement, with specific crystal structures lying within particular ranges [13]. Uranium presents significant opportunity here, as its unique crystal structure (Fig. 8.26) retains aspects of face centered tetragonal, simple cubic, and monoclinic structures.

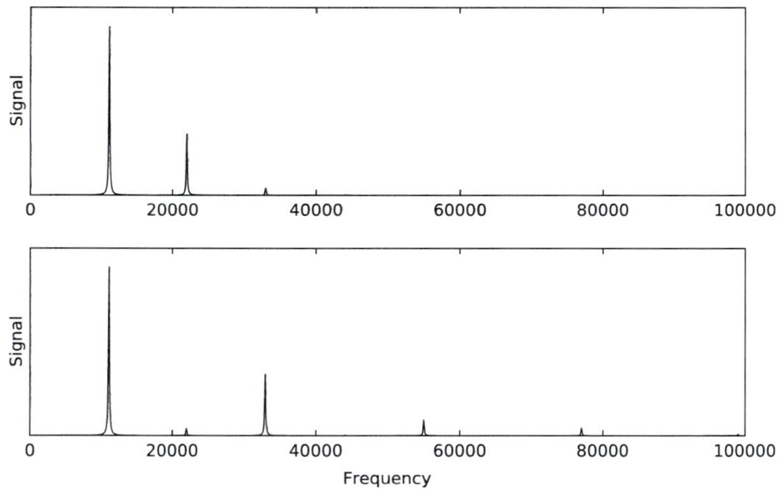

Fig. 8.25 Harmonics generated by a sinusoidal waveform at $f = 11$ kHz as applied to a nonlinear system exhibiting quadratic (*above*) and anomalous (*below*) behaviors

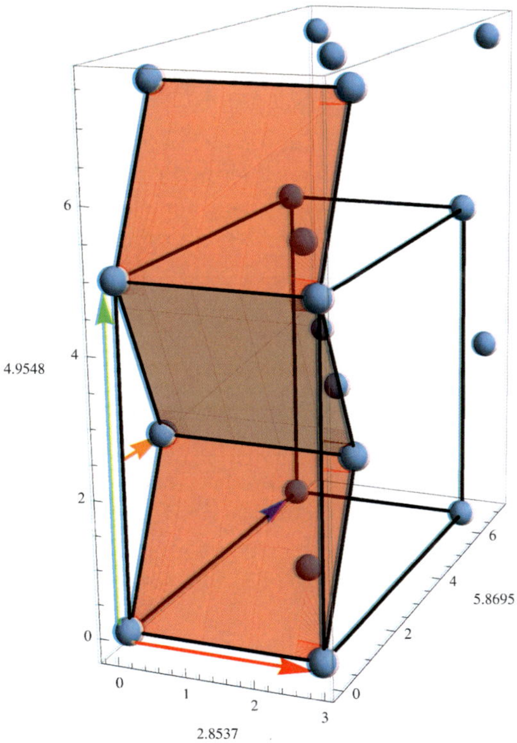

Fig. 8.26 Orthorhombic crystal structure of α-uranium. The highlighted sheets denote the "corrugated" atomic arrangement in the crystal

Table 8.3 Microstructure dimensional requirements for the necessity of including a nonlinear term for the chemical potential

Length scale	Example	Chemical potential
$\approx 10^{-8}$ m	Nanocomposites	Nonlinear
$\approx 10^{-7}$ m	Thin films, substrates	Linear
$\approx 10^{-3}$ m	Bulk samples	Linear

8.3.3 *Causes*

When dimension scales in the nano- to micrometer range describe the variation in the material microstructure, then nonlinear thermodynamics and kinetics are required to accurately describe the system. Tu [49] described the need for application of nonlinear terms in the expressions describing the evolution of nano-scale thin films, as indicated in Table 8.3. Tu showed that when the dimensions are very small and the potentials gradients are very steep, the transport equation needs nonlinear terms represented by higher order terms to properly describe the atomic transport in small dimensional materials.

The accumulation expression for nonlinear diffusion is

$$\frac{\mathrm{d}c}{\mathrm{d}t}=\bar{D}\frac{\mathrm{d}^2c}{\mathrm{d}x^2}\left[1+\frac{1}{8}\left(\frac{\lambda f_0}{kT}\frac{\mathrm{d}c}{\mathrm{d}x}\right)^2\right]-\frac{2D}{f_0''}\frac{\mathrm{d}c^4}{\mathrm{d}x^2}, \tag{8.76}$$

where it can clearly be seen that the first term represents Fick's 2nd Law of diffusion. The two additional nonlinear terms in (8.76) represent the nonlinear contribution of both the mobility and driving force terms and are required when the gradients become large. An example is provided in Sect. 8.5.3 for high current densities in metallic materials which are useful for reducing grain size.

8.3.3.1 Crystal Defects

The nonlinear interpretation can be applied to electron charge distribution of a defect in a crystal, such as the charge distribution of a dislocation, or line defect. The dislocation is represented as an extra half plane of atoms above the slip plane (such as in Fig. 8.13) but can also be described as an electric dipole along the dislocation line as shown in Fig. 8.27. Applying the Poisson equation to this charge distribution will allow the dislocations to be described as steep electric potential gradients, such as:

$$\phi=\phi_0\mathrm{e}^{-|x|/\lambda_0}. \tag{8.77}$$

The potential of this dislocation in a metal can be experienced only in extremely short distances due to the very small *Debye length* (i.e., the electron screening distance) for a metal which is given as:

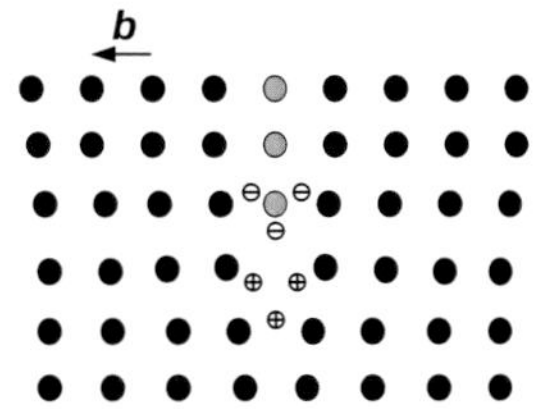

Fig. 8.27 Electronic description of an edge dislocation as an electronic dipole

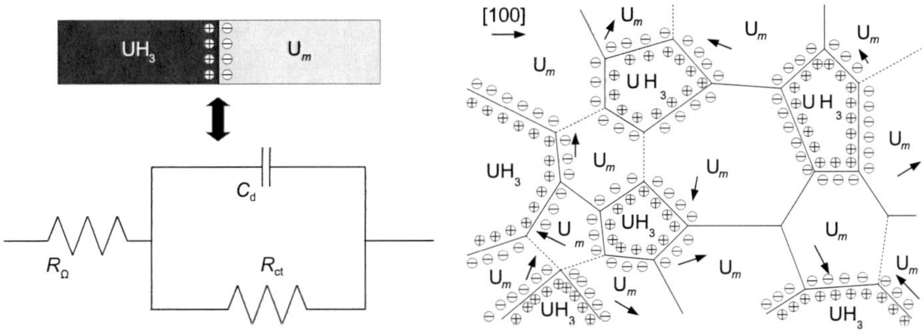

Fig. 8.28 (*Left*) A model circuit of the material sensed by low frequency impedance, comprised of capacitive interfaces between resistive grains. (*Right*) A proposed example of the charge distribution in fine grained uranium with hydrides present. The [100] direction is highlighted for the metallic uranium grains, with boundaries of significant mismatch indicated by *dashed lines*. With uranium's high degree of anisotropy in electrical conductivity, these locations develop significant impedance mismatch

$$\lambda_0 = \sqrt{\frac{\epsilon_0 kT}{e^2 n_\infty}}, \tag{8.78}$$

where k is Boltzman's constant, e is the electronic charge, and n_∞ is the electron concentration of the bulk material. Thus nonlinear behavior is observed when a cyclic perturbation of wavelength comparable to λ_0 is incident upon the material.

8.3.3.2 Secondary Phases

Inclusions in the metallic matrix have an electronic junction at their interfaces with metal and this junction can be treated also as a dipole. The interaction of a dislocation with an inclusion can be considered a dipole–dipole interaction which is a very short range interaction. Since these gradients are very steep and dimensionally small their charge distribution and electrical potential obtained by the Poisson equation should be treated by a nonlinear description. The nonlinear electrical potential can be assessed by measuring their higher order electric properties terms, such as with a higher harmonic-induced impedance measurement. This practice will offer new insight into dislocation behavior.

Advanced composite theory will need to describe the electronic charge gradient across the bonding interfaces based on electronic concepts. Figure 8.28 illustrates

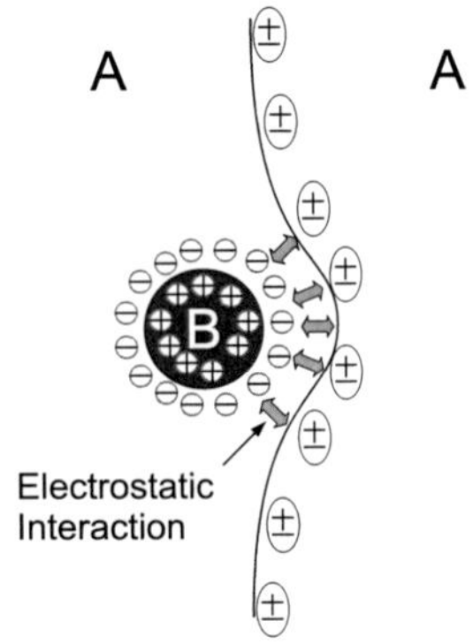

Fig. 8.29 Dipole–dipole interaction between a dislocation in phase *A* and inclusion (phase *B*)

an electronic junction resulting from a metal–metal hydride interface which will behave electronically as a local capacitor to an eddy current. These bonding interfaces will have steep electronic gradients in a metal described by the very short Debye length. The expressions describing such steep potentials will require additional nonlinear terms. Inclusions in the metallic matrix also have an electronic junction at their interfaces and this junction can be treated also as a charged dipole.

The interaction of a dislocation with an inclusion can be considered a dipole–dipole interaction which is a very short-range interaction. Since these potential gradients are very steep in a metal with its large electron concentration, the electrical potential obtained by the Poisson equation should be nonlinear. This nonlinear electrical potential can be assessed by measuring the higher order electric properties terms, such as through harmonic analysis of the induced voltage. This practice will offer new insight into dislocation behavior by measuring the higher harmonic frequencies (Fig. 8.29).

8.4 Solid State Physics of Uranium

Electronic band theory, Fermi energy, and Brillouin zones are all based on wave mechanics concepts. The use of the *free electron model* for a metal can be extended to consider electron–lattice potential interactions through the introduction of the *effective mass* of the electron, m^*.

The total energy (E) of an electron in a solid, based on wave mechanics is:

$$E = \frac{\hbar^2 k^2}{2m_e} + V, \tag{8.79}$$

where k is the electronic wave vector, and V is the potential that the nearly free electron is experiencing from the lattice, and $\hbar$ is the reduced Planck constant. The information of the potential of electron–lattice interactions (V) is incorporated into the electron mass factor m_e, then the total energy of electron in the lattice can be expressed in terms of effective mass m^* as:

$$E = \frac{\hbar^2 k^2}{2m^*} \tag{8.80}$$

These localized potentials represent the structural contributions that disturb the periodic lattice (Bloch function), such as dislocations, grain boundaries, areas of lattice strain, and phase changes. The effective mass of the electron is very sensitive to lattice changes in alloy and is derived to be

$$m^* = \frac{\hbar^2}{\frac{d^2 E}{dk^2}}. \tag{8.81}$$

Electronic property measurements offer a very sensitive NDE assessment tool for microstructural constituents and alloy stability, based on the second derivative factor in (8.81).

For example, when the Fermi energy surface contacts the Brillouin zone boundary, which represents an energy band gap that exists when the electron energy has a wave vector that will be diffracted, the conditions for a phase transformation occurs. After this event it quickly fills higher electronic energy states during further alloy additions, thus rapidly increasing the effective mass of the electron. To minimize energy, the lattice will select a different crystal structure, thus a new Brillouin zone, allowing for lower energy filling. This situation is the electronic explanation for phase transformations and is useful for understanding the charge density wave in Sect. 8.4.4.1.

8.4.1 Free and Nearly Free Electron Theory

Despite the failure to produce accurate predictions of properties for all but the simplest of systems, some features of rudimentary solid state theories offer qualitative insight into the electronic nature of metals.

Bloch's Theorem remains one of the fundamental achievements in the understanding of solid state physics as it lends tractability to the solution of the many-body problem encountered in electronic structure. In principle, the theorem is rooted in the translational symmetry of a crystal: the conditions at one point in a lattice are identical to those points anywhere else in the lattice when translated by an integral multiple of primitive lattice vectors. In addressing the "conditions" at a particular point $\mathbf{r}$, the Schrödinger equation finds appropriate use:

$$H\psi = \left(-\frac{\hbar^2}{2m}\nabla^2 + V(\mathbf{r})\right)\psi = \epsilon\psi, \tag{8.82}$$

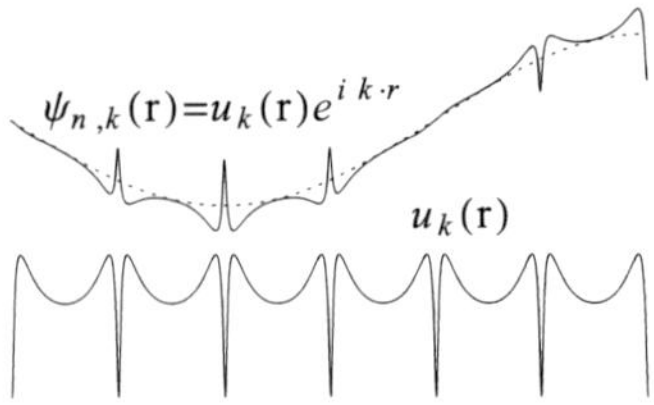

Fig. 8.30 The periodic potential (*below*) experienced by the electron as modulated by $e^{ik\cdot r}$ (*above*)

with ψ the eigenstate and $V(\mathbf{r})$ the potential encountered by an electron. As the Hamiltonian operator H, the eigenvalue ε is naturally energetic in nature. Bloch's theorem states that ψ can be expanded as a series of plane waves multiplied by a function which exhibits the same periodicity as the lattice, or,

$$\psi_{n\mathbf{k}}(\mathbf{r}) = e^{i\mathbf{k}\cdot\mathbf{r}} V_{n\mathbf{k}}(\mathbf{r}), \tag{8.83}$$

where $\mathbf{k}$ is a wavevector and

$$V_{n\mathbf{k}}(\mathbf{r}+\mathbf{R}) = V_{n\mathbf{k}}(\mathbf{r}) \tag{8.84}$$

given a translation vector $\mathbf{R} = n_1a_1 + n_2a_2 + n_3a_3$ $(n_i \in \mathbf{I})$ for primitive lattice vectors a_i (Fig. 8.30).

When the periodic potential is very weak, the eigenvalues of the free electron Schrödinger equation, ε° can be perturbed to first order by the inclusion of a Fermi energy ε:

$$(\varepsilon^{\circ}_{\mathbf{k}-\mathbf{K}} - \epsilon)c_{\mathbf{k}-\mathbf{K}} = \left(\frac{\hbar^2}{2m}q^2 - \epsilon\right)c_q = 0, \tag{8.85}$$

where c_q are the Fourier components of the (periodic) potential. The kinematic behavior of the electrons collapses to two distinct cases: (1) the perturbed energy differs significantly from the potential V, or (2) when the difference is on the order of V, i.e., when the electron wave vectors are degenerate. It can be shown[9] that in the degenerate case (1), the level ε varies as V^2 and in the second case the dominant terms vary as V. Thus for highly *symmetric* crystal systems, the degeneracies in k-space allow the electrons to behave in a manner that approaches that of the free electron model.

8.4.2 Departure from Ideality

In situations where the potential can no longer be thought of as "weak," or when the periodicity of the lattice is broken, the electrons can no longer be considered "free." While a thorn in the side of theory, it is these departures from the simple cases that provide substantial promise in the field of NDE.

[9] See, for example, Ashcroft and Mermin [1].

8.4.2.1 Disruptions to Periodicity

Innumerable features in the nano- to mesoscopic scale contribute to the collapse of an otherwise perfectly periodic potential. Point defects such as vacancies or interstitials act as charge sinks or sources, which locally alters the electron density and scatters the Bloch waves. Dislocations and planar defects (grain boundaries, secondary phases) effectively behave as electric dipoles, interacting with each other and with any elastic or electromagnetic wave that is effected upon the lattice.

8.4.2.2 Strong Potentials

Given the highly anisotropic nature of countless physical properties in uranium, it might be suspected that the periodic potential which the electron experiences can no longer be considered "weak." Indeed the electronic structure of uranium lends to the metal a number of unique properties which are as yet unseen in the other elements. Uranium exhibits three high temperature allotropes denoted α, β, and γ in order of stability for increasing temperatures corresponding to orthorhombic, tetragonal, and cubic crystal systems, respectively. It is the low temperature α phase (orthorhombic, stable below 940 K) which has attracted attention among those in the solid state physics community for a number of reasons.

Uranium is the only pure element with orthorhombic symmetry at room temperature, and as the temperature is decreased various anomalies in the physical properties emerge due to three additional phase transitions corresponding to charge density waves which then give way to a superconducting transition as high as 2 K.[10] The lattice dynamics and reduced symmetry of the crystal system contribute to extremely anisotropic physical properties including the elastic constants, conductivity tensor, and thermal expansion coefficients.

8.4.3 Role of f*-Orbitalsin Uranium*

The work of Fisher and McSkimmin [24] demonstrated anomalous behavior in the elastic constants—C_{11} in particular—at 43 K (see Fig. 8.31), echoing the findings of earlier experiments on thermal expansion [48], thermal conductivity [46], Hall coefficient [4], and other physical properties [35]. Early in the search for an explanation of uranium's unique behavior, the 5*f* electrons were thought to contribute to a magnetic transition at low temperatures in a manner similar to the 4*f* orbitals of the lanthanide series, and efforts would continue into the 1960s to prove that the spins of the 5*f* electrons were responsible.

---

[10] This text highlights various physical phenomena in uranium which present opportunities for the NDE community. For an excellent and much more thorough review of the experimental milestones in the field of solid state properties of uranium, see [35].

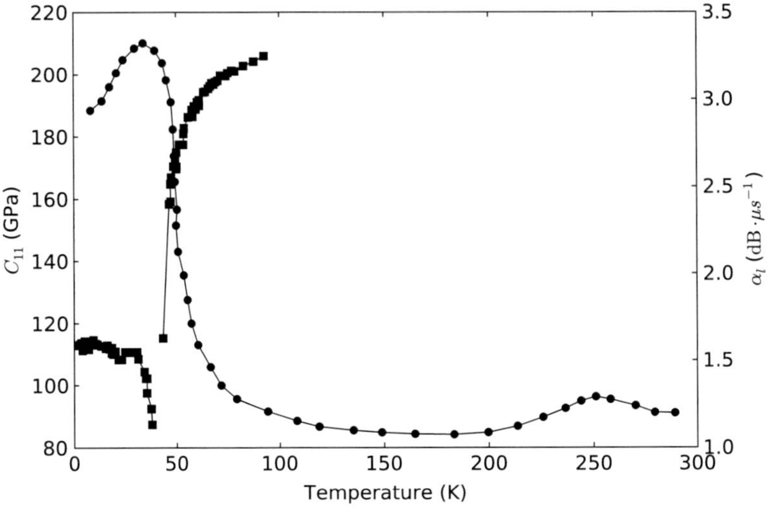

Fig. 8.31 Anomalous behavior of the C_{11} elastic constant (*filledsquare*) as determined by Fisher and McSkimin. The apparent discontinuity corresponds to the 43 K charge density wave transition and is accompanied by a dramatic increase in attenuation shown for polycrystalline uranium (*filledcircle*). Replotted from data obtained by [24] and [45]

Even by the 1970s, however, direct experimental evidence for a magnetic ordering phase transition (as observed for the lanthanides) could not be found. Magnetic susceptibility measurements by Ross and Lam [47] did not lend significant credibility to the theory of a magnetic transition or spin density wave (SDW) in α-uranium, but also couldn't rule out the possibility of their existence.

8.4.4 *Role of Phonons*

Early low temperature neutron scattering experiments [3] sought again to find evidence of magnetic ordering through the detection of additional peaks in the diffraction spectra. Later experimentation and interpretation of the extra peaks found in the data revealed disappointingly that their existence could not be due to magnetic order [36]. In light of these revelations, it became clear that the 5*f* electrons did not behave like those of the 4*f* shell in the lanthanides, and an alternative explanation for the anomalous behaviors would be necessary.

8.4.4.1 Charge Density Waves in Uranium

Work would continue on developing the phonon dispersion curves to elucidate the nature of what would eventually be recognized as a *charge density wave* (CDW) in uranium. Neutron scattering experiments performed by Crummett et al. [16]

Table 8.4 Atomic displacements and phase angles for the CDW in uranium at $T \leq 22$ K [37]

x_i	e_{x_i}	ϕ_{x_i}
1	0.027 Å	$99 \pm 3\,^\circ$
2	0.005 Å	$-\,24 \pm 5\,^\circ$
3	0.003 Å	$118 \pm 10\,^\circ$

revealed rapid fluctuations in the Σ_4 phonon mode (primarily optical), which involves atomic displacement in the [ζ00] direction. Thus the anomalous behavior of the C_{11} elastic constant could be explained by a *structural* transition arising from a periodic disruption to the equilibrium electron configuration: a CDW.

The charge density wave in uranium is characterized by atomic displacements of a specific periodicity given by (in reciprocal space)

$$\mathbf{q}_{T<22\,\mathrm{K}} = \frac{1}{2}\boldsymbol{a}^* + \frac{1}{6}\boldsymbol{b}^* + \frac{5}{27}\boldsymbol{c}^*, \tag{8.86}$$

where the term involving the reciprocal lattice vector $\boldsymbol{a}^*$ corresponds to lock-in of the transition for the q_x component at 37 K [8] to a commensurate value of 1 / 2. Thus below 37 K, the unit cell for α_1-uranium has essentially *doubled* in the x-direction; the unit cell below 22 K sees a six-fold increase in the y-direction and nearly as much in the z-direction, resulting in a unit cell volume for α_3 of nearly 6,000 Å^3.

Using a diffraction intensity-based model, the actual displacements and phase angles of the atoms at $(0, y, \frac{1}{4})$ were determined by Marmeggi et al. [38] and are listed in Table 8.4, for a displacement vector of the form

$$\mathbf{u}_{jk} = e_{x_1}\cos[2\pi\mathbf{q}\cdot(\mathbf{R}_j + \mathbf{r}_k) \pm \phi_{x_1}] \tag{8.87}$$

$$e_{x_2}\cos[2\pi\mathbf{q}\cdot(\mathbf{R}_j + \mathbf{r}_k) \pm \phi_{x_2}] \tag{8.88}$$

$$e_{x_3}\cos[2\pi\mathbf{q}\cdot(\mathbf{R}_j + \mathbf{r}_k) \pm \phi_{x_3}], \tag{8.89}$$

for lattice vector $\mathbf{R}_j$ and position $\mathbf{r}_k$ within the cell.

To elucidate the source of the CDW, it would be necessary to perform electronic structure calculations; by the late 1990s, computational infrastructure and ab initio methods had matured sufficiently to fulfill such a task. Curiously, it was found that the lack of correlation effects (which could have contributed to an SDW) could be ignored in favor of the local density approximation [15], significantly expediting the computation of the electronic structure of α-uranium.

The work of Fast et al. demonstrated that nesting of Fermi surface features was a necessary condition for the formation of the α_1 and (possibly) α_2 transitions by identifying the nesting vectors π / a and π / $6b$ and noting that the CDW did not occur when certain symmetry points associated with the nesting were omitted [21]. A decade later, computation of the phonon dispersion relations as a function of

Fig. 8.32 Nonlinear current–voltage relationship for a CDW-phased material

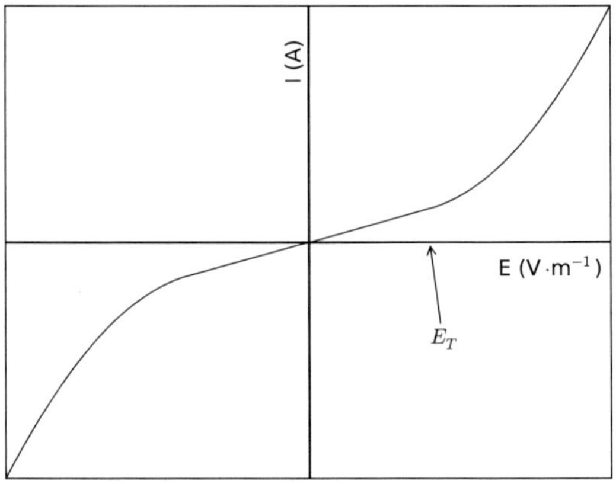

applied hydrostatic stress by Bouchet et al. would predict suppression of the CDW through the hardening of the Σ_4 mode to acoustic-like behavior [8]. Subsequent inelastic X-ray scattering experiments by Raymond et al. and computation of the Fermi surface at elevated pressures would benchmark the predictions by Bouchet et al. and confirm that while nesting of Fermi surface features was necessary, it was not sufficient. Rather, strong electron–phonon coupling was required to enable the formation of a CDW.

Just as defects (disruptions to the periodicity, e.g., impurities, dislocations, etc.) scatter single-particle electrons, they also serve to "pin" or otherwise inhibit the movement of the collective mode in the CDW, until a sufficient electric field E_T is applied. The critical field E_T serves to impart substantial nonlinearity to the conductivity relationship in a CDW material, as illustrated in Fig. 8.32. Pinned charge density waves can be transported through the lattice by electric fields and act as additional current carriers, contributing to the overall conductivity [29]

$$\sigma(\omega) = \frac{n_c e^2}{\mathrm{i}\omega m^*} \frac{\omega^2}{\omega_0^2 - \omega^2 - \mathrm{i}\omega/\tau}, \tag{8.90}$$

where m^* is once again the effective mass, n_c is the charge density in the CDW direction, and ω_0 is termed the pinning frequency. The relaxation time τ can be approximated [29]

$$\tau \approx \frac{\tau_N m^*}{m_{\mathrm{band}}}, \tag{8.91}$$

where the relaxation time for the uncondensed electrons $\tau_N = \sigma m^*/ne^2$.

Estimates for the pinning frequencies and critical fields based on excess specific heat are provided in Table 8.5.

Table 8.5 Values for the pinning frequencies and critical fields of the collective modes in the three CDW phases of uranium [42]

Phase	ω_0 (GHz)	E_T $(\mathrm{V \cdot m^{-1}})$
α_1	75	1.45×10^3
α_2	45	8.9×10^3
α_3	55	13.3×10^3

8.4.4.2 Superconductivity of Uranium

Another somewhat-contested feature of uranium is the superconducting transition, which occurs as high as 2 K at elevated pressures where the CDW phase is suppressed [35]. It was first supposed that superconductivity was filamentary, but later results would indicate that it was a bulk phenomenon. Recent de Haas–van Alphen experiments down to 20 mK, however, suggest that bulk superconductivity is actually impurity/defect driven, as Graf et al. were unable to observe a superconducting transition down to that temperature [27].

8.4.5 Opportunities in Uranium NDE

The unique properties and behaviors of uranium over a wide range of temperatures provide abundant prospects for NDE applications. Indeed, the nonlinear relationships between a host of physical quantities suggest that their careful assessment may progress the NDE community substantially. For high-value uranium products in particular, quantification of the temperature dependency of various properties will surely lead to significant improvements in understanding and qualifying microstructural features in this fascinating metal.

8.4.5.1 Anisotropy

Perhaps an otherwise undesirable characteristic of metallic uranium, the directional dependence of its physical properties lends to the NDE community a variety of quantities to measure—from which significant inferences can be made. In the case of polycrystalline uranium samples, the Zener intercrystalline thermoelastic effect of Sect. 8.2.5.7 can be observed in ultrasonic attenuation. Crystallographic texture and grain structure may be observed by careful measurement of the conductivity and Hall coefficient for a particular sample orientation due to the extreme variation observed in the conductivity and susceptibility of single crystal uranium—particularly in the [100] direction [9, 47]. The nonlinear electronic and elastic contributions may be significant, owing to the unique arrangement of atoms found in α-uranium.

8.4.5.2 Temperature-Dependent Properties

Numerous physical properties have been shown to change dramatically and exhibit anomalies when temperature is varied (with the exception of magnetic properties, rather surprisingly) [35]. Quantification of the elastic and attenuative behaviors as a function of temperature can reveal the interplay between defect structures and the lattice or impurities. Shifts in phase transitions may provide insight into the preferred orientation of polycrystalline specimens or qualify residual stresses in uranium. Conductivity measurements over a range of frequencies will surely cast light on the nature of the electron–phonon interactions which give uranium its unique properties.

8.4.5.3 Low Temperature Observations

If a specific application lends itself to cryogenic temperatures, the CDW and superconducting transitions below 43 K may offer significant opportunities for nondestructive evaluation of uranium. Here it is the nonlinear behavior of the current–voltage relationship in the CDW phase which is perhaps of greatest utility to the NDE community.

Application of an ultrasonic wave will compress and dilate individual crystallites, possibly leading to subdomains which are suitable for CDW or superconducting transitions, providing the nonlinear behavior associated with those phases. Different defects will pin the CDW to varying degrees, changing the harmonic-generating behavior in ways which may be reliably correlated. Further ultrasonic analysis at low temperatures will allow the observation of the electron- and phonon interactions addressed in Sects. 8.2.5.9 and 8.2.5.5. Finally, uranium forms a number of heavy-fermion compounds with elements of interest to the nuclear field [25] such that their precipitation may substantially alter the low temperature behavior of uranium.

8.5 Case Studies in UNDE

The principles of this chapter have been recently applied to the development of nondestructive evaluation techniques in a handful of uranium applications. Ultrasound has been used to detect low levels of carbon, residual stresses have been characterized by eddy current analysis, and work on nonlinear behavior resulting from high electrical current densities forms an ongoing research effort. Salient features and results from each work are provided herein.

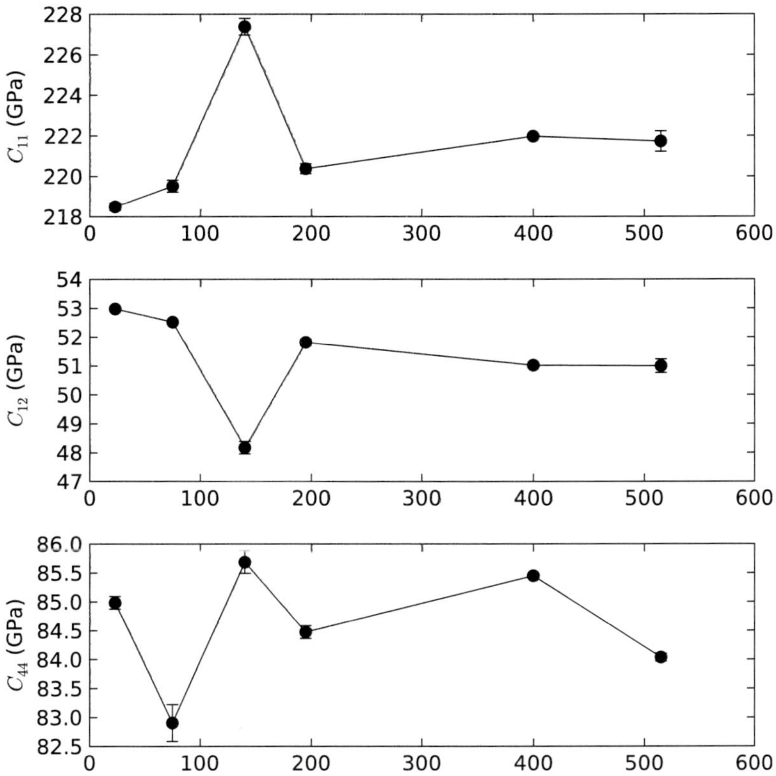

Fig. 8.33 Elastic constants of polycrystalline metallic uranium samples with varying carbon additions. The elastic tensor is assumed cubic to allow for variation in Poisson's ratio and determined by resonant ultrasound spectroscopy

8.5.1 Uranium Solute Analysis by Ultrasound

In the production of metallic uranium, commonly employed manufacturing technologies introduce various impurities to the detriment of subsequent processing and/or performance. Proper characterization of the extent to which certain elements are present is crucial, but current techniques are destructive and time consuming. Substantial concern over the carbon content in particular has initiated development of an analytical method which is sensitive to its presence irrespective of form: as interstitials in solution or as the secondary compound uranium monocarbide (UC), which forms the predominant carbon-containing phase in high purity uranium. The field of physical acoustics, as Sect. 8.2.5 briefly attests, offers a wealth of information encompassed by electronic, phononic, and defect-oriented properties—the potential for ultrasonic measurements to fulfill the required role is clear.

Initial focus was upon reliable quantification of the *elastic* properties of uranium as a function of carbon, which were obtained by resonant ultrasound spectroscopy (RUS). Figure 8.33 demonstrates the relationship between the moduli (C_{11}, C_{12},

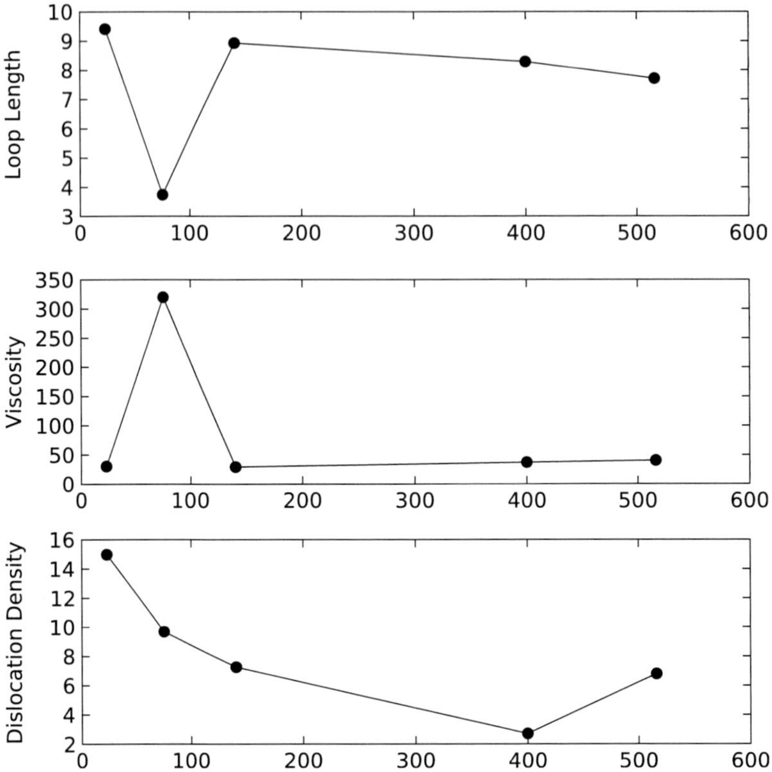

Fig. 8.34 Dislocation loop length (μm), viscosity (μPa s), and dislocation density (10^{15} m^{-2}) for additional carbon concentrations (ppm by weight) as determined by the model-based least-squares data reduction scheme outlined in the text

and C_{44}) and carbon content. The near-linear behavior is interrupted at a carbon content of about 140 ppm (by weight), which coincides with the limit of the γ-phase field in the uranium–carbon phase diagram [7]. Curiously, anomalous behavior is observed in the shear modulus C_{44} commensurate with the limit for the α phase to within 10 ppm (wt.) of the solubility quoted [7].

Analysis of the shear wave attenuation as a function of frequency formed the remainder of the study, leading to the development of a data reduction scheme for a dislocation-based model (c.f. Sect. 8.2.5.4 and (8.29) therein). Shear attenuation dispersion curves up to 5 MHz were fit to the Granato–Lücke equation by least-squares minimization using the Levenberg–Marquardt algorithm, yielding the dislocation loop length L, damping parameter or viscosity B, and density N for the assumed (010) slip system in the [100] direction (the most active slip system at room temperature [12]). Figure 8.34 highlights the features of the dislocation parameters as carbon is added; all parameters are of orders of magnitude consistent with available literature [6].

The dislocation loop length (Fig. 8.34, top) and damping parameter (center) exhibit behaviors consistent with the shear modulus (with discontinuities at carbon contents of 75 ppm by weight) but are otherwise linearly related to the carbon content. One may very well expect such divergence very near the solubility limit of a phase (i.e., prior to the precipitation of UC) as the number of dislocation pinning sites is at a maximum (decreasing the loop length) and the tension overdamps the vibration of the defects, manifest in the damping parameter (viscosity).

As carbon is added to the uranium, the dislocation density decreases by nearly an order of magnitude up to approximately 400 ppm (by weight). It is surmised that the carbon present in solution serves to lower the number of dislocations which are able to interact with the ultrasonic wave by providing additional pinning sites, resulting in lower values of dislocation density as determined by this technique. The subsequent rise in dislocation density may be attributable to the nucleation and growth of uranium monocarbide precipitates, increasing the overall lattice strain during solidification.

While the results are very interesting and show a great deal of promise, additional research and testing on the uranium–carbon and other systems will significantly improve understanding of such a complex material.

8.5.2 Strain Measurements of Machined Uranium

Residual stresses are considered internal stresses that exist in a body of material with no external applied force formed by nonuniform plastic deformation. Discrepancies exist in literature for a consistent description of residual stress classification. Dieter [17] claims that residual stress, although created by plastic deformation, should be considered to be strictly elastic. Moreover, the yield stress of the material is the maximum achievable residual stress [17]. McGonnagle [40], however, introduces two kinds of cold-working induced residual stresses: "elastic through a large portion of the material, and plastic microstresses within the deformed grains." Further specificity for the characterization of residual stress is found in the work by Fitzpatrick and Lodini [26], who classify the residual stress into three categories—first, second, and third order (or type I, type II, and type III, respectively), giving rise to the following definitions for residual stress.

- Residual stresses of the first order, or type I residual stresses, are homogeneous over very large number of crystal domains of the material. Such stresses are also termed *macrostresses*. The internal forces related to this stress are balanced on all planes. The moments related to these forces are equal to zero around all axes.
- Residual stresses of the second order, or type II residual stress, are homogeneous within small crystal domains of the material (e.g., a single grain or phase). The internal forces related to these stresses are in balance strictly between the different grains or phases.
- Residual stresses of the third order, or type III residual stress, are homogeneous on the smallest crystal domains of the material (over a few interatomic distances). The internal forces coupled to these stresses are in balance in very

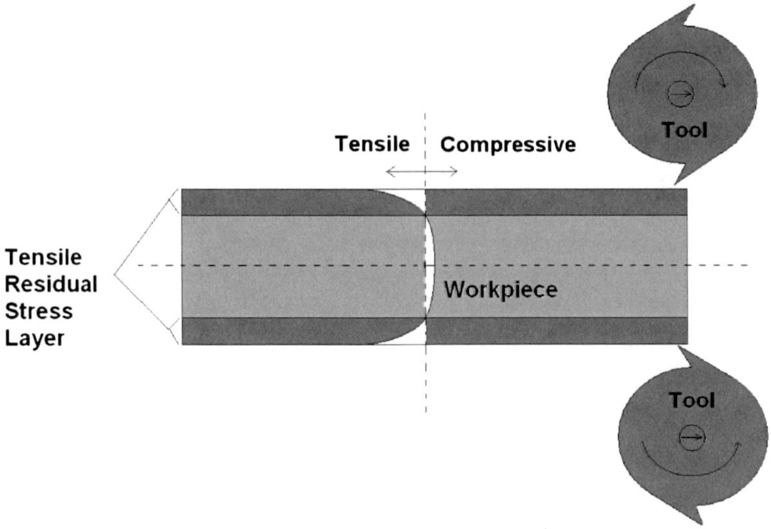

Fig. 8.35 This cross section schematic illustrates the residual stress layer that is formed from a metal removal process

small domains (such as around dislocations or point defects). Type II and III residual stresses are collectively termed *microstresses*.

Withers [51] has noted that type II (second order) stresses nearly always exist in polycrystalline materials due to the fact that the elastic and thermal properties are distinct between randomly oriented grains. In this discussion, milling and turning processes which produce plastic flow at the surface will be considered the sources of residual stress.

Deformation depth is dependent upon the metal removal process. The grains underneath the deformed surface remain unaffected while the surface grains are deformed. Thus, the unaffected grains have to accommodate for changed surface grains with residual strain opposing the surface traction. Typically, compressive residual stress will form with a tensile plastic strain deformation process [17]. However, once heat is produced by friction, tensile residual stresses will form. Equilibrium must be achieved in the workpiece with the residual stress system as shown in Fig. 8.35 with the counterbalance of the compressive and tensile stresses, residual stress manifests in a three-dimensional stress state, increasing the complexity of the system [17].

8.5.2.1 Nondestructive Strain Measurements

Strain measurements utilizing low-frequency-induced (eddy current) impedance and thermoelectric power were performed by VanHorn et al. [50]. Impedance measurements are a non-contact measurement and allow for strain to be quantified

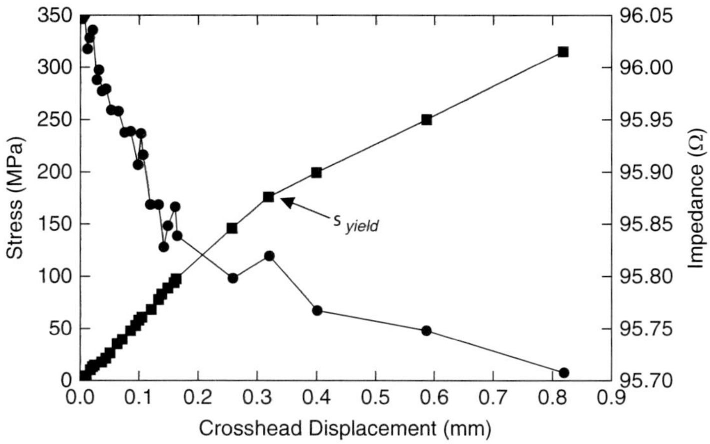

Fig. 8.36 Stress (*filledsquare*) and induced impedance (*filledcircle*) measurements taken at 20 Hz as a function of crosshead displacement for a uranium bar

at different depths into the material depending on the frequency utilized, as per (8.58). The thermoelectric power coefficient is measured via contact with the workpiece surface and is dependent on the strain field directly beneath the probe tip.

Figure 8.36 shows low frequency induced impedance measurements as a function of strain performed on a depleted uranium bar in a four-point bend. The impedance value decreases as the strain increases and was consistent with other data demonstrated by the investigators [50]. Figure 8.36 shows how an induced impedance nondestructive tool can be calibrated for different strains. Once a calibration curve is obtained on a specific composition, the tool can be utilized to obtain measurements on a machined surface to nondestructively determine how much strain the machining operation has introduced into the uranium.

Further demonstrating the process, Fig. 8.37 shows how induced impedance measurements were obtained over the convex surface of a plastically strained uranium bar (removed four point bend fixture, i.e., no applied load). The low-frequency-induced impedance measurements show that strain can be characterized nondestructively and when the uranium bar was stress relieved by vacuum anneal, the impedance measurement showed that strain was relieved. Since the applied load was removed from the bar, a residual compressive stress remains on the convex side of the bar (tensile residual stress on the concave side of the bar), and again this was consistent with the calibration curve initially obtained.

Similar to low-frequency-induced impedance, thermoelectric power coefficient strain measurements can be calibrated to a typical stress–strain curve then used on machined parts to determine residual stress on the surface nondestructively as shown in Fig. 8.38.

Based on the testing performed by VanHorn et al., it can be concluded that low-frequency-induced impedance can measure Type I residual stresses while the thermoelectric power coefficient is a qualifier of Type II and Type III residual stresses [50].

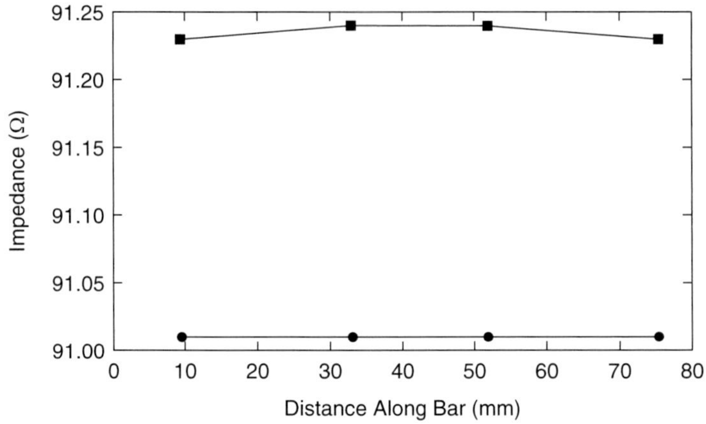

Fig. 8.37 Induced impedance measurements taken at 20 Hz along the outside radius of a bent uranium bar obtained before (*filledsquare*) and after (*filledcircle*) annealing

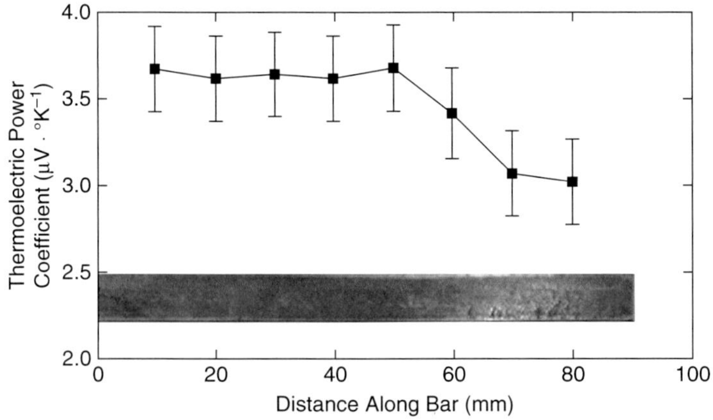

Fig. 8.38 Themoelectric power coefficient as a function of position along the uranium bar. The uranium bar is shown in the background of the plot while the pictures above the chart zoom in on the machining striations which differed from the rest of the bar and indicate the presence of higher residual stresses

8.5.3 Waves in Materials Processing

It has recently been observed that high current density electric pulses can be applied to metals to achieve grain refinement [55, 56]. The concept has been termed *electropulse processing* and has been applied to the more common metals. It is proposed in this section that electropulse processing is a viable option for grain refinement in uranium and that careful analysis of the pulse waveform can provide substantial insight into microstructural development in uranium.

8.5.3.1 Electropulse Processing

Grain refining mechanisms which occur commensurate with high current density pulses are fundamentally tied to atomic transport in the material. The pulse causes diffusion due to highly localized resistive heating, electromigration, and chemical potential gradients (c.f. (8.76)). These driving forces promote the nucleation and growth of new crystal grains, resulting in a decrease in the average grain size in the metal such that effectively the grains are refined.

During electropulse processing, an electrical potential is briefly applied to the metal. The electric field creates drag forces [22] which are typically coalesced into a single expression [31],

$$F_{\text{em}} = Z^* \frac{e\mathbf{E}}{kT}, \tag{8.92}$$

where Z^* is the effective valence and $\mathbf{E}$ is electric field.

The force given by (8.92) creates a transport phenomenon collectively termed *electrotransport*. Effects of the free electron "wind" and electrostatic forces are accounted for with the effective valence term.

Resistive heating occurs when an electric current is applied to a metal by an electropulse. Since most metals contain various inhomogeneities (e.g., grain boundaries or precipitates) with varying degrees of conductivity, the highly localized heating will result in steep thermal gradients which will provide another driving force for atomic diffusion [31],

$$F_{\text{th}} = -\frac{Q^*}{T} \nabla T, \tag{8.93}$$

where Q^* is the heat of transport for the particular process, taking into account phononic, electronic, and intrinsic contributions to heat flow.

The final electropulse driving force is promoted by the chemical potential gradients caused by the aforementioned electromigration and thermal effects. The chemical potential force is simply written as

$$F_{\mu} = -\mu_i, \tag{8.94}$$

for chemical potential μ of species i.

Collectively, the three atomic transport forces described can make solid state phase transformations thermodynamically favorable for short periods of time. Transformations are most likely to occur in regions where the Joule heating term dominates, as shown in Fig. 8.39.

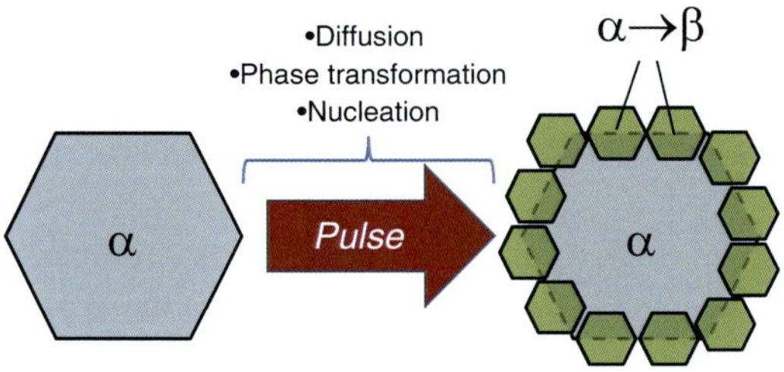

Fig. 8.39 A model of the grain refinement process caused by the delivery of high current density pulses to a metallic system. New grains nucleate and grow in regions of substantial variation in resistivity owing to the steep thermal gradients witnessed there

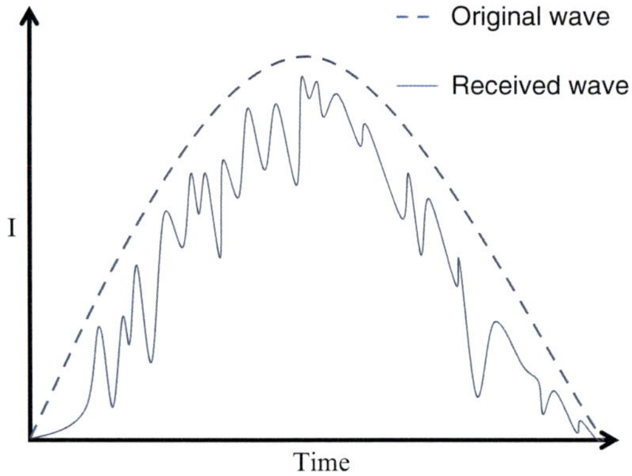

Fig. 8.40 An exaggerated view of a pulsed waveform deforming as it encounters various nonlinear electrical features within the microstructure

8.5.3.2 Electropulse Analysis

In response to its interactions with microstructural features, an applied current pulse will attenuate and, in the case of a sufficiently nonlinear system, distort as it propagates as shown in Fig. 8.40. Thus it stands to reason that careful analysis of the changes encountered by the pulse would allow inference of microstructure and its evolution. Such a procedure, referred to as *electropulse sensing* and illustrated in Fig. 8.41, would be of significant value in the processing of high-value metallic components—uranium in particular.

Among the exotic properties of metallic uranium which will play an important role in the success of electropulse processing/sensing for uranium, the single crystal anisotropy is perhaps the most dominant. Differences in electrical conductivity, thermal conductivity and expansion, and elastic constants between grains will develop substantial localized stresses to the lattice where significant thermal gradients are present. Dilatational stresses may induce nucleation of new grains directly owing to the pressure dependency of phase stabilities in uranium, wherein

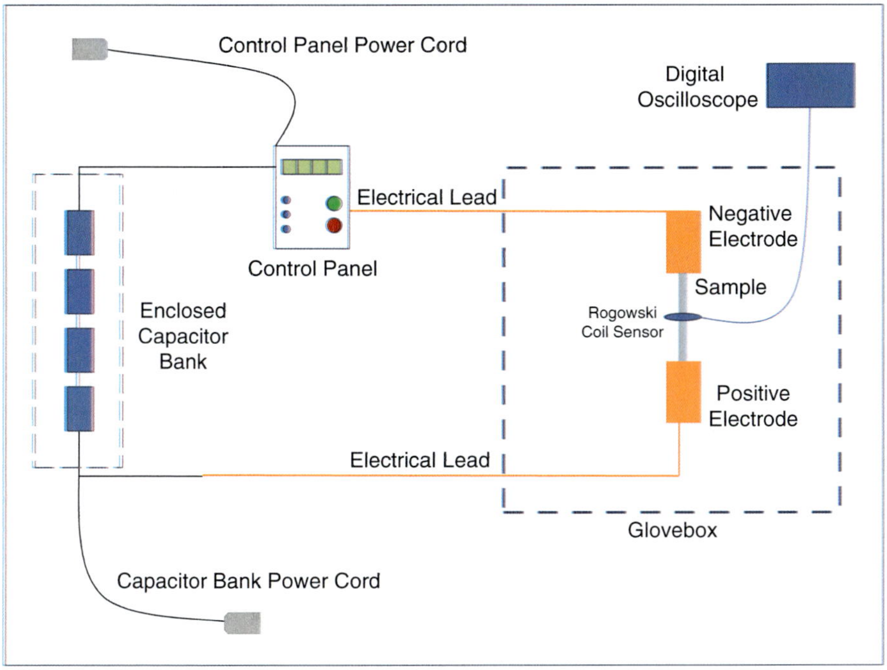

Fig. 8.41 Experimental apparatus for electropulse processing/sensing

the γ-phase field is expanded to lower temperatures. Alternatively, the limited solubility of a number of elements in uranium will lend to the metal a propensity for developing inhomogeneities—at grain boundaries in particular. Such processes will surely impart nonlinear behavior to the electropulse sensing scheme, dramatically improving its efficacy as an online nondestructive technique.

8.6 Future Perspective

Much work remains to be done before a text such as this can perform the role of a *review* of uranium NDE, as such a feature would surely be deemed impotent in light of the scarcity of current NDE applications in uranium. It is possible that nondestructive techniques will develop naturally in the course of the trickling down of scientific progress from basic research to practical application, but it is ultimately the necessity, not the interest, which gives birth to invention.

It has been the goal of this text to inspire the reader with a vision for the tremendous opportunity and challenges that uranium offers to the field of nondestructive evaluation and to offer pertinent insight into the methods by which the field may be extended to such a fantastic material. In particular, the related fields of physical acoustics, nonlinear wave mechanics, and solid state physics have progressed substantially in the past half century and are maturing to the point where their application to uranium NDE is judicious, perhaps necessary.

References

1. Ashcroft NW, Mermin DN (1976) Solid state physics. Brooks/Cole, Cengage Learning, Belmont, CA
2. Balanis CA (1989) Advanced engineering electromagnetics. Wiley, Hoboken, NJ
3. Barrett C, Mueller M, Hittermann R (1963) Crystal structure variations in alpha-uranium at low temperatures. Phys Rev 129(2):625–629
4. Berlincourt TG (1959) Hall effect, resistivity, and magnetoresistivity of Th, U, Zr, Ti, and Nb. Phys Rev 114(4):969–977
5. Beyer RT, Letcher SV (1969) Physical ultrasonics. Pure and applied physics, vol 32. Academic Press, New York, NY
6. Bhatia A (1967) Ultrasonic absorption an introduction to the theory of sound absorption and dispersion in gases, liquids and solids. Dover Publications, New York
7. Blumenthal B (1960) Constitution of low carbon u-c alloys. J Nucl Mater 2(3):197–208
8. Bouchet J (2008) Lattice dynamics of α uranium. Phys Rev B 77:024,113–1–7
9. Brodsky M, Griffin N, Odie M (1969) Electrical resistivity of alpha uranium at low temperatures. J Appl Phys 40:895–897
10. Cahn J, Hilliard J (1958) Free energy of a non-uniform system I. Interfacial free energy. J Chem Phys 28:258
11. Cahn J, Hilliard J (1959) Free energy of a non-uniform system II. Thermodynamic basis. J Chem Phys 30:1121–1124
12. Cahn R (1953) Plastic deformation of alpha-uranium; twinning and slip. Acta Metall 1:49–70
13. Cantrell J, Salama K (1991) Acoustoelastic characterisation of materials. Int Mater Rev 36(4):125–145
14. Cantrell JH (2004) Ultrasonic nondestructive evaluation engineering and biological material characterization, chap. 6. CRC Press, Boca Raton, FL
15. Chantis AN, Albers R, Jones M, Schilfgaarde Mv, Kotani T (2008) Many-body electronic structure of α-uranium. Phys Rev B 78:081,101–1–4
16. Crummett W, Smith H, Nicklow R, Wakabayashi N (1979) Lattice dynamics of α-uranium. Phys Rev B 19(12):6028–6037
17. Dieter G (1986) Mechanical metallurgy, 3rd edn. McGraw-Hill, New York
18. Dobmann G, Meyendorf N, Schneider E (1997) Nondestructive characterization of materials: A growing demand for describing damage and service-life-relevant aging processes in plant components. Nucl Eng Design 171(1–3):95–112
19. Du P, Kibbe WA, Lin SM (2006) Improved peak detection in mass spectrum by incorporating continuous wavelet transform-based pattern matching. Bioinformatics 22(17):2059–2065
20. Elmore WC, Heald MA (1969) Physics of waves. McGraw-Hill Book, New York
21. Fast L, Eriksson O, Johansson B, Wills J, Straub G, Roeder H, Nordström L (1998) Theoretical aspects of the charge density wave in uranium. Phys Rev Lett 81(14):2978–2981
22. Fiks V (1964) Ion entrainment by electrons in metals. Phys Lett 9:299–300
23. Fisher E, McSkimin H (1958) Adiabatic elastic moduli of single crystal alpha-uranium. J Appl Phys 29(10):1473–1484
24. Fisher E, McSkimin H (1961) Low-temperature phase transition in alpha uranium. Phys Rev 124(1):67–70
25. Fisk Z, Sarrao J, Smith J, Thompson J (1995) The physics and chemistry of heavy fermions. Proc Natl Acad Sci USA 92:6663–6667

26. Fitzpatrick M, Lodini A (2003) Analysis of residual stress by diffraction using neutron and synchrotron radiation. Taylor and Francis, London
27. Graf D, Stillwell R, Murphy T, Park J, Kano M, Palm E, Schlottmann P, Bourg J, Collar K, Cooley J, Lashley J, Willit J, Tozer S (2009) Fermi surface of α-uranium at ambient temperature. Phys Rev B 80(24):241,101–1–4
28. Granato A, Lücke K (1956) Theory of mechanical damping due to dislocations. J Appl Phys 27(6):583–593
29. Grüner G (1988) The dynamics of charge-density waves. Rev Mod Phys 60(4):1129–1181
30. Hart E (1959) Thermodynamics of inhomogeneous systems. Phys Rev 113:412–416
31. Huntington H (1972) Electro- and thermomigration in metals. In: Diffusion. Seminars presented at an ASM conference, Metals Park, OH, pp. 155–183. ASM
32. Kittel C (1955) An electron transfer mechanism for ultrasonic attenuation in metals. Acta Metall 3:295–297
33. Landau L, Lifshits E, Pitaevskii L (1984) Electrodynamics of continuous media, Landau and Lifshitz course of theoretical physics, vol 8, 2nd edn. Butterworth-Heinemann (Elsevier), Oxford, UK
34. Landau L, Lifshitz E (1986) Theory of elasticity, Course of theoretical physics, vol 7, 3rd edn. Butterworth-Heinemann (Elsevier), Oxford, UK
35. Lander G, Fisher E, Bader S (1994) The solid-state properties of uranium a historical perspective and review. Adv Phys 43(1):1–111
36. Lander G, Mueller M (1970) Neutron diffraction study of α-uranium at low temperatures. Acta Crystall B 26:129–136
37. Marmeggi J, Delapalme A, Lander G, Vettier C, Lehner N (1982) Atomic displacements in the incommensurable charge-density wave in alpha-uranium. Solid State Comm 43(7):577–581
38. Marmeggi J, Lander G, Smaalen S, Brückel T, Zeyen C (1990) Neutron-diffraction study of the charge-density wave in α-uranium. Phys Rev B 42(15):9365–9376
39. Mason W (1958) Physical acoustics and the properties of solids. D. Van Nostrand Company, Princeton, NJ
40. McGonnagle W (1961) Nondestructive testing. McGraw-Hill, New York
41. Migliori A, Sarrao JL (1997) Resonant ultrasound spectroscopy: applications to physics, materials measurements, and nondestructive evaluation. Wiley, New York, NY
42. Mihaila B, Opeil C, Drymiotis F, Smith J, Cooley J, Manley M, Migliori A, Mielke C, Lookman T, Saxena A, Bishop A, Blagoev K, Lashley J, Lang B, Boerio-Goates J, Woodfield B, Schmiedeshoff G (2006) Pinning frequencies of the collective modes in α-uranium. Phys Rev Lett 96:076,401–1–4
43. Morse R (1955) Ultrasonic attenuation in metals by electron relaxation. Phys Rev 97(6):1716–1717
44. Randall R, Rose F, Zener C (1939) Intercrystalline thermal currents as a source of internal friction. Phys Rev 56:343–349
45. Rosen M (1968) Elastic moduli and ultrasonic attenuation of polycrystalline uranium from 4.2 to 300 ° K. Phys Lett 28A(6):438–439
46. Rosenberg H (1955) The thermal conductivity of metals at low temperatures. Phil Trans R Soc Lond A 247:441–497
47. Ross J, Lam D (1968) Magnetic susceptibility of single-crystal alpha-uranium. Phys Rev 165(2):617–620
48. Schuch A, Laquer H (1952) Low temperature thermal expansion of uranium. Phys Rev 86(5):803
49. Tu K (1985) Interdiffusion in thin films. Ann Rev Mater Sci 15:147–176
50. VanHorn C, Olson D, Mishra B (2008) Stress corrosion cracking analysis on machined pure uranium utilizing thermoelectric power coefficient and low frequency induced impedance advanced nondestructive tools. Master's thesis, Colorado School of Mines
51. Withers P, Bhadeshia H (2001) Residual stress: Part 1 - measurement techniques. Mater Sci Tech Ser 17:355–365

52. Zener C (1938) II. General theory of thermoelastic internal friction. Phys Rev 53:90–99
53. Zener C (1939) Intercrystalline thermal currents as a source of internal friction. Phys Rev 56:343–349
54. Zener C (1948) Elasticity and anelasticity of metals. University of Chicago Press, Chicago
55. Zhang W, Sui M, Zhou Y, Li D (2003) Evolution of microstructures in materials induced by electropulsing. Micron 34:25–35
56. Zhou Y, Zhang W, Wang B, Guo J (2003) Ultrafine-grained microstructures in a Cu-Zn alloy produced by electropulsing treatment. J Mater Res 18:1991

Chapter 9
High-Density, Low-Enriched Uranium-Based Target for Radioisotope Production

Gary L. Solbrekken, Kyler K. Turner, and Srisharan G. Govindarajan

Abstract Demand for a reliable, economic, steady supply of radioisotopes for medical diagnostic imaging is growing globally. The current process most commonly used for Molybdenum-99 production, the most widely used radioisotope, is through fission product recovery from uranium targets that have been neutron irradiated in a reactor. Current targets use high-enriched uranium (HEU) as the seed material. Nonproliferation efforts made by the United States and other partner nations are leading to the elimination of HEU. A replacement target strategy is therefore needed. Fission-product recovery using high-density, low-enriched uranium targets is explored in this chapter.

The development of a new target for reactor-based irradiation production requires that safety constraints imposed by the reactor be met. Those safety constraints are typically in the form of a maximum temperature limit and a fission-product release limit. They effectively imply that a target be structurally sound while providing effective heat transfer paths from the fissioning uranium and the reactor coolant. A unique feature of the high-density target described in this chapter is the monolithic structure of the uranium—basically that it is a piece of metal foil as opposed to a dispersion compact. The monolithic structure provides challenges as far as evaluating the thermal and mechanical design of the target. Analysis suggests that safe irradiation of a high-density target is possible. Further coupled analysis with the target assembly process is needed to provide robustness to the analysis described in this chapter.

G.L. Solbrekken (✉) • S.G. Govindarajan
Mechanical and Aerospace Engineering, University of Missouri-Columbia,
E2411 Lafferre Hall, Columbia, MO, USA
e-mail: solbrekkeng@missouri.edu; sgkff@mail.missouri.edu

K.K. Turner
Mechanical and Aerospace Engineering, University of Missouri-Columbia,
Columbia, MO, USA
e-mail: Kyler.K.Turner@gmail.com

J.S. Morrell and M.J. Jackson (eds.), *Uranium Processing and Properties*,
DOI 10.1007/978-1-4614-7591-0_9, © Springer Science+Business Media New York 2013

Keywords LEU foil • Dispersion • Annular • Plate • Target • Thermal-mechanical • Nonproliferation • Radioisotope • Thermal contact resistance • Medical Isotopes • Molybdenum-99

9.1 Introduction

9.1.1 A Brief History of Nuclear Medicine

Radiation has been used since its discovery in medicine to both diagnose and treat various ailments. In 1895 the German physicist Wilhelm Konrad Roentgen observed crystals of barium platinocyanide luminescing when placed near a highly evacuated electrical discharge tube. Experimentation indicated that the density of the object greatly affected the attenuation of this luminescent energy and would ultimately lead to the development of X-ray imaging technology, the first use of radiation in the medical field [1].

In 1932 Cockcraft and Walton produced the first man-made neutron-deficient isotopes using a particle accelerator. Shortly after, in 1938, Dr. Robley D. Evans of the Massachusetts General Hospital and members of the Massachusetts Institute of Technology used iodine-138 to observe isotope uptake in the thyroid gland of a rabbit. In 1950 the first localized radiation scanner was developed and designed specifically for the study of iodine-138 uptake in the thyroid gland. Soon after the first whole-body scanner was developed by Donner Laboratory. In the 1960s technology continued to advance developing to the point where nuclear medicine was considered a medical specialty in its own right [1].

9.1.2 Diagnostic Radiopharmaceuticals

Approximately 95 % of all radiopharmaceuticals are used for diagnostic purposes. Diagnostic radiopharmaceuticals are designed to develop internal images of the patient while minimizing tissue damage. In contrast, therapeutic radiopharmaceuticals are designed to damage the target tissue. Radiopharmaceuticals consist of two components, the radionuclide which is the radioactive isotope, and the pharmaceutical which is the delivery agent. The characteristics of both will determine how the radiopharmaceutical is administered and used. Designing of a radiopharmaceutical starts by finding a nontoxic pharmaceutical that demonstrates an affinity for the target organ such as the heart. Next the radionuclide is paired (tagged) with this pharmaceutical and injected into the patient. The uptake of the radiopharmaceutical and the emission of radiation that follows allow organ imaging [2].

The ideal diagnostic radiopharmaceutical will have several characteristics making it both safe and effective. First, a reliable and readily available supply should exist.

Second, the effective half-life, a combination of the physical half-life and the biological half-life, should be short minimizing patient dose. Finally, a diagnostic radiopharmaceutical must emit a gamma ray allowing for the radiation to escape the patient's body for imaging [2].

9.2 Reactor-Based Isotopes and Production: Technetium-99m

Technetium-99m is only one example of the many medical man-made isotopes utilized today. Technetium-99m is the product of molybdenum-99 decay, has an atomic number of 43, and a half-life of 6-h, and is in a metastable state [3]. The nuclear decay equations, (9.1), (9.2), and (9.3), show the complete decay chain of molybdenum-99 into stable ruthenium-99:

$$^{99}\mathrm{Mo} \rightarrow {}^{99m}\mathrm{Tc} + \beta \tag{9.1}$$

$$^{99m}\mathrm{Tc} \rightarrow {}^{99}\mathrm{Tc} + \gamma \tag{9.2}$$

$$^{99}\mathrm{Tc} \rightarrow {}^{99}\mathrm{Ru}_{\mathrm{Stable}} + \beta^- \tag{9.3}$$

where Mo is molybdenum, Tc is technetium, Ru is ruthenium, γ is a gamma ray, and β^- is a beta particle. The decay process is shown schematically in Fig. 9.1.

In 1965 Brookhaven National Lab produced the first technetium-99m generator that allowed for the collection, or "milking," of the technetium-99m from molybdenum-99 decay [4]. Technetium-99m's short half-life makes stockpiling and long-distance transportation difficult; thus molybdenum-99's longer half-life is used to increase stockpiling and transportation distance. The parent–daughter relationship between molybdenum-99 and technetium-99m is a nonequilibrium relationship meaning, the daughter's half-life is shorter than the parent's half-life. The half-lives of molybdenum-99 and technetium-99m allow for the milking or removal of technetium-99m from the generator, once a day. Column chromatography is used to separate the two isotopes during the milking process. This process separates the isotopes into negatively charged particles of TcO_4^- and MoO_4^{2-}. The single negative charge on the TcO_4^- adheres to the alumina sorbent in the generator

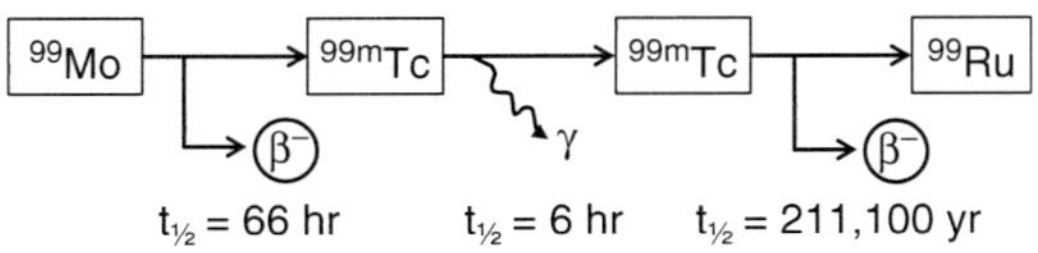

Fig. 9.1 Molybdenum-99 decay process schematic. Note that Tc process has a radioactive half-life of 6 h, making it ideal for diagnostic applications. The decay time to Ru-99 far exceeds the biological half-life

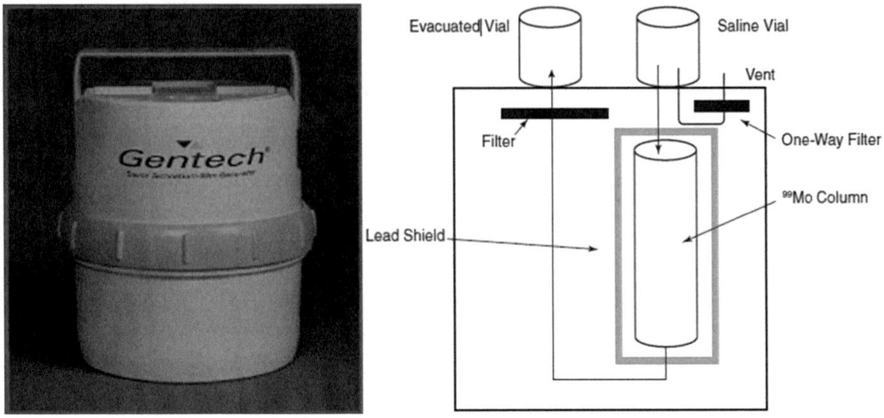

Fig. 9.2 Technetium-99m generator and diagram illustrating the internal structure and its function [5]

far less than the double-negative charge on the MoO_4^{2-}. A saline solution is used to flush the free-floating TcO_4^- from the generator producing a saline solution full of technetium-99m [4]. An image of a technetium-99m generator and a diagram illustrating the internal layout of a generator can be seen in Fig. 9.2.

Technetium-99m, from a medical standpoint, is the ideal radiopharmaceutical for several reasons. First, the technetium-99m generator is easy to transport and store with limited exposure. Shielding required for the average activity transported is approximately 2 cm of lead. Second, a 6-h half-life is short, minimizing the patient's radiation dose. On average a person injected with technetium-99m will return to background radiation levels within 2.5 days. Finally, technetium-99m only emits gamma rays at 140 keV, allowing the gamma rays to leave the patient's body while limiting the radiation dose administrated.

9.3 Reactor-Based Techniques in the Context of Nonproliferation Issues

9.3.1 Current Fission-Based Molybdenum-99 Production Techniques with High-Enriched Uranium

The most common process used for molybdenum-99 production throughout the world uses high-enriched uranium (HEU) powder in a dispersion target that is irradiated in a conventional reactor. In this process HEU is mixed in powder form with aluminum powder. The two powders are then heated and compressed between two aluminum plates creating a solid plate target. The completed target has a uranium–aluminum core and an aluminum picture frame and cladding [6].

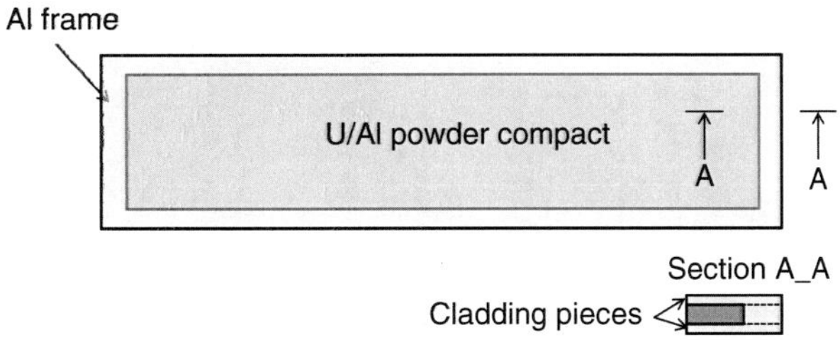

Fig. 9.3 Conventional HEU dispersion plate target

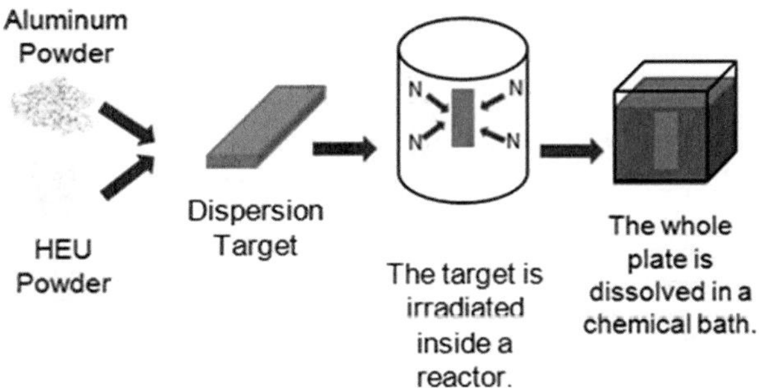

Fig. 9.4 Process overview for HEU dispersion target

A diagram showing the aluminum frame and uranium–aluminum core can be seen in Fig. 9.3.

The solid nature of the HEU dispersion target ensures that there is no gas gap between the uranium and the aluminum leading to a low thermal contact resistance through the target. The molybdenum-99 is collected by dissolving the entire plate through chemical processing [7]. A simplified diagram showing this process can be seen in Fig. 9.4.

An annular geometry dispersion target is another currently used HEU target type. In essence, a flat dispersion plate target is curved to form an annular cylinder. The annular geometry is optimized for high production rates in certain reactors. After irradiation the target is dissolved in a process similar to the dispersion plate target [8]. A diagram of the annular target is shown in Fig. 9.5.

Another HEU target type used in molybdenum-99 production is a pin target. In this design the uranium–aluminum core is arranged in pins similar to pressurized water reactor (PWR) and boiling water reactor (BWR) fuel elements. The target core is surrounded by a finned outer aluminum shell to improve target cooling. After irradiation the target is dissolved in a process similar to the dispersion plate and annular dispersion targets [9]. An image of the pin target can be seen in Fig. 9.6.

9.3.2 Current Molybdenum-99 Reactors and Producers

There are currently six reactors used by high-volume molybdenum-99 producers, most of which use HEU [7, 10]. Table 9.1 provides a summary of location and

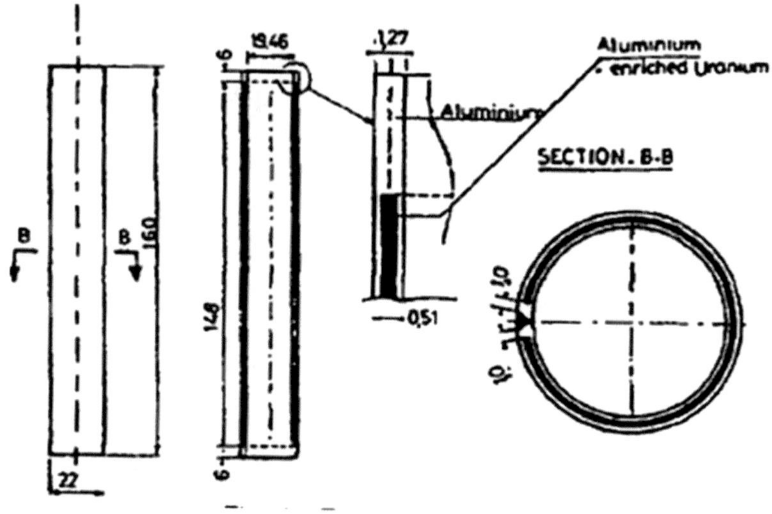

Fig. 9.5 Drawing of HEU annular dispersion target [8]

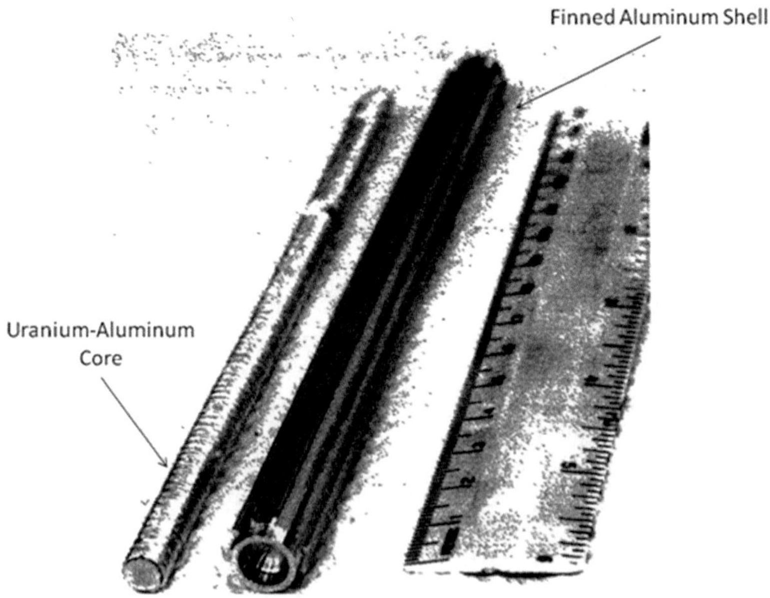

Fig. 9.6 HEU pin-type target design [9]

Table 9.1 Reactors used by high-volume producers [7, 10]

Reactor name	Location	Owner	Target type
NRU	Chalk River, Canada	AECL	HEU pin
HFR	Petten, The Netherlands	European Commission	HEU dispersion plate
BR2	Mol, Belgium	SKC-CEN	HEU dispersion, annular
Osiris	Saclay, France	CEA	HEU dispersion plate
SAFARI-1	Pelindaba, South Africa	NECSA	HEU and LEU dispersion
OPAL	Lucas Heights, Australia	ANSTO	LEU dispersion plate

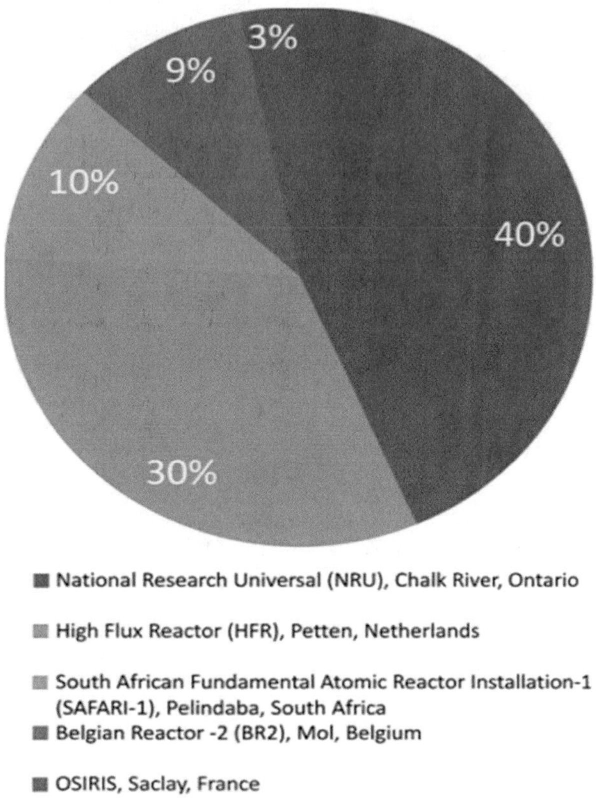

Fig. 9.7 Percentage of molybdenum-99 supplied by global reactors [11]

target type for each reactor. The global molybdenum-99 production fractions are shown in Fig. 9.7, without considering the contribution from OPAL.

A few reactors also create small amounts of molybdenum-99 for domestic and limited regional use with both RA-3 and GA SIWABESS Y MPR using LEU targets. Table 9.2 summarizes these regional reactors.

After irradiation, targets are shipped to the molybdenum-99 processers where they are dissolved, extracting the molybdenum-99. The bulk molybdenum-99 is then shipped to generator manufacturers where it is packaged into technetium-99m

Table 9.2 Reactors used by regional producers [7, 10]

Reactor name	Location	Owner	Target type
RA-3	Buenos Aires, Argentina	CNEA	LEU dispersion plate
GA SIWABESS Y MPR	Batan, Indonesia		LEU dispersion plate

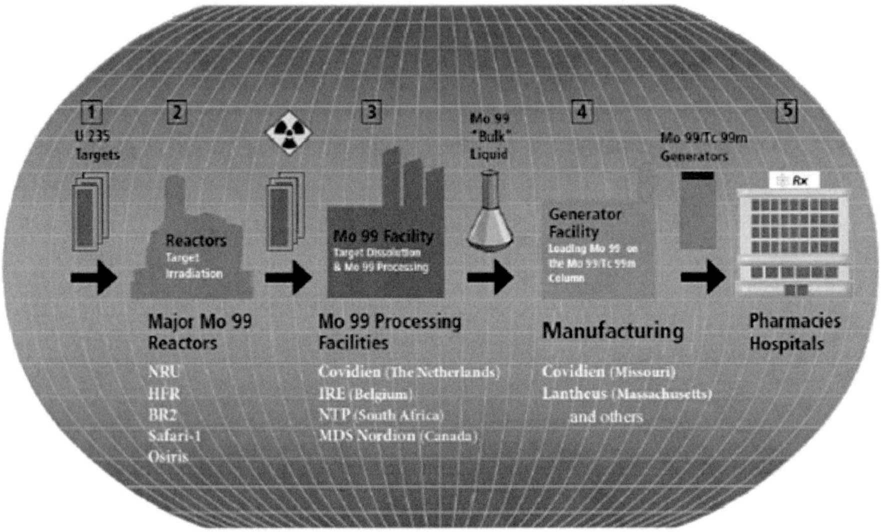

Fig. 9.8 Global supply chain of molybdenum-99 [10]

generators before finally being shipped to a radiopharmacy or a hospital. Figure 9.8 illustrates, in brief, the global molybdenum-99 supply chain.

There are three high-volume producers that dissolve irradiated targets. Covidien is located in the Netherlands and produces approximately 40 % of the US supply and about 25 % of the global supply from targets supplied by HFR and BR2. IRE, located in Belgium, produces about 20 % of the global supply from targets supplied by the HFR, BR2, and Osiris reactors. MDS Nordion located in Canada provides 60 % of the US supply and 40 % of the global supply from targets irradiated at the NRU reactor. NTP Radioisotopes located in South Africa produces 10 % of the global supply and can assist with supply to the United States during shortages with targets coming from the SAFARI-1 reactor [7].

9.3.3 *The US Government Molybdenum-99 Efforts: The Office of Global Threat Reduction*

The office of the Global Threat Reduction Initiative (GTRI), within the US Department of Energy, exists to "reduce and protect vulnerable nuclear and radiological materials located at civilian sites worldwide [12]." As part of their efforts, GTRI has

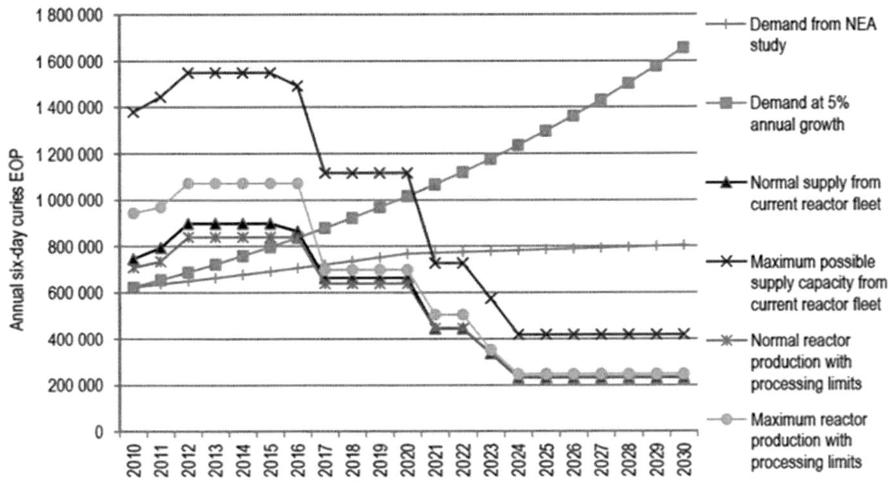

Fig. 9.9 Market projections for the current molybdenum-99 production rate [15]

been an international leader in the development of alternate technologies to ensure a safe, reliable source of molybdenum-99 [13]. Their strategy has been to fund cooperative agreements with interested providers for developing non-HEU methods. Approaches include neutron capture, solution reactor, accelerator systems [14], and fission-based solutions. If the technical targets for the new technologies are achieved, a sustainable supply of molybdenum-99 will be made available.

9.3.4 Economic Issues with the Current Molybdenum-99 Production Market

The current molybdenum-99 production industry is a complex and fragile system that is susceptible to any change in market conditions. Following the large supply shortages in 2008 and 2010 the Organization for Economic Co-Operation and Development's Nuclear Energy Agency (OECD-NEA) analyzed the current molybdenum-99 market and provided recommendations for its improvement [15].

The OECD-NEA established the High-Level Group on the Security of Supply of Medical Radioisotopes (HLG-MR) in 2009 to specifically address this issue. The HLG-MR includes representatives from the molybdenum-99 production industry and governments where molybdenum-99 is produced. The OECD-NEA began by first examining the current molybdenum-99 production market and examining the ability of the current market structure to provide molybdenum-99 in the future. Their findings showing a range of scenarios can be seen in Fig. 9.9.

OECD-NEA projections indicate that the current molybdenum-99 production community will be unable to support demand beginning in 2021 time frame. When including the current market with new entrants, the projections on supply appear

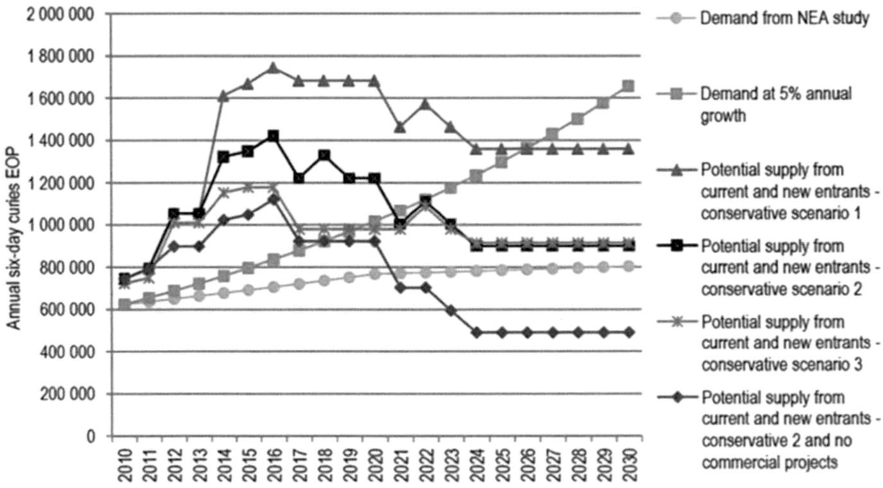

Fig. 9.10 Market projections for molybdenum-99 production rate with new producers [15]

much more sustainable. Figure 9.10 shows that meeting the demand will be easier with new entrants. However the first HLG-MR concluded and developed a set of six principles needed to establish this market condition in 2011:

- First, all of the participants involved in the molybdenum-99 production chain should implement full-cost recovery including cost related to capital replacement. The OECD-NEA concluded that a removal of subsidies and a move to a full-recovery market is the only way to create an economical sustainable production market.
- Second is management of the production reserve capacity, which addresses the ability of the production community to supply molybdenum-99 during reactor outages. OECD-NEA recommends reactors coordinate their schedules more closely to ensure that waste in the supply chain is minimized.
- Third is that government should not directly supply molybdenum-99 to the market and that its role should be in the establishment of proper infrastructure for full-cost recovery. The role of government has been extremely detrimental to the molybdenum-99 market and has been one of the greatest deterrents for new entrants in the market through government subsidization.
- Fourth, the government through its international commitments to nuclear non-proliferation and security should provide appropriate support to the reactors and processers in the conversion to LEU. Governments can implement this by supporting R&D on new LEU-based targets providing an economical way of producing molybdenum-99 with LEU. It is also recommended that the International Atomic Energy Agency (IAEA) be used as the conduit to share technologies developed with the global community.

- Fifth, international collaboration should continue as molybdenum-99 is an international issue and implementation of this policy will ensure a level playing field for all producers.
- Sixth, the OECD-NEA's final recommendation is a periodic review of the market and supply chain insuring that producers are implementing full-cost recovery, thus ensuring that an economically sustainable market exists [15].

The implementation of the HLG-MR's principles is extremely important in ensuring the establishment of a fair and full-cost recovery market favorable to new entrants and non-HEU-based production. It is imperative for the current supply chain and production community transition to ward off the possibility of another extended shortage comparable to those in 2008 and 2010.

9.4 LEU Reactor-Based Targets

9.4.1 Advantages of the LEU Foil Target

While alternate molybdenum-99 production solutions are being explored, it is likely that international molybdenum-99 production will continue to be dominated by fission-based production for the foreseeable future. A major drawback with converting to LEU-based production using the current dispersion-based target design is that the total uranium density (U-235 and U-238) is relatively low, on the order of 2 g/cm^3. When one transitions from HEU to LEU-based targets, the ratio of fissile U-235 to U-238 decreases from about 9:1 to below 0.25:1. For a target of a given geometry and mass, this suggests that the production of molybdenum-99 will be correspondingly reduced.

The reduction in production can be estimated by using a simple activity model that assumes linear behavior and does not consider reactor-specific effects. Such a model can be written as [8]

$$A_{\mathrm{Mo99}} = \frac{1}{k}\left(\frac{N_{\mathrm{A}}}{M_{\mathrm{U235}}}\right)\left(\gamma_{\mathrm{Mo99}}\right)\left(\sigma_{\mathrm{f,\,U^{235}}}\right)\left(\phi_{\mathrm{th}}\right)\left(1 - \mathrm{e}^{-\frac{\mathrm{Ln}(2)}{t_{1/2\mathrm{Mo99}}}t_{\mathrm{irr}}}\right) \tag{9.4}$$

The model estimates the activity of the target (A_{Mo99}) after irradiation with units of $\mathrm{Ci_{Mo99}/grams_{U\text{-}235}}$. $M_{\mathrm{U\text{-}235}}$ is the molar mass of U-235, N_{A} is Avagadro's number, γ_{Mo99} is fission yield of molybdenum-99, $\sigma_{\mathrm{f,U\text{-}235}}$ is fission cross section, ϕ_{th} is thermal neutron flux, t_{irr} is irradiation time and $t_{½\ \mathrm{Mo99}}$ is molybdenum-99 half-life. The parameter k is a unit conversion constant depending on what units are used for all the other parameters. While the results may not account for all production parameters, including irradiation shielding, they do allow the effect of U-density to be explored. Figure 9.11 shows that by increasing the U-density of the target, the activity rate of molybdenum-99 increases linearly. The figure also shows that when converting from 90 % enriched HEU material to 19.75 % enriched LEU

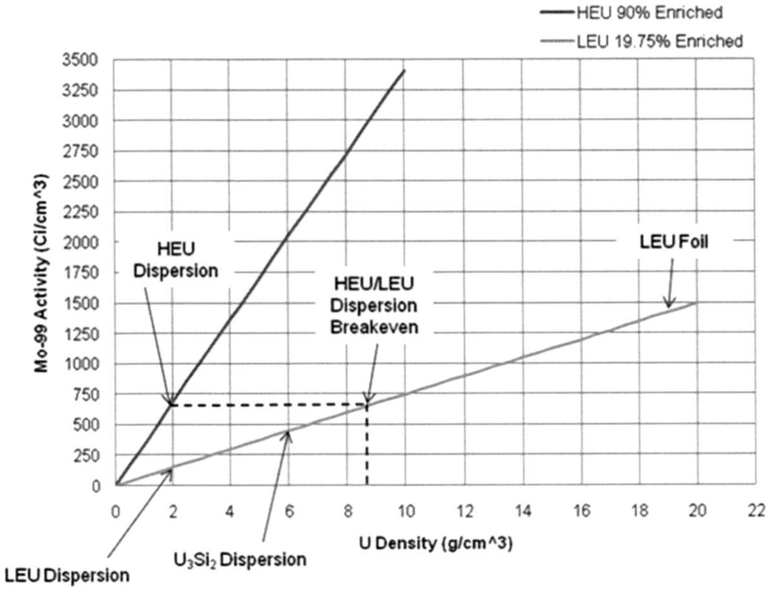

Fig. 9.11 Estimate of Mo99 activity assuming a 7-day irradiation time, 6 % fission yield for Mo99, and a thermal neutron flux of 2.0×10^{14} n/s cm^2

material, there will be a significant decrease in molybdenum-99 activity for a given target U-density. If one assumes that a dispersion target can be fabricated with a 2 g/cm^3 U-density, the molybdenum-99 production rate will be reduced by about a factor of 4 when transitioning from HEU to LEU. In order to recover the original production activity using LEU material it would be necessary to increase the U-density to about 8.5 g/cm^3. This is a significant jump in current target fabrication technology, and therefore unlikely to occur.

The density of U-foil is around 19 g/cm^3. If one considers using a target based on a foil of U-metal, it is apparent from Fig. 9.11 that the molybdenum-99 activity increases beyond the point of production with an HEU dispersion target. While it is not suggested that actual production gains based on LEU foil will be realized, the results from (9.11) do warrant further study into the potential of high-density, LEU foil targets.

9.4.2 High-Density Target Concepts

There are three nominal target designs based on high-density LEU foil considered here. They are the annular, flat plate, and curved plate designs.

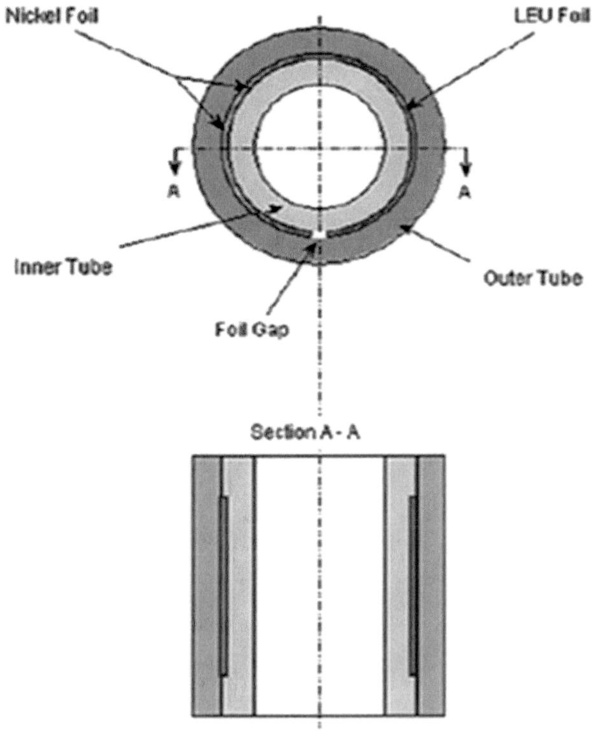

Fig. 9.12 Cross-sectional view of the LEU annular target concept [18]

9.4.2.1 Annular

The Argonne National Lab (ANL) pioneered the development of the annular geometry target [16]. It is a design that consists of two concentric aluminum cylinders, welded at their ends, containing the LEU foil and a nickel recoil barrier. The nickel recoil barrier prevents the LEU foil and the aluminum cladding from fusing together during irradiation. A sketch of the approach can be seen in Fig. 9.12 with an image of a mock target in Fig. 9.13. The LEU foil and nickel recoil barrier are placed between two aluminum cylinders which are drawn together using a plug die, or some other expansion technique. The ends of the cylinders are welded to completely seal the LEU foil within the cladding.

The ANL annular target was not necessarily designed with high-volume production or the existing reactor infrastructure in mind. It was, however, designed to reduce the amount of liquid waste that results after post-irradiation chemical dissolution. The monolithic structure of the LEU, coupled with the Ni recoil barrier, allows the target to be cut open and the LEU foil removed for chemical dissolution using the modified Cintichem process [17]. The solid aluminum cladding does not have to be dissolved, and thus can be disposed of as solid waste. Pakistan Research

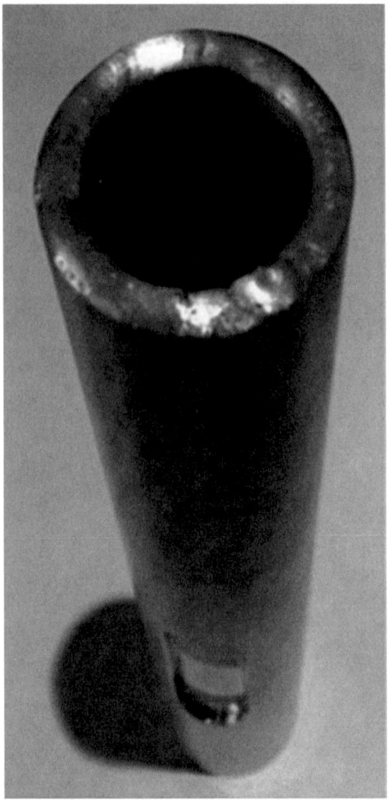

Fig. 9.13 Image of a mock annular target [19]

Reactor-1 (PARR-1) has performed several safety test irradiations on the Argonne annular design. The results from various computer codes indicate safe operation in PARR-1 [10].

9.4.2.2 Plate/Curved

There is no reason to believe that a plate-based target design based on the Argonne annular target could not be successful. In fact, one can consider the annular target to be a special case of a more general plate design concept. The plate design can be curved to the point that its ends meet reproducing the annular geometry. Further, current irradiators use plate designs for their current dispersion targets, and thus have supporting infrastructure, like irradiation rigs, designed for plate geometries.

A nominal flat plate target design has been developed at the University of Missouri (MU) and is designed for economical high-volume molybdenum-99 production. The nominally flat plate LEU foil-based concept is illustrated in Fig. 9.14, with photographs of flat and curved plate mock-ups shown in Fig. 9.15.

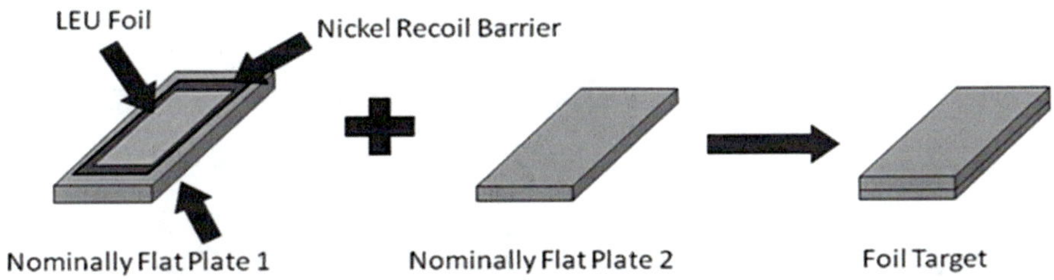

Fig. 9.14 LEU foil-based, nominally flat plate target design concept [19]

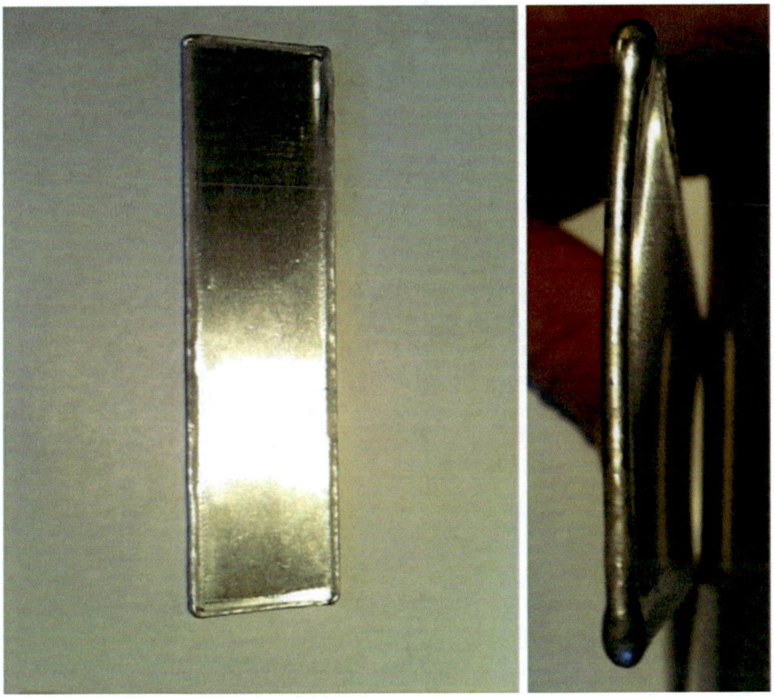

Fig. 9.15 Mock-ups of a flat plate (*left*) and curved plate (*right*) target

The MU concept involves placing the LEU foil between two pieces of aluminum cladding and sealing the edges. Sealing, in the form of a weld, prevents the escape of fission products into the reactor cooling system. After irradiation, the welds may be removed allowing for easy retrieval of the LEU foil, which is then chemically processed using the modified Cintichem process, obtaining molybdenum-99 [17].

The laminated structure of all the LEU foil target designs suggests that there is an opportunity for gaps to open between the fissioning LEU foil and the cladding. The annular design is expected to mitigate gap opening more effectively than the flat plate. As a means to improve the plate geometry resistance to 'pillowing', curvature can be introduced to the plate. As illustrated in the photo mock-up in Fig. 9.15, the

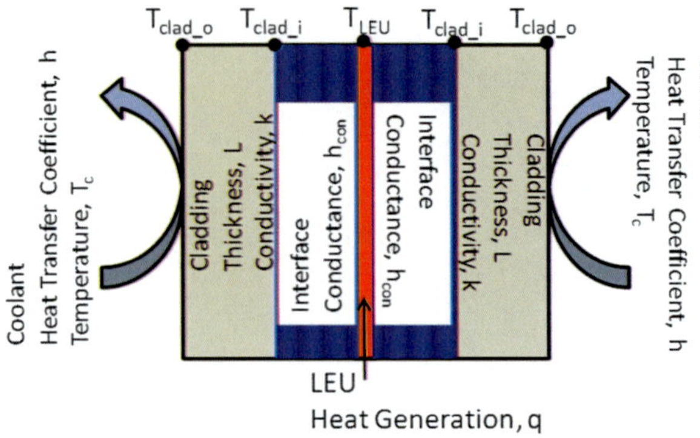

Fig. 9.16 Cross-sectional view of LEU foil-based target. Note that the components are not drawn to scale, especially the interface. Temperatures labeled at the *top* of the figure are assumed to prevail over the entire vertical plane based on the 1-D assumption

amount of curvature is slight, sufficient to bias any potential "pillowing" to one side of the target, or the other. But even as with the annular target design the thermal/mechanical behavior of the high-density, LEU foil targets must be evaluated.

9.4.3 Technical Considerations for Target Design

Safe irradiation of a fission-based target requires that the structure be mechanically sound and not have any component exceed the prescribed maximum temperature noted in a reactor safety case. Mechanical soundness can often be described in terms of how close the materials used in the target structure approach some fail criteria when undergoing irradiation. Neither the mechanical soundness nor the maximum temperature achieved by the target can be easily measured during irradiation, so both need to be evaluated using modeling tools to demonstrate the likeliness that any violations will be possible.

One of the major concerns in using high-density foil targets is the interface between the target meat and the cladding. The interface provides a point of weakness in the target structure where imperfect bonding causes a rise in temperature due to thermal contact resistance between the heat-generating foil undergoing fission and the cladding. Indeed, for targets that are to be disassembled in order to reduce the amount of mass that needs to be dissolved, an imperfect bond is required. The question then becomes one of how much thermal contact resistance can be tolerated.

A generalized thermal resistance model can be used to estimate the temperature distribution within a target, including the temperature of the LEU foil. For first-order estimates, a 1-D resistance network can be used for both annular and plate geometries to provide the temperature distribution estimates. If one looks at either the geometric cross section of a plate geometry shown in Fig. 9.14 or a cross section of an annular geometry shown in Fig. 9.12 a similar profile can be observed as illustrated in Fig. 9.16. The resulting thermal resistance network for either geometry with assumed bidirectional cooling is shown in Fig. 9.17. It is assumed in the

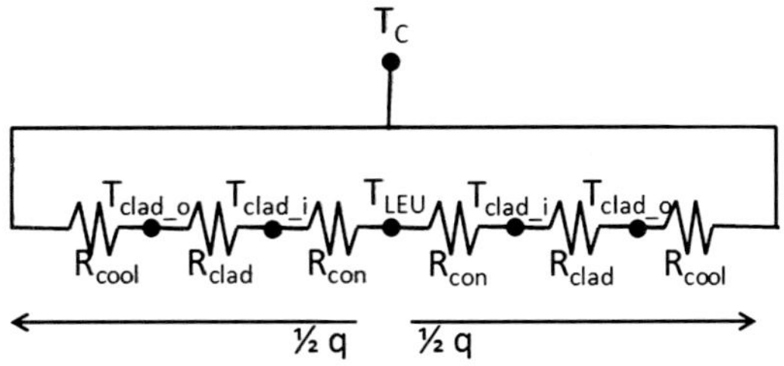

Fig. 9.17 Thermal resistance network for LEU foil-based target. The resistors line up with the spatial sketch in Fig. 9.16

resistance network that the thermal resistance between the LEU and the coolant is the same in both directions such that the heat flow, q, is evenly split between the two paths as suggested in Fig. 9.17. If the resistance between the two paths is different from one another then the heat flow split must be modified appropriately.

In order to estimate each of the thermal resistances it will be necessary to determine the heat transfer mode represented by each resistor. The R_{cool} resistances represent the convective cooling provided by the reactor coolant. R_{clad} is the thermal resistance through the solid clad material, and will be assumed to be by conduction. The contact thermal resistance, R_{con}, is the thermal resistance at the interface between the LEU foil and the clad material. It is known that there is a temperature difference between two mating materials and that the magnitude of the temperature difference is caused by the geometric structure of the interface and the thermal properties of the mating materials. Therefore it is necessary to understand the geometric structure of the mating materials, especially during irradiation when temperature gradients have a tendency to distort the materials. As the materials distort there is a potential for gaps to open between the heat-generating LEU foil and the cladding. The distortion of the materials also creates mechanical stresses within the cladding and at the welds that need to be withstood in order to prevent any fission-product release to the reactor coolant. The rest of this chapter explores the thermal-mechanical behavior of foil-based targets, starting with simplified models to help estimate the operating envelope and then evolving to more complicated models that can be calibrated with experimental data to provide more precise target behavior estimates.

9.4.3.1 Convective Thermal Resistance

In-depth analysis of the convection thermal resistance, R_{cool}, is beyond the scope of this chapter. The general expression for the convection thermal resistance is

$$R_{\text{cool}} = \frac{1}{h_{\text{cool}}A} = \frac{(T_{\text{clad\_o}} - T_{\text{C}})}{\frac{1}{2}q} \tag{9.5}$$

The area, A, is the nominal surface area on one side of the target that is exposed to the coolant, assumed to be the same as the LEU foil footprint. Keep in mind that the total surface area exposed to the coolant on both sides of the target will be $2A$.

The coolant heat transfer coefficient, h_{cool}, can be found using a variety of techniques described in texts like those written by Bejan [20] or Burmeister [21] or Kays and Crawford [22]. The specific value for h_{cool} to be used in a given reactor will depend on the geometry of the irradiation position, the coolant flow velocity, and the properties of the fluid. For the analysis presented here, a range of heat transfer coefficients will be used as part of the parametric studies. A typical value of heat transfer coefficient for water flow with velocities of 5 m/s is about 10,000 W/m^2 K [23].

9.4.3.2 Cladding Thermal Resistance

The thermal resistance for the cladding material, R_{clad}, can be estimated by assuming 1-D conduction through the clad material. The form of the resistance equation depends on whether the target geometry is nominally planar (plate) or cylindrical (annular). For a planar-type target the thermal resistance can be found from

$$R_{cond} = \frac{L}{kA} = \frac{(T_{clad\_i} - T_{clad\_o})}{\frac{1}{2}q} \tag{9.6}$$

The thermal resistance for a cylindrical target is

$$R_{cond} = \frac{\ln(r_{clad\_o}/r_{clad\_i})}{2\pi kH} = \frac{(T_{clad\_i} - T_{clad\_o})}{\frac{1}{2}q} \tag{9.7}$$

The outer radius of the annular target is r_{clad_o} and the inner radius of the annular target is r_{clad_i}. The LEU height is H. It should be noted that for targets with very thin cladding (on the order of 1 mm), it is possible to use the planar form of the conduction thermal resistance with minimal error. Figure 9.18 shows that using the planar equation to calculate the conduction thermal resistance results in less than 10 % error in thermal resistance estimate for tubes that are of an inner diameter that is larger than 2.54 cm. Typically the cladding thermal resistance will be small compared to the convection and contact thermal resistances.

9.4.3.3 Contact Thermal Resistance

For any two materials with different material constants (atomic lattice structure) that are in contact with one another, there will always be a thermal contact resistance. Swartz and Pohl [24] developed the diffuse mismatch model that describes how phonons are transported across an interface. Their expression for the intrinsic thermal contact resistance is developed in terms of the mating

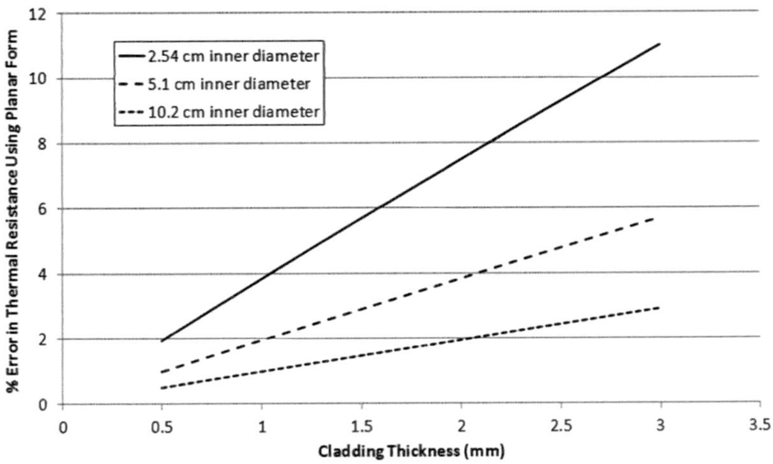

Fig. 9.18 Error associated with using planar thermal resistance expression as opposed to annular resistance expression

materials' heat capacity and provides reasonable trends for materials that are sufficiently different. However for materials which are similar in atomic structure a more accurate approach is needed. Chen and Zeng [25] developed a model that provides a more rigorous accounting of phonon origin, and therefore a more accurate model of the intrinsic thermal contact resistance of similar materials at elevated temperatures. The magnitude of the intrinsic thermal contact resistance is usually much smaller than the overall contact thermal resistance for materials that are larger than 1 μm in thickness. For materials of larger dimensions the thermal contact resistance due to surface mismatch tends to dominate.

Real interfaces will always have a thermal contact resistance that exceeds the intrinsic thermal contact resistance. The difference is due to the fact that the mating materials do not have perfectly aligned lattices, nor are the mating surfaces finished to allow perfect contact. The imperfect contact, suggested in Fig. 9.19, traps gas between the points of solid contact. So instead of heat being conducted between the two solid materials with relatively large thermal conductivities, heat is conducted from solid to solid at the points of material contact in parallel with heat being conducted through the trapped gas. The analysis of the thermal resistance across such an interface is generally conducted by assuming 1-D heat transfer through the two conduction paths.

There are a variety of expressions that have been developed to estimate the thermal contact resistance given surface properties of the mating materials. The general expression used for estimating the thermal contact resistance is typically done in terms of the effective conductance, h_{con}:

$$R_{\mathrm{con}} = \frac{\Delta T}{q} = \frac{1}{h_{\mathrm{con}} A} \tag{9.8}$$

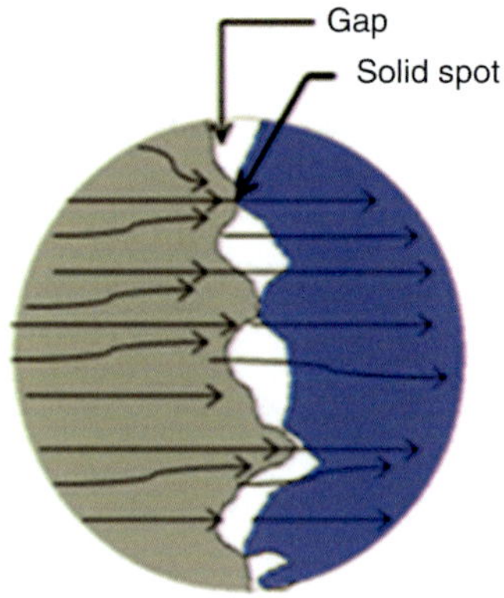

Fig. 9.19 Sketch of real solid surfaces with periodic contact

When both of the mating surfaces are assembled with plastic deformation, an estimate for the effective conductance was derived by Mikic and Rohsenow [26] to be

$$h_{\mathrm{con}} = \left(1.13k\frac{\tan\theta}{\sigma}\right)\left(\frac{P}{H}\right)^{0.94} \tag{9.9}$$

The thermal conductivity used in the expression, k, is the harmonic mean of the thermal conductivity for the two mating materials, k_1 and k_2, and is defined as

$$k = \frac{k_1 + k_2}{2k_1k_2} \tag{9.10}$$

The pressure term, P, is the apparent pressure that is based on the total contact area and the hardness, H, is the microhardness of the softer of the two materials. Finally, the $\tan\theta$ and σ terms are the average slope of the surface asperities and the average surface roughness of the mating materials.

While (9.10) is not always accurate in predicting the thermal contact conductance, it does illustrate the influencing parameters. In particular, the ratio of the applied pressure to microhardness effectively characterizes the ratio of the actual contact area to the apparent contact area, and how much the mating materials can be expected to deform under pressure. The existence of the contact pressure in (9.10) provides a hint that knowledge of the thermal-mechanical structure of the target is important for ensuring as large a contact conductance as possible.

It should be noted that the thermal contact conductance analysis described so far has assumed that the mating surfaces are nominally flat or matched with one another. In reality the mating surface could have any arbitrary waviness or warpage that will influence the overall thermal contact resistance. A study by Chiu et al. [27] illustrated that if it is not possible to have nominally flat surfaces in contact with one another, then it is preferred to have the mating surfaces be convex so that there is a relatively small thermal resistance in the central area of the heated region.

From a practical standpoint it is not easy to characterize the mating surface materials at a level to use (9.10). Experimental characterization is many times a more practical approach to characterize thermal contact resistance. There are a variety of experimental techniques that can be used to evaluate the thermal contact resistance [28–31]. But it is clear from (9.10) that surface finish and contact pressure conditions need to be matched as closely as possible. Thus experiments that have the same heating condition as the real application are needed. For LEU-based targets, however, it is virtually impossible to replicate the internal heat generation due to the fissioning LEU. Therefore the only real practical solution to establish if an LEU foil-based target is indeed safe to irradiate is to conduct studies of bounding scenarios, and then assess the risk that a target will ever be subjected to that set of circumstances. An alternate approach is to create a target design that will be able to withstand the conservatively bounding scenarios.

9.4.3.4 Limiting Thermal Contact Resistance

The limiting behavior for thermal contact resistance is if a gap opens between the LEU foil and the cladding. If it is assumed that the gap is filled with a gas, then a simple assumption is that heat conducts through the gap. Heat conduction through the gap can be estimated using either (9.6) or (9.7), depending on whether the target is of a plate or an annular design. For a sufficiently thin gap the planar model will suffice. If it is assumed that the temperature of the clad material is 100 °C, or 373 K, and that the gap is filled with a gas with thermal conductivity the same as air, then it is possible to determine the LEU temperature for a range of heat flux values and gap thicknesses. Figure 9.20 shows how both higher heat flux values and larger gap openings result in higher LEU temperatures. The figure shows that even though it is not preferred, it is likely possible to tolerate small gaps within the target.

9.5 Thermal/Mechanical Analysis for High-Density, LEU Foil-Based Target

9.5.1 Nondimensional Simply Supported Plate

Understanding the natural thermal-mechanical tendencies of plates subjected to heat loading will give guidance for more complex models. A simple plate with a uniform temperature gradient through the thickness was chosen as a representative case as a closed-form solution to an analytical model can be obtained. The edges of the plate are simply supported meaning they were not allowed to translate in the *X*-, *Y*-, or *Z*-directions. Simply supported also means there is no moment applied to the edge and a point on that edge is free to rotate as if on a hinge. A sketch illustrating the simply supported boundary condition can be seen in Fig. 9.21.

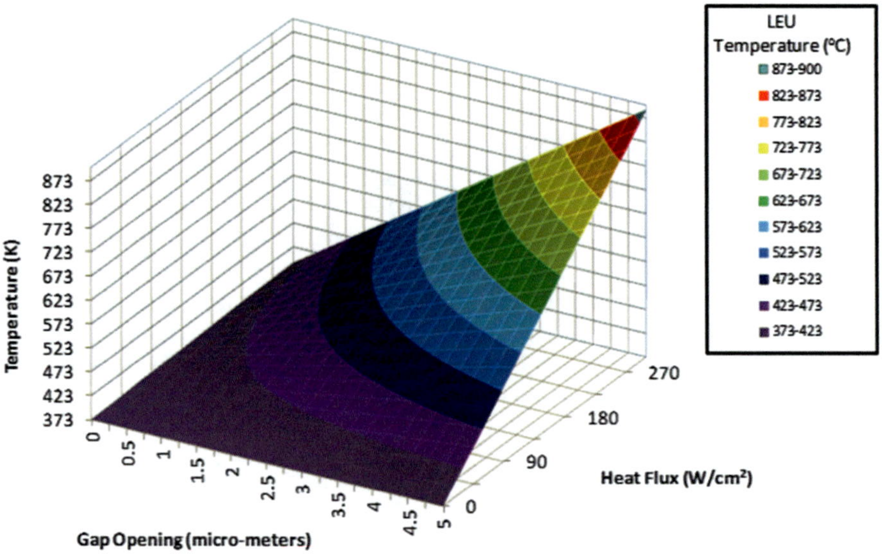

Fig. 9.20 Estimate of thermal contact resistance if a uniform gap opens between the LEU foil and cladding. This is a highly unlikely scenario as the foil will most likely remain in contact with at least periodic contact with the cladding

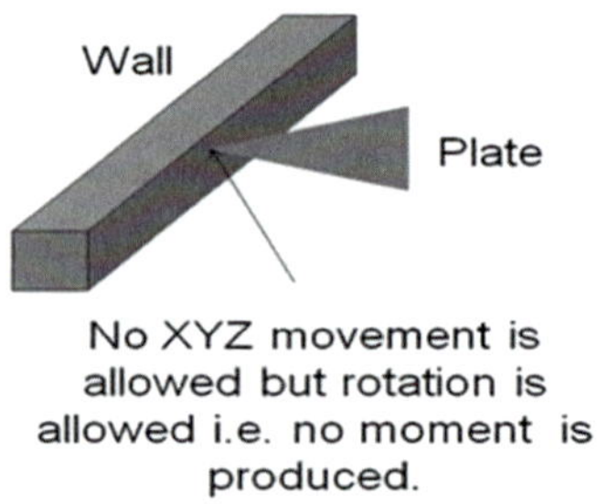

Fig. 9.21 Sketch of the boundary conditions assumed for a simply supported plate model

The focus of this analysis is on the development of nondimensional analytic model describing the thermal-mechanical behavior of a simply supported plate providing direction to the length, width, and thickness selection for a molybdenum-99 production target. Based on the simply supported boundary condition assumptions, it is expected that the deflections obtained from the model will overestimate the amount of "pillowing" that might be expected in a flat plate target with welded edges.

Nondimensional analytical modeling started with a simply supported model based on work done by Noda et al. [32]. A sketch with a labeled plate dimensions can be seen in Fig. 9.22.

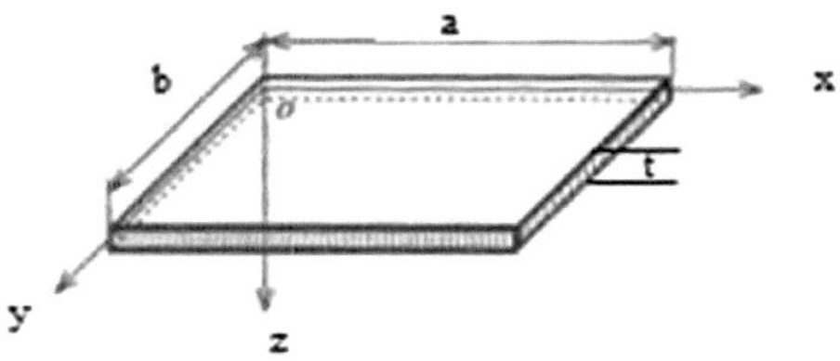

Fig. 9.22 Sketch of plate with labels used for analytic model solution with simply supported edges

The governing equation for a simply supported plate is [32]

$$\nabla^2\omega = -\frac{1}{(1-\upsilon)D}M_{\mathrm{T}} \tag{9.11}$$

The deflection of the plate in the z-direction is ω, the material Poisson's ratio is ν, and the plate stiffness D is defined as

$$D = \frac{Et^3}{12(1-\upsilon^2)} \tag{9.12}$$

In the definition of the plate stiffness, E is the Young's modulus of the plate material and t is the plate thickness. The driving force for the plate deflection in (9.11) is the thermally induced moment described by

$$M_{\mathrm{T}} = \alpha E\int_0^t Tz\mathrm{d}z \tag{9.13}$$

The thermal diffusivity is α, and T is the temperature distribution through the thickness of the plate. For 1-D heat transfer with a uniformly applied heat flux on one side of the plate, the temperature distribution will be a linear function of z.

In order to generalize the results of the solution to the above analytic model, (9.11), (9.12), and (9.13) were non-dimensionalized. The parameters defined in terms of the labels shown in Fig. 9.22 are the following:

$$x^* = \frac{a}{L_{\mathrm{c}}} \tag{9.14}$$

$$y^* = \frac{b}{L_{\mathrm{c}}} \tag{9.15}$$

$$t^* = \frac{t}{L_{\mathrm{c}}} \tag{9.16}$$

$$\omega^* = \frac{\omega - \omega_{\mathrm{o}}}{L_{\mathrm{c}}} \tag{9.17}$$

where all of the star values represent the nondimensional variables, L_c represents the characteristic length taken to be the length of the plate in the x-direction, and ω_o is the initial deflection of the plate. The nondimensional boundary conditions for the model are then

At $x^* = 0$ and 1:

$$\omega^* = \frac{0 - \omega_o}{L_c} = 0 \tag{9.18}$$

At $y^* = 0$ and b/a,

$$\omega^* = \frac{0 - \omega_o}{L_c} = 0 \tag{9.19}$$

Solving the governing equation using Helmholtz's method gives a solution as a function of two infinite summations [33],

$$\omega^*(x, y) = -\sum_{m=1}^{\infty}\sum_{n=1}^{\infty} F_{mn} \frac{\sin\left(\frac{m\pi x^*}{1}\right)\sin\left(\frac{n\pi y^*}{b/a}\right)}{\left(\frac{m\pi}{1}\right)^2 + \left(\frac{n\pi}{b/a}\right)^2} \tag{9.20}$$

where F_{nm} is given by

$$F_{mn} = -\left(\frac{0.014177}{(b/a)t^*}\right)\left(-\frac{1}{m\pi}\cos(m\pi) + \frac{1}{m\pi}\right)\left(-\frac{b/a}{n\pi}\cos(n\pi) + \frac{b/a}{n\pi}\right) \tag{9.21}$$

The general deflection profile for the nondimensional solution for a square plate is shown in Fig. 9.23. The contour plot shows that the edges do indeed appear to hinge while the plate tends to deform into a basket shape. It should be noted that the heat flux is applied on the convex, or exterior, part of the plate, while the concave, or interior, part of the plate is kept at a lower temperature. Thus it appears that the plate has a natural tendency to want to bend INTO the heat source. This is an important behavior to understand as it suggests that the edge welds will see significant stress if the natural tendency of the plate bending is to be overcome.

A parametric study was conducted by varying the aspect ratio and plate thickness of the target and evaluating the maximum plate deflection. A plot of the nondimensional maximum deflection, located in the center of the plate, versus the nondimensional thickness and length and width aspect ratio can be seen in Fig. 9.24. The temperature difference through the thickness of the plate is held constant.

The plot clearly shows that the greatest deflection will occur when the thickness of the plate is small and when the aspect ratio of the plate is equal to 1. The plot also indicates that the thermal-mechanical behavior is nonlinear when both the thickness and aspect ratios are adjusted. The behavior indicates that there is a point when a change in the aspect ratio will not greatly affect the thermal-mechanical deflection.

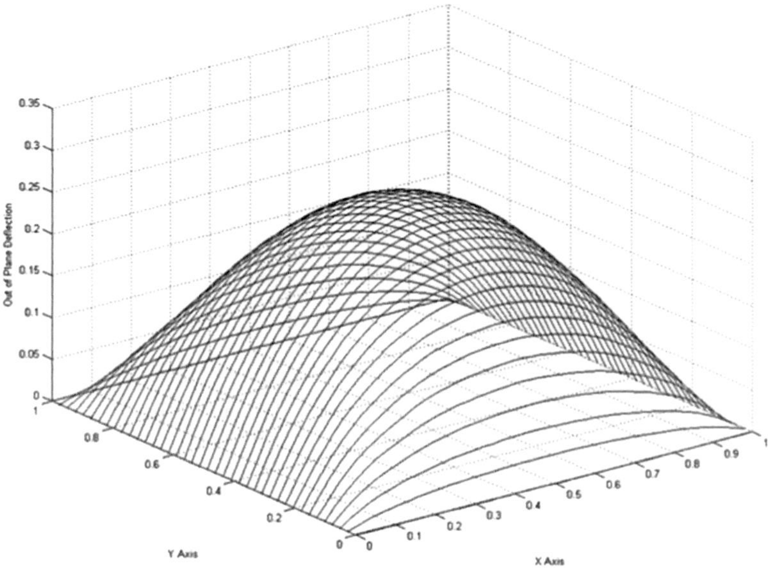

Fig. 9.23 Nondimensional simply supported plate profile for an *a*/*b* aspect ratio of 1 and a nondimensional thickness of 0.001

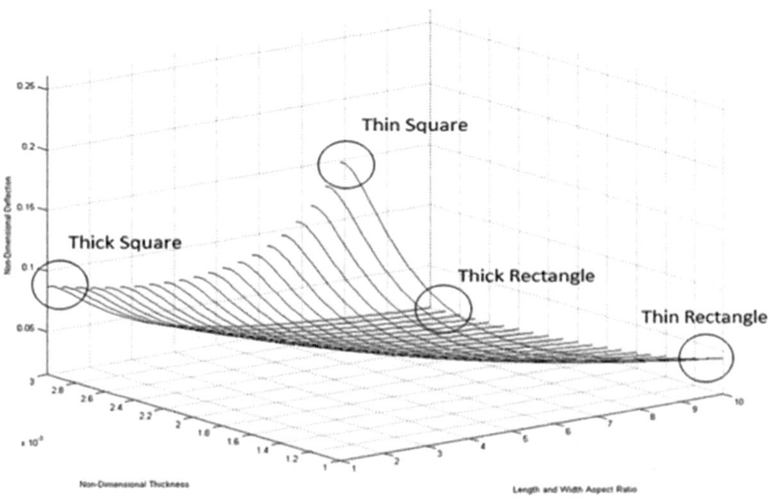

Fig. 9.24 Nondimensional deflection results from parametric study of plate thickness and plate aspect ratio. The temperature difference through the plate thickness is held constant

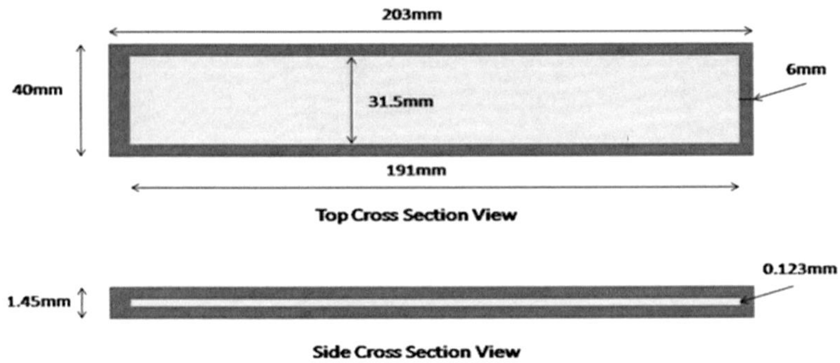

Fig. 9.25 Cross-sectional views of the LEU foil plate target used for numeric modeling using Abaqus. The overall dimensions match the HFR dispersion target currently used for molybdenum-99 production

The plot also shows that increasing the thickness will decrease the deflection. This occurs because the thermal moment, M_T, will remain the same while the stiffness value D will increase.

9.5.2 *Nominally Flat and Curved LEU Foil-Based Target*

Extensive thermal-mechanical analysis work on plate targets has been carried out. A small summary of results is presented here. The reader interested in more details is referred to publications 34–40.

Commercial finite element software (Abaqus) was used to conduct the analysis on more realistic target perform the primary analysis for more realistic target models. In particular, edge welds and nonuniform heating profiles have been examined. Previous studies determined that free-edge holding conditions produced the least amount of stress in the cladding material and the least amount of deflection. Therefore only the free-edge boundary condition will be considered here. For bracketing purposes, heat generation rates on the order of 30 kW are used. The heat generation rate within a production target is expected to be less. The dimensions of the target analyzed were chosen to match the current HFR dispersion target as shown in Fig. 9.25. Lighter areas correspond to where the LEU foil would be placed within the cladding, and thus correspond to the location of applied thermal load.

Welded edges are simulated by using "tie" constraints. The nodes on each plate in contact around the edges of the plates are bonded together. For model convergence reasons one end of the target was not allowed to translate or rotate, using the ENCASTRE boundary condition in Abaqus. Symmetry was assumed in these models and reported values taken from the edge opposite of the fully constrained edge. This assumption is effectively the same as using Saint Venant's principle. The boundary conditions used are shown schematically in Fig. 9.26.

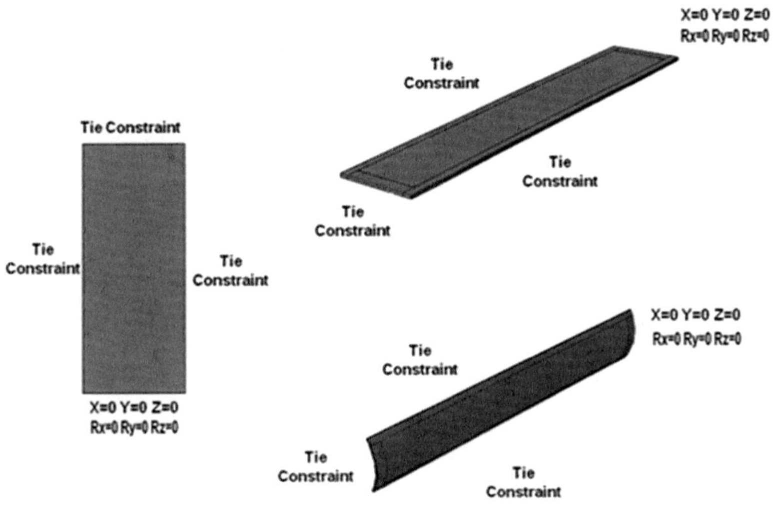

Fig. 9.26 Illustration of the mechanical boundary conditions applied to the numeric models for the simulations presented here

Table 9.3 Angles of curvature studied and the corresponding radii of curvatures

Angle of curvatures used in models (°)	Radius (m)
0 (Flat)	∞
20	~0.115
40	~0.057

The purpose of this study is to develop an understanding of prototypical target behavior during irradiation. It should be noted that assembly residual stresses in the cladding material are not comprehended in the study. Future work clearly must include an evaluation of the assembly process and the corresponding mechanical state of the materials before irradiation. The current study does allow the natural tendency of the materials to be clearly understood in such a way as to guide assembly process design.

One of the goals of the present study is also to understand the influence of plate curvature on the thermal-mechanical behavior of targets. Three different levels of curvature, ranging from perfectly flat to an angle of 40°, were examined. The angles of curvature and the corresponding radii of curvature are shown in Table 9.3.

A range of thermal loads and cooling conditions were also applied simulating heating/cooling conditions produced during irradiation. The thermal load was applied to the foil as volumetric heat generation and cooling conditions were simulated by applying a uniform heat transfer coefficient to the outer surface of the target. The heating/cooling conditions used in the study can be seen in Table 9.4. They cover the expected range of heating and cooling conditions that are expected for target irradiators.

Table 9.4 Thermal loads, heat generation values, and heat transfer coefficients used for numeric simulations

Thermal load (W)	Heat generation applied to the foil (W/cm^3)	Heat transfer coefficients (W/m^2 K)
22,500	3.040e4	50,000; 40,000; 30,000
27,500	3.716e4	
32,500	4.392e4	

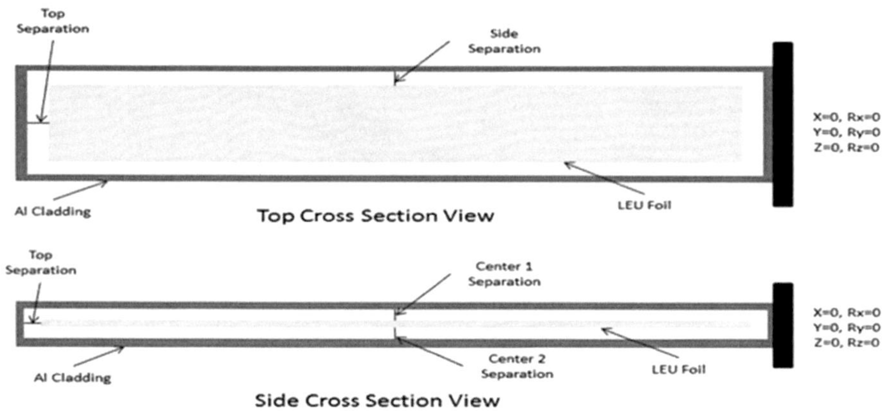

Fig. 9.27 Separation monitoring points for plate numeric simulations

Separation between the uranium foil and the aluminum cladding during irradiation was examined in the simulation results. Various locations were used to monitor separation between the foil and cladding. The location of the monitoring points and a diagram of the separations in a target cross section are identified in Fig. 9.27.

Target failure could also occur from thermal stress development in the target. Von Mises stress was used to characterize the risk of failure. Monitoring points were located along the weld and the center of the target cladding. The locations of the von Mises monitoring points are identified in Fig. 9.28.

Separation at various conditions was directly compared determining the effect of different curvatures and heating/cooling conditions. The center 1, center 2, side, and top separations can be seen in Figs. 9.29, 9.30, 9.31, and 9.32. In each of those plots, it appears that the magnitude of the plate separation, noted as the gap that tends to want to open between the LEU foil and the cladding, is on the order of 1 μm in the central area of the target. When this is coupled with the allowable gap that can be tolerated from Fig. 9.20, it is evident that a plate geometry has the potential to be a viable design option. Larger gaps that tend to want to open near the edges suggest that assembly residual stress may be needed to reduce the tendency for a gap to open. It should be clearly stated, however, that the gaps forming at the side and top monitor points are LATERAL gaps (look closely at Fig. 9.27), and will thus represent a very small increase to the overall thermal resistance of the target.

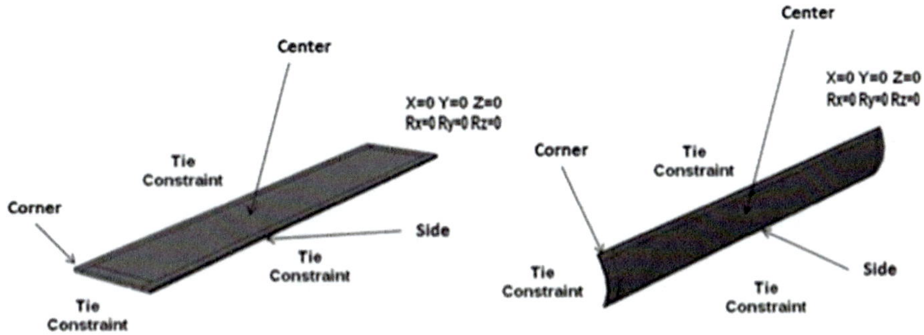

Fig. 9.28 Von Mises stress monitoring points for plate numeric simulations

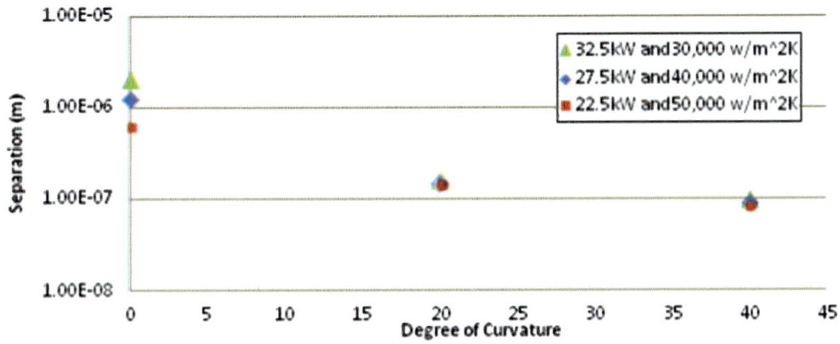

Fig. 9.29 Separation gap for range of curvature at Center 1 monitor point

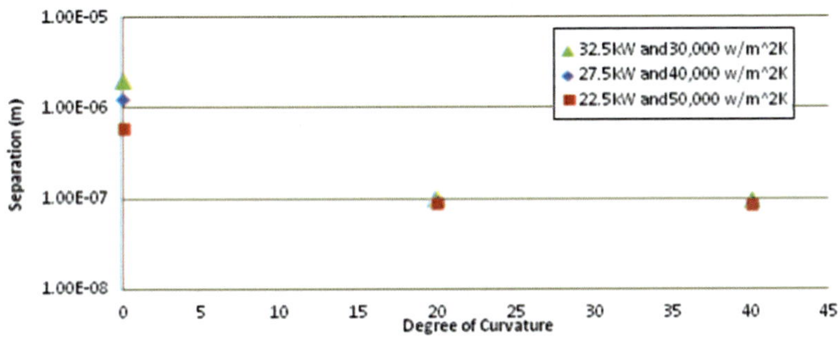

Fig. 9.30 Separation gap for range of curvature at Center 2 monitor point

It is evident that curvature has an effect on the foil/cladding separations in the target. Separation at both center locations drastically decreases non-linearly as curvature is increased. The separation at the side monitor location also decreases with added curvature but not as drastic as with the center monitoring points. The top separation increases non-linearly with added curvature, suggesting that moderate curvature is probably preferred over a large amount of curvature.

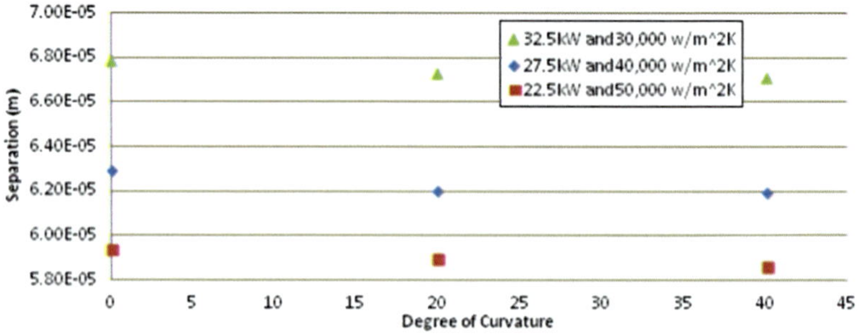

Fig. 9.31 Separation gap for range of curvature at Side monitor point

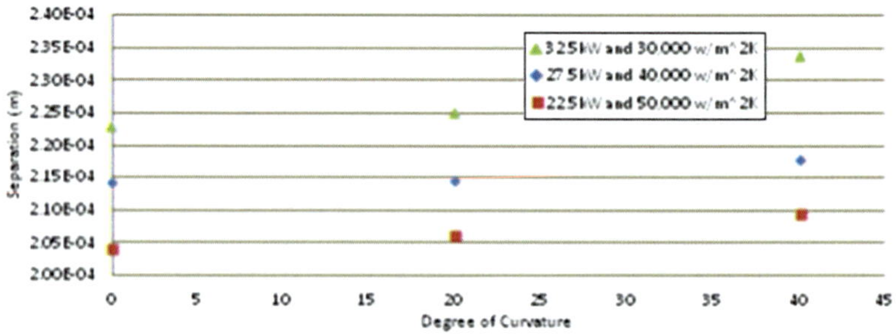

Fig. 9.32 Separation gap for range of curvature at Top monitor point

Table 9.5 Percent of heat passing through three foil faces

Single center face (%)	Single side face (%)	Single top face (%)
49.7	0.4	0.07

To illustrate more clearly that the gap-opening tendency on the side and top faces is probably not of significant concern, the amount of heat dissipated through each of the six faces of the foil and target was obtained from the numeric solution. The percentage of the total heat flow is shown in Table 9.5 for three of the edges. Note that the remaining three faces provide the exact same percentage of the corresponding face shown in Table 9.5.

Based on the percentages in Table 9.5 the majority of the heat generated in the foil will pass through the center faces where the separation is the lowest (total percentage through the center faces is $2 \times 49.7\ \% = 99.4\ \%$). A very small amount will pass through the top and side faces where separation is the greatest, reducing the risk of failure from the separations.

Von Mises stress was examined for its magnitude and distribution. The calculated values were plotted and compared to the yield strength of several aluminum alloys. The alloys chosen were Al-1100, Al-3003, and Al-6061. The yield strength for each of these alloys can be seen in Table 9.6.

Table 9.6 Aluminum alloy yield strengths

Al-1100 yield strength (Pa)	Al-3003 yield strength (Pa)	Al-6061 yield strength (Pa)
1.05e8	1.25e8	2.75e8

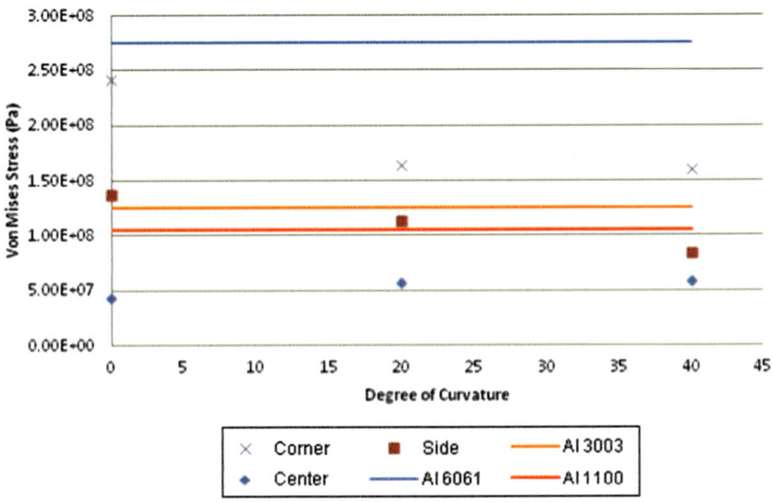

Fig. 9.33 Von Mises comparison at 32.5 kW and 30,000 W/m^2 K

The comparison shown here was for the worst-case heating and cooling conditions, 32.5 kW and 30,000 W/m^2 K. The stresses at these conditions can be seen in Fig. 9.33.

Figure 9.33 indicates that von Mises stress on the corner and side will decrease as curvature is increased and the center will increase slightly and then level off between 20° and 40° of curvature. These trends indicate that increasing curvature lowers or levels out the von Mises stress in the target and can be used to influence how stress develops during thermal loading. The magnitude of stress that is developed in the cladding suggests that softer grades of aluminum will be at risk for yielding without considering any assembly residual stress. However the stress levels at all locations within the target fall below the yield stress value for Al-6061.

9.5.3 LEU Foil-Based Annular Target

A study similar to that which was just conducted for plate target geometries was also conducted on the annular target, using numeric simulations to analyze a reference target design. As with the plates, the heat generated by the LEU foil during irradiation increases the target temperature causing thermal expansion-based deformation of the target. Since the LEU and the aluminum have different thermal

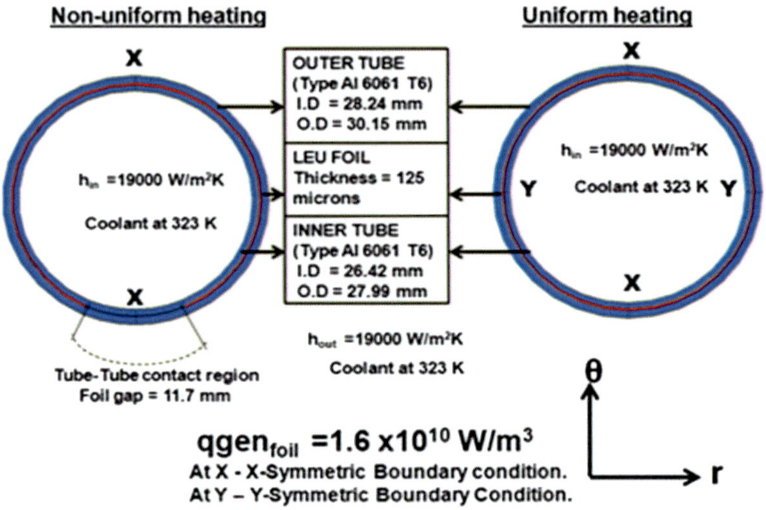

Fig. 9.34 Uniform and nonuniform heating model parameters

expansion coefficients, their magnitudes of thermal expansion will differ. Further, there will be a temperature gradient that is established in the aluminum cladding that will tend to induce a difference in thermal expansion of a single tube. The thermal expansion mismatch creates a tendency for gaps to form between the LEU and cladding, as well as stresses to develop in the target. The advantage of using an annular target geometry is the robust structural integrity.

As suggested in Fig. 9.12, the LEU foil does not fully wrap around the tube cladding, leaving an open gap (not to be confused with the natural tendency for a gap between the LEU and cladding to form during irradiative heating). The purpose of the gap is to provide a cutting street that facilitates the target opening process. From a thermal-mechanical standpoint, the open gap implies that there will be circumferential nonuniform heating within the target and this makes it difficult to develop a closed-form solution to an analytical model. Previous work has been conducted to compare a simplified analytic model to numeric simulations in order to establish the numeric modeling settings to be used [41, 42]. The learning from those studies is applied to the study presented here.

The numerical model in Abaqus FEA consists of the annular target modeled as a 2-D plane strain problem with a low-enriched uranium foil sandwiched between two aluminum tubes (Al-6061 T6). A uniform and a nonuniform heating model were built in Abaqus FEA as illustrated in Fig. 9.34. The residual stresses from the assembly of the target were not considered in the fully coupled thermal stress model and perfect contact was assumed at the interface of the parts before thermal loading was applied. Since the uniform heating model in Fig. 9.34 is symmetric about the x-axis and y-axis, x-symmetric and y-symmetric boundary conditions are applied to the nodes.

Fully coupled quadratic reduced integration elements (CPE8RT) were used to obtain the finite element mesh and the assembled annular target for the UNIFORM

Table 9.7 Analytical and numerical comparison for uniform heating

Parameter	Inner tube maximum error (%)	Outer tube maximum error (%)
Temperature	0.03	0.04
Radial stress	3.61	2.74
Hoop stress	3.08	0.89

heating case had 108,000 elements and 300,000 nodes with 20 elements through the thickness of each part. While the outer tube of the circumferential NONUNIFORM heating model has a uniform circular cross section, the inner tube has a recess cut on its surface to accommodate the LEU foil. This is illustrated in Fig. 9.34. Fully coupled quadratic reduced integration elements (CPE8RT) were used to obtain the finite element mesh and the assembled annular target had 120,000 elements and 370,481 nodes with 20 elements through the thickness of each tube and foil.

To validate the uniform heating numerical model analytical models were constructed. It has been shown [43] that in a composite model, due to the thermal expansion mismatch and direction of heat flow, contact is either contact can be established or withdrawn. Based on this, it is expected that the contact between the foil and inner tube will be reinforced while a gap would open up between the foil and the outer tube. A thermal resistance network similar to that shown in Fig. 9.17 was used to obtain the temperature distribution through the inner and outer tubes. The formulation for thermal stresses in a single tube subject to a heat flux [44] can be applied to the outer tube as it separates from the foil. Hence the properties of the outer tube alone need to be accounted for in calculating the stresses and displacement. For the inner tube–foil composite, the radial stresses on the inner surface of the inner tube and outer surface of the foil are zero. The interface compatibility dictates stress and displacement continuity. These four conditions can be used to solve for the constants in [45] and hence determine the stresses and displacements on the inner tube and the foil.

The purpose behind building a uniform heating numerical model was to validate it and use it to calibrate the nonuniform heating model. Table 9.7 presents the analytical and numerical comparison for the uniform heating analysis, where the error is defined as the difference between the analytic and numeric models.

The radial displacement contours for the non-uniformly heated target subjected to the conditions shown in Fig. 9.34 are illustrated in Fig. 9.35. Beginning at the top of the target (0°), there is no separation between the foil and the inner tube. A gap between the outer tube and the foil exists due to the larger thermal expansion coefficient for the aluminum cladding compared to the LEU foil. At 90° there is no separation between the foil and the inner tube. However, the magnitude of separation between the foil and the outer tube is smaller than at 0°. This shows that the widest gaps would tend to want to form at 0°. Between 90° and 180° a transition region exists as the ends of the foil move away from the recess and flap out to make contact with the outer tubes as the LEU foil expands. This causes the foil to pull away from the inner tube and make contact with the outer tube. A gap tends to form

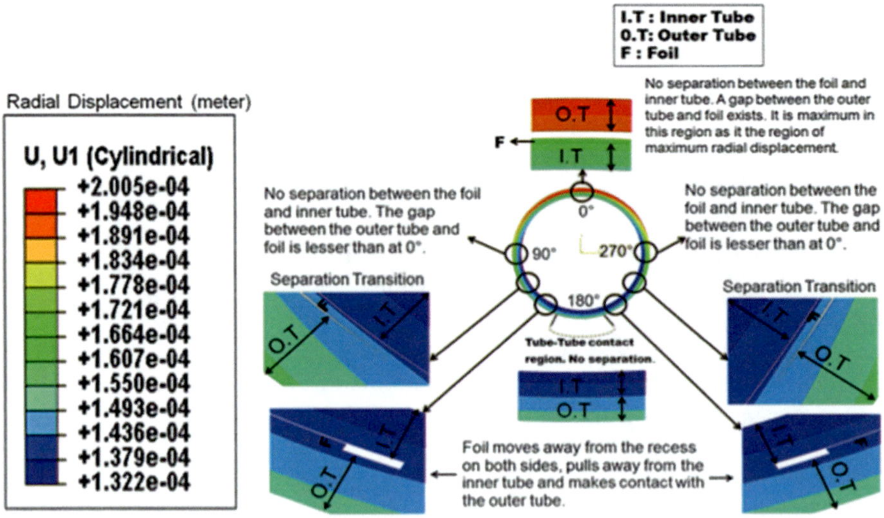

Fig. 9.35 Contour plot showing radial displacement of annular target where there is initially no residual stress and perfect contact is assumed

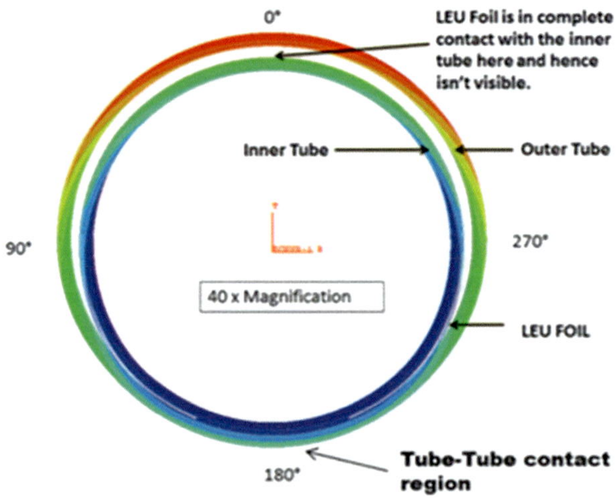

Fig. 9.36 Final deformed shape of the annular target. Note the teardrop shape as the un-foiled region has a smaller radius of curvature than the heated, foiled region

between the foil and the inner tube. The final geometry of the deformed target is shown in Fig. 9.36.

It is to be reiterated here that the current analysis does not take into account assembly process residual stresses. The current analysis effectively characterizes the qualitative separation that has a natural tendency to form during irradiative heating. That tendency can be used to develop an assembly process that will ensure that gaps actually do not form between the LEU foil and the cladding.

9.6 Summary and Conclusion

A stable, low-cost supply of the radioisotope molybdenum-99 is critical to ensure that the needed diagnostic medical procedures are allowed to take place as scheduled. This demand coupled with the drive to comply with nonproliferation agreements requires that a new, high-density target be developed for use in current reactor-based production operations. An LEU foil-based target is one such option.

It has been suggested here that multiple geometric arrangements for an LEU foil-based target could be viable from a thermal-mechanical safety standpoint. Annular-based designs have been successfully irradiated and have been in existence for over 20 years. Plate-based designs are not as mature but there is really no technical study that is currently known to dispute their potential. Thus, it is recommended that additional effort be expended to gain confidence in using LEU foil-based targets for molybdenum-99 production where economic and yield constraints exist.

One key item that needs to be considered more completely for annular and plate target designs is the assembly residual stress. It is hypothesized that the residual stress can be tailored to help mitigate the natural tendency of gaps to form as well as to help alleviate stresses that develop in the cladding material.

Acknowledgements Charlie Allen at the University of Missouri Research Reactor (MURR) has been a valuable source of information regarding the history of molybdenum-99 production. He has also provided valuable insight into how different target designs would integrate into a reactor irradiation space.

References

1. Williams RW (1976) Technetium-99m production from uranium-235 fission. University of Missouri, Columbia, MO
2. Saha GB (2004) Fundamentals of nuclear pharmacy, 5th edn. Springer, New York
3. Sonzogni A (n.d.) Chart of nuclides. Web. 27 Feb 2013. http://www.nndc.bnl.gov/chart/
4. Zolle I (2007) Technetium-99m pharmaceuticals: preparation and quality control in nuclear medicine. Springer, New York
5. National Research Council of the National Academies (2009) Medical isotope production without highly enriched uranium. The National Academies Press, Washington, DC
6. Turner K (2009) Thermal-mechanical analysis of targets for high volume production of molybdenum-99 using low-enriched uranium. Mechanical and Aerospace Engineering Department, University of Missouri, Columbia, MO
7. National Nuclear Security Administration (2012) United States, Belgium, France, and The Netherlands announce joint statement on HEU minimization and the reliable supply of medical

isotopes, 26 Mar 2012. Web. 27 Feb 2013. http://nnsa.energy.gov/mediaroom/pressreleases/usbfnmediso32612
8. Malouch F, Wohleber X, Durande-Ayme P (2011) Enhancement of irradiation capabilities of moly targets in the osiris reactor. Presented at the European Research Reactor Conference (RRFM), Rome, Italy, 20–24 Mar 2011
9. Mushtaq MIA, Bokhari IH, Mahmood T, Mahmood T, Ahmad Z, Zaman Q (2008) Neutronic and thermal hydraulic analysis for production of fission molybdenum-99 at Pakistan Research Reactor-1. Ann Nucl Energy 35:345–352
10. Salacz J (1987) Production of fission Mo-99, I-131, and Xe-133. In: Fission molybdenum medical use, Karlsruhe, Germany, 13–16 Oct 1987, pp 129–132
11. Solbrekken GL, Turner KK, Allen CW (2010) Thermal-mechanical analysis of varying boundary conditions on a LEU foil based molybdenum-99 plate processing target. Presented at the ASME International Mechanical Engineering Congress and Exposition, Vancouver, British Columbia, 2010
12. National Nuclear Security Administration (n.d.) Office of global threat reduction. Web. 27 Feb 2013. http://nnsa.energy.gov/aboutus/ourprograms/nonproliferation/programoffices/officeglobalthreatreduction
13. National Nuclear Securing Administration (2011) NNSA works to minimize the use of HEU in medical isotope production, 1 Nov 2011. Web. 27 Feb 2013. http://www.nnsa.energy.gov/mediaroom/factsheets/factsheet20100125
14. Ruth T (2011) Direct production of Tc-99m including a Mo-100 supply. In: Mo-99 Topical Meeting, Santa Fe, NM
15. OECD Nuclear Energy Agency (2011) Nuclear energy in perspective: the path to a reliable supply of medical isotopes, June 2011. Web. 27 Feb 2013. http://www.oecd-nea.org/press/in-perspective/2011-reliable-supply-medical-radioisotopes.pdf
16. Snelgrove JL, Hofman GL, Wiencek TC, Wu CT, Vandegrift GF, Aase S, Buchholz BA, Dong DJ, Leonard RA, Srinivasan B, Wu D, Suripto A, Aliluddin Z (1995) Development and processing of LEU targets for 99Mo production—overview of the Anl program. Presented at the 1995 International Meeting on, Reduced Enrichment for Research and Test Reactors, Paris, France, 18–21 Sept 1995
17. Srinivasan BH, Johnson JC, Vandegrift GK (1994) Development of dissolution process for metal foil target containing low enriched uranium. Presented at the Reduced Enrichment for Research and Test Reactors, Williamsburg, VA
18. Allen CW, Solbrekken GL (2007) Feasibility study—Part 2: Production of fission product Mo-99 using the LEU-modified Cintichem process. MURR TDR-0104, University of Missouri, Columbia, MO, March 2007
19. Allen C, Solbrekken G, Jollay L, Creasy J (2011) Equivalent fission Mo-99 target without highly enriched uranium. In Mo-99 Topical Meeting, Santa Fe, NM, 2011
20. Bejan A (2004) Convection heat transfer, 3rd edn. Wiley, New York
21. Burmeister LC (1993) Convective heat transfer. Wiley, New York
22. Kays WM, Crawford ME (2004) Convective heat and mass transfer, 4th edn. McGraw-Hill, New York
23. Ozisik MN (1985) Heat transfer: a basic approach. McGraw-Hill, New York
24. Swartz ET, Pohl RO (1989) Thermal boundary resistance. Rev Mod Phys 61:605–668
25. Chen G, Zeng T (2001) Nonequilibrium phonon and electron transport in heterostructures and superlattices. Microscale Thermophys Eng 5:71–88
26. Mikic BB, Rohsenow WM (1996) Thermal contact resistance. MIT Report No. 4542-41, Department of Mechanical Engineering, MIT, Sept 1966
27. Chiu CP, Solbrekken GL, Young T (2000) Thermal modeling and experimental validation of thermal interface performance between non-flat surfaces. Proceedings of Itherm 2000, Las Vegas, NV, 24–26 May 2000, pp 55–62
28. Prasher R (2006) Thermal interface materials: historical perspective, status, and future directions. Proc IEEE 94(8):1571–1586

29. Solbrekken GL, Chiu CP, Byers B, Reichenbacher D (2000) The development of a tool to predict package level thermal interface material performance. In: Proceedings of ITHERM 2000, vol 1. Las Vegas, NV, pp 48–54
30. Park JJ, Taya M (2006) Design of thermal interface material with high thermal conductivity and measurement apparatus. J Electronic Packaging 128:46–52
31. Madhusudana CV (1996) Thermal contact conductance. Springer, NewYork
32. Naotake Noda RBH, Tanigawa Y (2003) Thermal stresses, 2nd edn. Taylor & Francis, New York
33. Asmer NH (2005) Partial differential equations with Fourier series and boundary value problems. Prentice Hall, New York
34. Turner KK, Solbrekken GL, Allen CW (2011) Thermal-mechanical response of non-uniformly heated LEU foil based target. Proceedings of IMECE11, Denver, Colorado, USA, 11–17 Nov 2011, Paper IMECE2011-64734
35. Turner KK, Solbrekken GL (2011) Comparison of uniform and non-uniform heating in the plate type LEU foil based molybdenum-99 production target. Proceedings of 2011 ANS Winter Meeting and Nuclear Technology Expo, Washington, DC, 30 Oct–3 Nov 2011
36. Solbrekken GL, Allen C, Turner K, Makarewicz P, Govindarajan S (2011) Development, qualification, and manufacturing of Leu-Foil targetry for the production of Mo-99. Proceedings of the European Research Reactor Conference, Rome, Italy, 20–24 Mar 2011
37. Solbrekken GL, Turner K, Makarewicz P, Govindarajan S, Allen C (2011) Thermal/mechanical development of a Leu-Foil based target for production of Mo-99. Proceedings of the European Research Reactor Conference, Rome Italy, 20–24 Mar 2011
38. Turner KK, Solbrekken GL (2010) Thermal-mechanical analysis of varying boundary conditions on a LEU foil based molybdenum-99 plate processing target. Proceedings of ASME 2010 International Mechanical Engineering Congress & Exposition, Vancouver, British Columbia, Canada, 12–18 Nov 2010, Paper IMECE2010-39393
39. Turner KK, Solbrekken GL, Morrell JS (2010) Non-dimensional analytical analysis of a simply supported molybdenum-99 production target. Proceedings of 2010 ANS Winter Meeting and Nuclear Technology Expo, Las Vegas, NV, 7–11 Nov 2010
40. Makarewicz P, Turner KK, Solbrekken GL, Morrell JS (2010) Thermal/mechanical/hydraulic experimental tools for molybdenum-99 production target analysis. Proceedings of 2010 ANS Winter Meeting and Nuclear Technology Expo, Las Vegas, NV, 7–11 Nov 2010
41. Turner K, Solbrekken GL, Allen CW (2009) Thermal-mechanical analysis of annular target design for high volume production of molybdenum-99 using low-enriched uranium. Proceedings of ASME 2009 International Mechanical Engineering Congress & Exposition, Lake Buena Vista, FL, 13–19 Nov 2009, Paper IMECE2009-13238
42. Govindarajan SG, Solbrekken GL, Allen CW (2012) Thermal-mechanical analysis of a low-enriched uranium foil based annular target for the production of molybdenum-99. Proceedings of IMECE 2012, Houston, TX, 9–15 Nov 2012, Paper IMECE2012-86921
43. Madhusudhana CV (1999) Thermal conductance of cylindrical joints. Int J Heat Mass Transf 42:1273–1287
44. Boley BA, Weiner JH (1960) Theory of thermal stresses. Dover Publications Inc., New York, pp 272–281
45. Timoshenko S, Goodier JN (1951) Theory of elasticity. McGraw Hill Book Company, New York, pp 410–413

Editors Biography

Dr. Jonathan S. Morrell is a senior chemist and technical manager at the Y-12 National Security Complex in Oak Ridge, Tennessee, and has led the Compatibility and Surveillance Section of the Development Division since 2005. As a technical manager, Dr. Morrell oversees projects related to nuclear programs supporting core surveillance with nondestructive, destructive, and component testing. This involves testing of nuclear reactor fuels, cladding, and structural materials for performance evaluation, as well as long-term compatibility issues. Further, he conducts fundamental studies of material property variations and performance due to the direct effects of environment-induced degradation in the presence of temperature gradients, mechanical stresses, reactive and moist ambiances, and chemical and radioactive conditions for national security applications. Dr. Morrell currently serves on a coordinated research project at the International Atomic Energy Agency in Vienna, Austria on lifetime extension of aging research reactors. He is also an adjunct faculty professor in the Department of Chemistry at the University of Tennessee and in the Natural and Behavioral Science Department at Pellissippi State Community College in Knoxville. Dr. Morrell has ten issued patents, authored and coauthored more than 30 publications in archived journals and refereed

J.S. Morrell and M.J. Jackson (eds.), *Uranium Processing and Properties*,
DOI 10.1007/978-1-4614-7591-0, © Springer Science+Business Media New York 2013

conference proceedings, authored over 80 formal reports, and edited three technical books. He is currently a member of the editorial boards of the *International Journal of Molecular Engineering*, *International Journal of Nano and Biomaterials*, and *International Journal of Nanoparticles*.

Dr. Morrell received his bachelor degrees with honors from King College in chemistry and in biology. His honors research was conducted at the J. H. Quillen College of Medicine on the campus of East Tennessee State University, Johnson City, Tennessee, and was funded under a grant from the Department of Biochemistry and Molecular Biology at Johns Hopkins Medical Institute. The topic of this research was the antibiotic inhibition of ribosomal subunit formation in *Escherichia coli* cells. His doctor of philosophy in inorganic chemistry was obtained in 2000 from the University of Tennessee in Knoxville under the guidance of Professors David B. Beach and Ziling B. Xue. For his doctoral dissertation, Dr. Morrell studied the heteroepitaxial growth of complex oxide thin films via solution chemical techniques using novel alkoxide precursors. This research involved both the synthesis of precursor solutions and the processing and characterization of epitaxial thin films for ferroelectric and superconductor applications. The research was conducted at the Oak Ridge National Laboratory (ORNL), Oak Ridge, Tennessee, under a grant from the Office of Basic Energy Sciences, United States Department of Energy. Dr. Morrell, briefly employed by ORNL as a post-doctoral fellow, joined the Y-12 National Security Complex, Oak Ridge, Tennessee, in the summer of 2000. As a principal investigator in the Development Division, he was involved in experimental testing and small-scale verification of operating conditions associated with industrial production issues related to national defense. Dr. Morrell later initiated research and development programs in pyrochemistry, materials processing, forensic analysis, machining, and environmental aging studies. Dr. Morrell currently resides in his native home of East Tennessee with his wife, Holly, and twin sons, Graham and Parker.

Dr. Mark J. Jackson began his engineering career in 1983 when he studied for his O.N.C. part I examinations and his first-year apprenticeship-training course in mechanical engineering. After gaining his Ordinary National Diploma in Engineering with distinctions and I.C.I. prize for achievement, he read for a degree in mechanical and manufacturing engineering at Liverpool Polytechnic and spent periods in industry working for I.C.I. Pharmaceuticals, Unilever Industries, and Anglo Blackwells. After graduating with a Master of Engineering (M. Eng.) degree with Distinction under the supervision of Professor Jack Schofield, M.B.E., Doctor Jackson subsequently read for a Doctor of Philosophy (Ph.D.) degree at Liverpool in the field of materials engineering focusing primarily on microstructure–property relationships in vitreous-bonded abrasive materials under the supervision of Professor Benjamin Mills. He was subsequently employed by Unicorn Abrasives' Central Research & Development Laboratory (Saint-Gobain Abrasives' Group) as materials technologist, then technical manager, responsible for product and new business development in Europe, and university liaison projects concerned with abrasive process development. Doctor Jackson then became a research fellow at the Cavendish Laboratory, University of Cambridge, working with Professor John Field, O.B.E., F.R.S., on impact fracture and friction of diamond before becoming a lecturer in engineering at the University of Liverpool in 1998. At Liverpool, Dr. Jackson established research in the field of micromachining using mechanical tools, laser beams, and abrasive particles. At Liverpool, he attracted a number of research grants concerned with developing innovative manufacturing processes for which he was jointly awarded an Innovative Manufacturing Technology Center from the Engineering and Physical Sciences Research Council in November 2001. In 2002, he became associate professor of mechanical engineering and faculty associate in the Center for Manufacturing Research, and Center for Electric Power at Tennessee Technological University (an associated university of Oak Ridge National Laboratory), and a faculty associate at Oak Ridge National Laboratory. Dr. Jackson was the academic adviser to the Formula SAE Team at Tennessee Technological University. In 2004 he moved to Purdue University as Professor of Mechanical Engineering in the College of Technology.

Doctor Jackson is active in research work concerned with understanding the properties of materials in the field of microscale metal cutting, micro- and nanoabrasive machining, and laser micromachining. He is also involved in developing next-generation manufacturing processes. Doctor Jackson has directed, codirected, and managed research grants funded by the Medical Research Council, Engineering and Physical Sciences Research Council, The Royal Society of London, The Royal Academy of Engineering (London), European Union, Ministry of Defense (London), Atomic Weapons Research Establishment, National Science Foundation, N.A.S.A., U. S. Department of Energy (through Oak Ridge National Laboratory), Y-12 National Security Complex at Oak Ridge, Tennessee, and Industrial Companies, which has generated research income in excess of $15 million. Dr. Jackson has organized many conferences and served as General Chairman of the International Surface Engineering Congress and is Deputy President of the World Academy of Materials and Manufacturing Engineering. He has authored and coauthored over 250 publications in archived journals and refereed conference proceedings, has written a book on "micro and nanomanufacturing," is guest editor to a number of refereed journals, and has edited a book on "commercializing micro- and nanotechnology products." He is the editor of the *International Journal of Nanomanufacturing*, *International Journal of Nano and Biomaterials*, *Journal of Manufacturing Technology Research,* associate editor of the *International Journal of Molecular* Engineering, and is on the editorial board of the *International Journal of Machining and Machinability of Materials*, *International Journal of Computational Materials Science and Surface Engineering,* and the *International Journal of Manufacturing Research.*

Index

J.S. Morrell and M.J. Jackson (eds.), *Uranium Processing and Properties*,
DOI 10.1007/978-1-4614-7591-0,

9781461475903